Hak-Keung Lam and Frank Hung-Fat Leung

Stability Analysis of Fuzzy-Model-Based Control Systems

Studies in Fuzziness and Soft Computing, Volume 264

Editor-in-Chief

Prof. Janusz Kacprzyk
Systems Research Institute
Polish Academy of Sciences
ul. Newelska 6
01-447 Warsaw
Poland
E-mail: kacprzyk@ibspan.waw.pl

Further volumes of this series can be found on our homepage: springer.com

Vol. 249. Bernadette Bouchon-Meunier, Luis Magdalena, Manuel Ojeda-Aciego, José-Luis Verdegay, Ronald R. Yager (Eds.)
Foundations of Reasoning under Uncertainty, 2010
ISBN 978-3-642-10726-9

Vol. 250. Xiaoxia Huang
Portfolio Analysis, 2010
ISBN 978-3-642-11213-3

Vol. 251. George A. Anastassiou
Fuzzy Mathematics: Approximation Theory, 2010
ISBN 978-3-642-11219-5

Vol. 252. Cengiz Kahraman, Mesut Yavuz (Eds.)
Production Engineering and Management under Fuzziness, 2010
ISBN 978-3-642-12051-0

Vol. 253. Badredine Arfi
Linguistic Fuzzy Logic Methods in Social Sciences, 2010
ISBN 978-3-642-13342-8

Vol. 254. Weldon A. Lodwick, Janusz Kacprzyk (Eds.)
Fuzzy Optimization, 2010
ISBN 978-3-642-13934-5

Vol. 255. Zongmin Ma, Li Yan (Eds.)
Soft Computing in XML Data Management, 2010
ISBN 978-3-642-14009-9

Vol. 256. Robert Jeansoulin, Odile Papini, Henri Prade, and Steven Schockaert (Eds.)
Methods for Handling Imperfect Spatial Information, 2010
ISBN 978-3-642-14754-8

Vol. 257. Salvatore Greco, Ricardo Alberto Marques Pereira, Massimo Squillante, Ronald R. Yager, and Janusz Kacprzyk (Eds.)
*Preferences and Decisions,*2010
ISBN 978-3-642-15975-6

Vol. 258. Jorge Casillas and Francisco José Martínez López
Marketing Intelligent Systems Using Soft Computing, 2010
ISBN 978-3-642-15605-2

Vol. 259. Alexander Gegov
Fuzzy Networks for Complex Systems, 2010
ISBN 978-3-642-15599-4

Vol. 260. Jordi Recasens
Indistinguishability Operators, 2010
ISBN 978-3-642-16221-3

Vol. 261. Chris Cornelis, Glad Deschrijver, Mike Nachtegael, Steven Schockaert, and Yun Shi (Eds.)
35 Years of Fuzzy Set Theory, 2010
ISBN 978-3-642-16628-0

Vol. 262. Zsófia Lendek, Thierry Marie Guerra, Robert Babuška, and Bart De Schutter
Stability Analysis and Nonlinear Observer Design Using Takagi-Sugeno Fuzzy Models, 2010
ISBN 978-3-642-16775-1

Vol. 263. Jiuping Xu and Xiaoyang Zhou
Fuzzy-Like Multiple Objective Decision Making, 2010
ISBN 978-3-642-16894-9

Vol. 264. Hak-Keung Lam and Frank Hung-Fat Leung
Stability Analysis of Fuzzy-Model-Based Control Systems, 2011
ISBN 978-3-642-17843-6

Hak-Keung Lam and Frank Hung-Fat Leung

Stability Analysis of Fuzzy-Model-Based Control Systems

Linear-Matrix-Inequality Approach

Springer

Authors

Dr. Hak-Keung Lam
Division of Engineering,
King's College London,
Strand, London,
United Kingdom, WC2R 2LS
E-mail: hak-keung.lam@kcl.ac.uk
http://www.kcl.ac.uk/schools/nms/diveng/research/cdspr/hkl

Dr. Frank Hung-Fat Leung
Centre for Signal Processing,
Department of Electronic and
Information Engineering,
The Hong Kong Polytechnic University,
Hong Kong
E-mail: enfrank@inet.polyu.edu.hk

ISBN 978-3-642-42296-6 ISBN 978-3-642-17844-3 (eBook)

DOI 10.1007/978-3-642-17844-3

Studies in Fuzziness and Soft Computing ISSN 1434-9922

Softcover re-print of the Hardcover 1st edition 2010

Typeset & Cover Design: Scientific Publishing Services Pvt. Ltd., Chennai, India.

Printed on acid-free paper

9 8 7 6 5 4 3 2 1

springer.com

To my family and my wife, Esther Wing See Chan, for standing by me.

H.K. Lam

To Hypatia of Alexandria
F.H.F. Leung

Preface

Fuzzy logic control has been studied and developed for decades. It is known to be an effective control approach to some ill-defined and complex control processes. Thanks to fuzzy logic, expert knowledge about the control processes can be employed to heuristically design fuzzy controllers with some linguistic IF-THEN rules. Practically, human knowledge can be represented as linguistic statements and incorporated into the fuzzy logic controller. As a result, the design of the control action can be understood easily, and the fuzzy logic controller can operate with intelligence.

With years of investigation, the analysis and design of fuzzy logic control systems have been developing into a broad paradigm of active research. Instead of following the conventional heuristic approach, the design of fuzzy logic controllers can be realized by adopting a fuzzy-model-based (FMB) approach with engineering concerns such as the system stability and performance. The FMB control approach has received great attention of the researchers in the fuzzy control community owing to the systematic mathematical platform it offers to carry out rigorous analysis on general nonlinear control systems. The Takagi-Sugeno/Takagi-Sugeno-Kang (TS/TSK) fuzzy model is generally accepted as a powerful mathematical tool to represent/model/identify nonlinear plants in a systematic way to facilitate the stability analysis and controller synthesis. Based on the Lyapunov stability theory, the stability and performance of the FMB control systems can be investigated. Stability and performance conditions in terms of linear matrix inequalities (LMI) have been reached to aid the design of stable and well-performed FMB control systems. Convex programming techniques can be applied to find the solution of the LMI-based conditions numerically.

In this book, the state-of-the-art FMB based control approaches are covered. A comprehensive review about the stability analysis of type-1 and type-2 FMB control systems using the Lyapunov-based approach is given, presenting a clear picture to researchers who would like to work on this field. A wide variety of continuous-time nonlinear control systems such as state-feedback, switching, time-delay and sampled-data FMB control systems, are covered.

In short, this book summarizes the recent contributions of the authors on the stability analysis of the FMB control systems. It discusses advanced stability analysis techniques for various FMB control systems, and founds a concrete theoretical basis to support the investigation of FMB control systems at the research level. The analysis results of this book offer various mathematical approaches to designing stable and well-performed FMB control systems. Furthermore, the results widen the applicability of the FMB control approach and help put the fuzzy controller in practice. A wide range of advanced analytical and mathematical analysis techniques will be employed to investigate the system stability and performance of FMB-based control systems in a rigorous manner. Detailed analysis and derivation steps are given to enhance the readability, enabling the readers who are unfamiliar with the FMB control systems to follow the materials easily. Simulation examples, with figures and plots of system responses, are given to demonstrate the effectiveness of the proposed FMB control approaches.

This book is organized in 10 chapters. Chapter 1 offers a review on the state-of-the-art stability analysis techniques of FMB control systems. Various techniques under the categories of adaptive, sampled-data, state-feedback, switching, and time-delay fuzzy control schemes, are presented. The stability of the FMB control systems under the different categories of fuzzy controllers are discussed.

In in the first half of Chapter 2, the basic concept and mathematical background of the FMB control systems are given. The technical details of the TS fuzzy model that represents the nonlinear plant and facilitates the stability analysis and controller synthesis are given. A state-feedback fuzzy controller is used to control the nonlinear plant represented by the TS fuzzy model, forming an FMB control system. Some published membership-function-shape-independent (MFSI) stability conditions and performance conditions in terms of LMIs are given in the second half of the chapter.

In chapter 3, the membership-function-shape-dependent (MFSD)-based stability of FMB control systems with perfectly/imperfectly matched membership functions is investigated based on quadratic Lyapunov function. The first approach employs the membership function boundary (MFB) information of both the TS fuzzy model and the fuzzy controller to facilitate the stability analysis. The MFB information is carried by some slack matrices in the stability analysis for relaxing the stability conditions. LMI-based stability and performance conditions are derived to realize a stable and well-performed FMB control system. The second approach employs the staircase membership functions (SMF) to approximate the membership functions of the TS fuzzy model and fuzzy controller for conducting the stability analysis. Using this approach, the membership functions of the TS fuzzy model and fuzzy controller are able to help reaching the stability conditions. Consequently, the LMI-based stability conditions are dedicated to the FMB control system under consideration, which makes the approach to be more suitable for dealing with practical control problems.

In Chapter 4, a state-feedback fuzzy controller with time-varying feedback gains are proposed to strengthen the feedback compensation capability. Based on a quadratic Lyapunov function, stability conditions in terms of bilinear matrix inequalities (BMIs) are derived. The BMI conditions demonstrate a potential to further relax the stability conditions thanks to their nonlinear nature. However, efficient numerical methods are lacked to look for the solution. In this chapter, in order to make the proposed fuzzy control scheme feasible, a genetic-algorithm (GA)-based convex programming technique is developed to search for the solution to the BMI stability conditions. The proposed fuzzy control scheme offers a mathematical analysis platform that extends the LMI-based analysis approach in Chapter 2 to BMI-based ones for the further relaxation of stability conditions.

In Chapter 5, a parameter-dependent Lyapunov function (PDLF) is employed to investigate the system stability of the FMB control systems. Compared with the quadratic Lyapunov function, which is a kind of parameter-independent Lyapunov function (PILF), the PDLF is a nonlinear function that facilitates the stability analysis in different operating regions governed by the membership functions. The results show that the PDLF offers a nice property for the relaxation of stability conditions in some cases.

In Chapter 6, a regional switching fuzzy controller is proposed. The fuzzy controllers described in Chapter 2 to Chapter 5 are for nonlinear plants working in full operating region. As the nonlinear plant demonstrates stronger nonlinearity when it works at the full operating region than the sub-operating region, the idea of regional switching control is motivated. The full operating region is divided into a number of sub-operating regions. Corresponding to each sub-operating region, a regional fuzzy controller is designed. As the nonlinearity of the sub-operating regions is relatively weak, it is more likely to come up with stable regional fuzzy controllers than a fuzzy controller for the full operating region. According to the working conditions, the corresponding regional fuzzy controller is employed to control the nonlinear plant. On investigating the system stability, the regional information of the operating domain is employed for the relaxation of stability conditions.

In Chapter 7, a fuzzy combination control technique is employed to combine the state-feedback fuzzy controller and switching controller under the consideration of the system stability. A local and global fuzzy models are proposed to represent the nonlinear plant operating in local and full operating domains, respectively. A local fuzzy controller and global switching fuzzy controller are then proposed to stabilize the nonlinear plant based on the local and global fuzzy models. The nice properties of both types of controllers are integrated by fuzzy logic to offer an effective control scheme. It forms a theoretical background to support further research on fuzzy combination of various types of controllers.

In Chapter 8, the system stability of time-delay FMB control systems with system states subject to time delay is considered. Time delay is one of the causes for system instability that can be found in most domestic and industrial applications. Time-delay independent/dependent analysis approaches are adopted to investigate the system stability. It offers a systematic approach to study the nonlinear systems with time delay under the consideration of the system stability. LMI-based stability conditions for both analysis approaches are given.

In Chapter 9, the model reference tracking control for sampled-data FMB control systems is considered. A sampled-data fuzzy controller is proposed to drive the system states of the nonlinear plant to following those of a stable reference model. As the sampled-data fuzzy controller can be implemented by a microprocessor or a digital computer that is available at low cost, the implementation cost and time can be reduced. However, due to the zero-order-hold unit, control signal is kept constant during the sampling period. It causes discontinuity introduced by the sampling activity that complicates the system dynamics and thus makes the analysis difficult. A Lyapunov-based analysis approach is proposed to handle the discontinuity under the consideration of system stability. The tracking performance is guaranteed by applying the H_∞ control theory.

From Chapter 2 to Chapter 9, the system stability analyses reported are for type-1 FMB control systems only. In Chapter 10, we extend the stability analysis approach to interval type-2 FMB control systems. By using the interval type-2 fuzzy logic, the parameter uncertainties of the nonlinear plant can be captured and the system dynamics can be described with some simple fuzzy rules. An interval type-2 TS fuzzy model is then proposed to describe the system dynamics of the nonlinear plant. An interval type-2 fuzzy controller is proposed to close the feedback loop. Some MFSI and MFSD matrices are introduced to facilitate the stability analysis and controller synthesis. LMI-based stability conditions are derived to guarantee the stability of the interval type-2 FMB control system.

Hak-Keung Lam
Frank Hung-Fat Leung

Acknowledgements

I wish to sincerely thank everyone who has made possible the publication of this book.

I would like to thank the publisher Springer for the publication of this book and the staff offering support for the preparation of the manuscript.

I would like to thank King's College London for providing me a nice research environment and my nice colleagues in the Division of Engineering, King's College London for their invaluable support and comments on this book. Their cooperation and assistance have made my life much easier.

Also, I would also like to thank the co-author, Dr F.H.F. Leung for providing input and ideas to this work. Thanks are also delivered to my PhD student Mr Mohammad Narimani who has offered his support on the investigation of fuzzy-model-based.

In particular, I am greatly indebted to my wife, Esther Wing See Chan, for her patience, understanding, support and encouragement that make this work possible.

The work described in this book was substantially supported by grants from King's College London and from the Research Grants Council of the Hong Kong Special Administrative Region, China (Project No. PolyU 5224/08E).

H.K. Lam

Contents

Acronyms

BMI Bilinear matrix inequality
LMI Linear matrix inequality
FLF Fuzzy Lyapunov function
FMB Fuzzy-model-based
IT2 Interval type-2
MFB Membership-function boundary
MFSD Membership-function-shape-dependent
MFSI Membership-function-shape-independent
MRAC Model reference adaptive control
PDC Parallel distributed compensation
PDLF Parameter-dependent Lyapunov function
PILF Parameter-independent Lyapunov function
PLF Piecewise Lyapunov function
RS Regional switching
SWLF Switching Lyapunov function
SMF Staircase membership function
TD Time delay
TS Takagi-Sugeno
TSK Takagi-Sugeno-Kang

Chapter 1
Introduction

1.1 Introduction

Inspired by the fuzzy set theory established by Zadeh in 1965, Mamdani proposed fuzzy controllers to tackle nonlinear systems [86, 87]. Since then, fuzzy control has become a promising research platform. Despite the lack of a concrete theoretical basis, many successful applications of fuzzy control were reported in various areas such as sludge wastewater treatment [115], control of cement kiln [39], etc. These successes show that fuzzy controllers are capable of handling ill-defined plants with significant parameter uncertainties. As pointed out by Mamdani, stability of fuzzy systems is an important issue, and the remarked disadvantage of fuzzy control is the lack of appropriate tools for the analysis of controller performance [47]. In the absence of an in-depth analysis, fuzzy control systems may come with no guarantees of stability, good robustness and satisfactory performance; even some guidelines or rules-of-thumb for designing fuzzy controllers may not be available. In view of these limitations, a lot of research work had been done during the past two decades.

Fuzzy-model-based (FMB) control [27] is a powerful approach for tackling mathematically ill-defined nonlinear systems. To investigate the system stability, the Takagi-Sugeno (TS) fuzzy model (which is also known as Takagi-Sugeno-Kang (TSK) model) [101, 102] was proposed to provide a general and systematic framework to represent the nonlinear plant as a weighted sum of some linear sub-systems. Each linear sub-system effectively models the dynamics of the nonlinear plant in a local operating domain. As the linear and nonlinear parts of the nonlinear plant are extracted, the TS fuzzy model exhibits a semi-linear characteristic in favour of doing stability analysis and controller synthesis. Based on the TS fuzzy model, a fuzzy controller can be designed to close the feedback loop and form a FBM control system as shown in Fig.1.1.

There are in general two approaches to obtain the TS fuzzy model. The first approach is to construct the TS fuzzy model by using some system

H.-K. Lam and F.H.F. Leung: FMB Control Systems, STUDFUZZ 264, pp. 1–11.
springerlink.com

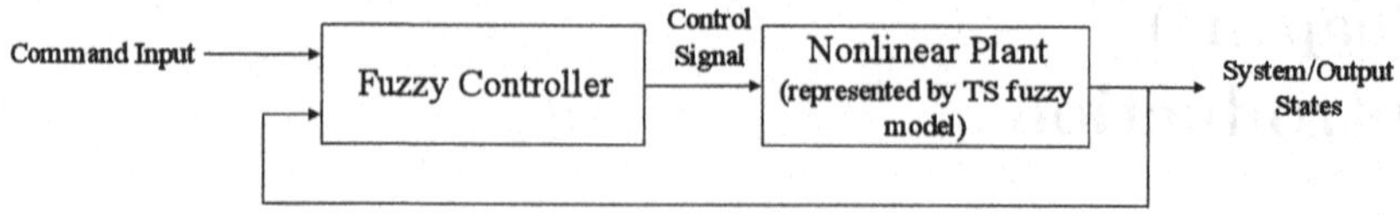

Fig. 1.1 A block diagram of the FMB control system.

identification algorithms [101, 102] based on the input-output data. This approach is suitable for nonlinear systems without mathematical models but with input-output data available. The second approach assumes that the mathematical model of the nonlinear system is available. The TS-fuzzy model can be derived from the mathematical model using the concept of sector nonlinearity or local approximation [110, 122].

This book mainly focuses on the stability analysis of continuous-time FMB control systems. Various types of FMB control systems such as time-delay and sampled-data FMB control systems are considered in the latter chapters. Synthesis of various types of fuzzy controllers using LMI (linear matrix inequality) or BMI (bilinear matrix inequality) approaches are covered. In this chapter, a review of the previous work is given. Various approaches of analyzing the stability of FMB control systems are described in the following sections.

1.2 Review of FMB Control

In this section, some FMB control schemes are briefly reviewed. Figure 1.2 shows a general framework for various FMB control schemes.

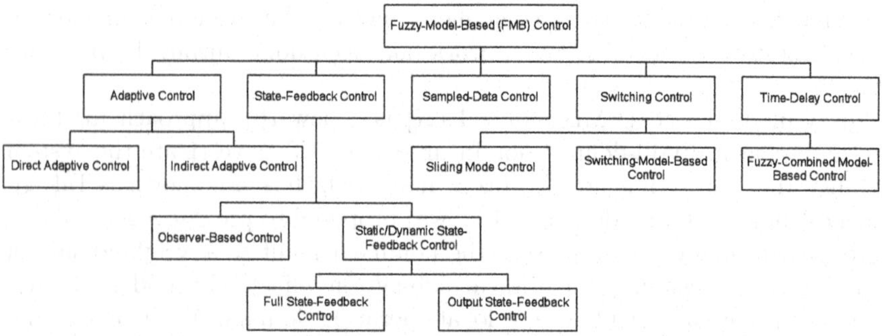

Fig. 1.2 A diagram showing various fuzzy control approaches.

1.2.1 FMB Adaptive Control

FMB adaptive control is good at handling nonlinear systems subject to parameter uncertainties or disturbances which are assumed to be bounded. To facilitate the stability analysis and controller synthesis, the TS fuzzy model is employed to represent the nonlinear plant in which the values of the membership grades or system parameters are uncertain or unknown. In some cases, external disturbances are included in the TS fuzzy model. Based on the TS fuzzy model, stability analysis and design of an adaptive fuzzy controller can be carried out. The parameters of the adaptive fuzzy controller are then updated by some online update laws.

In general, there are two classes of adaptive control schemes, namely indirect and direct adaptive control schemes [41]. The block diagrams of these two adaptive fuzzy control schemes are shown in Fig. 1.3 and Fig. 1.4, respectively. Referring to Fig. 1.3, which shows the indirect adaptive fuzzy control scheme [95], the adaptive fuzzy controller is characterized by the system parameters of the TS fuzzy model. As the values of the membership grades or system parameters are uncertain or unknown, their values are online estimated by a fuzzy estimator with some update laws. The estimated system parameters are then employed by the fuzzy controller to generate the control signal for the control process.

Figure 1.4 shows the direct adaptive fuzzy control scheme [50, 77, 132]. It is assumed that there exists a stable and well-performed FMB control system formed by the TS fuzzy plant model and an ideal fuzzy controller connected in a closed loop. Under this assumption, the parameters of the TS fuzzy model can be written in terms of the parameters of the fuzzy controller; e.g. the feedback gains. The values of the controller parameters are then online

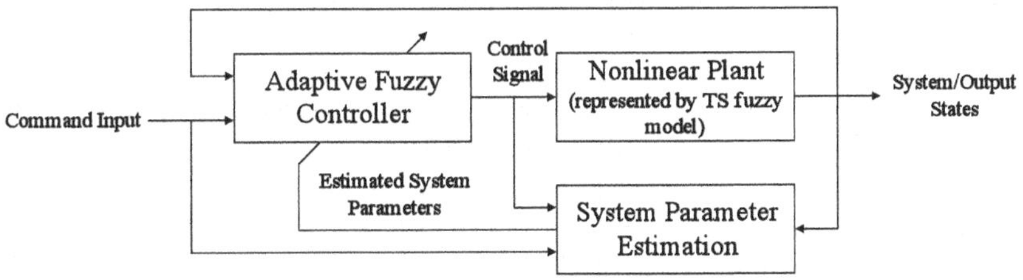

Fig. 1.3 Indirect adaptive fuzzy control.

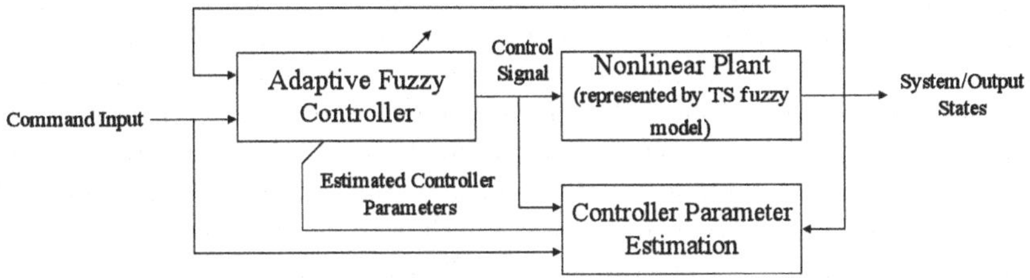

Fig. 1.4 Direct adaptive fuzzy control.

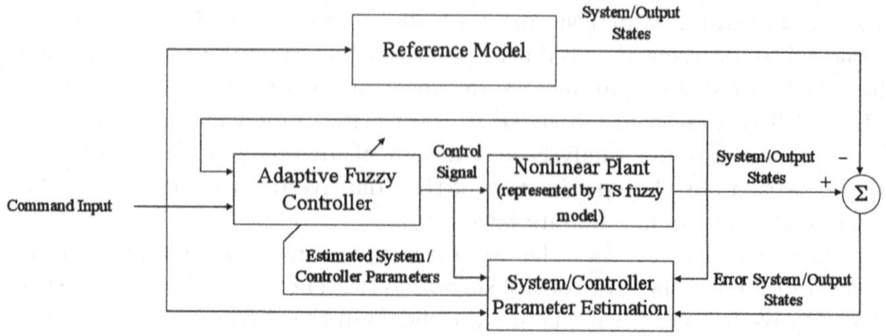

Fig. 1.5 Indirect/Direct fuzzy model reference adaptive control.

estimated with some update laws. The estimated controller parameters are directly applied to the fuzzy controller to perform the control process.

The indirect/direct adaptive fuzzy control schemes were extended to indirect/direct fuzzy model reference adaptive control (MRAC) schemes [5, 94], which can be represented by the block diagram in Fig. 1.5. Under this control scheme, a stable reference model is included in the system. The adaptive fuzzy controller is designed to drive the system/output states following those of the reference model. The parameters of the fuzzy controller are online updated according to the indirect or direct update laws.

1.2.2 State-Feedback FMB Control

On applying the state-feedback FMB control scheme, the fuzzy controller generates the control signal based on the system states. Referring to Fig. 1.2, there are in general two categories of FMB control schemes that can be found in the literature, namely fuzzy static/dynamic state-feedback fuzzy control scheme [15, 17, 24, 40, 45, 48, 76, 81, 82, 97, 105, 112, 122, 125] and fuzzy observer-based control scheme. The static/dynamic state-feedback fuzzy control scheme can be further divided into full state-feedback and output feedback fuzzy control schemes.

The FMB static/dynamic state-feedback control system can be depicted by the block diagram in Fig. 1.1. The state-feedback fuzzy controller is governed by some fuzzy rules with the consequents being some linear static/dynamic state-feedback sub-controllers. The firing strength of each rule (grade of membership) is governed by some membership functions. The state-feedback fuzzy controller is thus a weighted sum of some linear static/dynamic state-feedback sub-controllers, where the weight corresponding to each sub-controller is the grade of membership. In the full state-feedback control scheme, all the system states will be used by the state-feedback fuzzy controller to produce the control signal for the control process. When only the output states of the system are employed for the control process, it is called the output feedback fuzzy control [17, 24, 40, 45, 76]. When a static state-feedback fuzzy control scheme is considered, the values of the feedback gains are determined and kept constant during the control process. When a dynamic state-feedback fuzzy control scheme is considered, a dynamic compensator [76] inside the fuzzy controller adds dynamics, which is governed by a set of first-order differential equations, to its output. Unlike the static state-feedback fuzzy controller, the dynamic state-feedback fuzzy controller makes use of the system states and the extra states from the dynamic compensator to realize the control process. This type of fuzzy controller is good at dealing with the reference tracking control and disturbance rejection problems. In case the system states of the nonlinear plant are not measurable, an observer-based control scheme [3, 14, 16, 33, 85, 105, 112, 118, 119] can be applied. A fuzzy observer is employed to estimate the system states which will be used by the fuzzy controller to perform the control process.

1.2.2.1 Stability Analysis of State-Feedback FMB Control Systems

Lyapunov-based approach is the most popular approach to investigate the system stability of the FMB control systems. There are different stability analysis approaches as shown in Fig. 1.6 that can be found in the literature to investigate the stability of FMB control systems based on the Lyapunov stability theory. In general, the stability analysis are based on two types of TS fuzzy models, namely type-1 [101, 102] and type-2 [4, 70, 78] fuzzy models. In the type-1 TS fuzzy model, the nonlinearity of the plant is described by type-1 fuzzy sets. Type-2 fuzzy sets [18, 88, 89] are extended from the type-1 fuzzy sets, which are good at handling system uncertainties. Compared to the type-1 TS fuzzy model, the type-2 TS fuzzy model is able to describe the system dynamics of the nonlinear plant subject to parameter uncertainties captured by the type-2 fuzzy sets. The type-2 TS fuzzy model can be regarded as a model formed by an infinite number of type-1 TS fuzzy models. In this

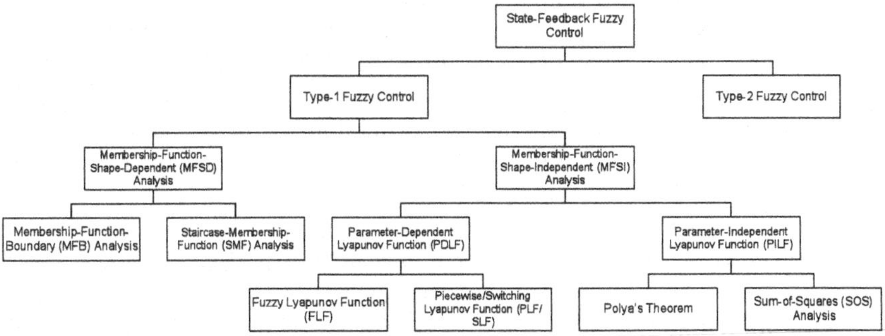

Fig. 1.6 Stability analysis approaches for state-feedback FMB control systems.

chapter, we will focus on the type-1 FMB control systems and the type-2 ones will be discussed in Chapter 10.

For type-1 FMB control systems, referring to Fig. 1.6, there are in general two streams of analysis approaches, namely the membership-function-shape-independent (MFSI) and membership-function-shape-dependent (MFSD) analysis approaches based on the Lyapunov stability theory. Under the MFSI analysis approach, the shapes of the membership functions are not taken into account in the stability analysis. Once the FMB control system is guaranteed to be stable by some stability conditions, the stability is not affected by the shapes of the membership functions.

MFSI Stability Analysis

Under the MFSI analysis approach, parameter-independent and parameter-dependent Lyapunov functions (PILF and PDLF) can be employed to investigate the stability of the FMB control system. A PILF is a Lyapunov function which is independent of the system parameters and/or states. In [15, 24, 48, 81, 82, 97, 105, 112, 122, 125], a quadratic Lyapunov function was employed. It was shown in [15, 122] that the FMB control system is guaranteed to be asymptotically stable if there exists a common solution to a set of linear matrix inequalities (LMI) [44]-[45], which are regarded as the stability conditions. It was shown in [122] that the stability conditions can be relaxed by using the parallel distributed compensation (PDC) design technique of which the fuzzy controller shares the same premise membership functions as those of the TS fuzzy model. Under the PDC design, the stability conditions were further relaxed in [24, 48, 81, 82, 97, 105, 112, 125] based on the Polya's theorem. In [108, 111], a sum-of-squares (SOS) analysis approach was proposed. The TS fuzzy model is modified to a polynomial TS fuzzy model of which the system and input matrices (which are constant in the traditional TS fuzzy model) are allowed to be polynomials in terms of the system states. The polynomials are able to model the nonlinearity more effectively than some constants. Consequently, the polynomial TS fuzzy model is able to alleviate some limitations of system modelling suffered from traditional TS fuzzy models. Unlike the quadratic Lyapunov function used in [15, 24, 48, 81, 82, 97, 105, 112, 122, 125], which is a polynomial of degree 2 in terms of system states, the Lyapunov function in the SOS analysis approach depends on a sum of higher-degree monomials of the system states. The stability of the polynomial FMB control system is guaranteed if there exists a Lyapunov function which is an SOS.

Under the PDLF approach, a Lyapunov function that depends on the system parameters is employed for conducting the stability analysis. Referring to Fig. 1.6, there are two types of Lyapunov functions, namely fuzzy Lyapunov function (FLF) and piecewise/switching Lyapunov function (PLF/SLF), found in the literature. A FLF is an average weighted sum of some local quadratic Lyapunov functions [103], where the weights are the grades of membership. The membership functions define the working domain of the local Lyapunov functions. Hence, the time derivative of the FLF is possible to be negative (which implies the system stability) even if not all time derivatives of the local Lyapunov functions are negative. In contrast, the time derivative of the quadratic Lyapunov function in the PILF approach is required to be negative for the full working domain. Consequently, the PDLF analysis approach has the potential to relax the stability conditions. It has been shown in [19, 36, 67, 96, 103, 107] by simulations that the stability conditions derived based on the FLF are more relaxed than the PILF in some cases. As the time derivatives of the membership functions are generated in the stability analysis, in general, the PDLF approach are more complex than the PILF approach to derive the stability conditions.

Apart from the FLF, PLF/SLF [25, 26, 28, 29, 43, 93] were also proposed to realize the PDLF approach for the stability analysis of FMB control systems. PLF/SLF consists of some local Lyapunov functions. On using FLF, the local Lyapunov functions are smoothly stepped over among themselves. The transition is governed by the membership functions. On using PLF/SLF, however, the local Lyapunov functions are switched among themselves and only one single local Lyapunov function is activated at a time. Owing to this hard switch property, it is required that the local Lyapunov functions at the boundaries are continuous to make sure that the PLF/SLF is a valid Lyapunov function. Some design methodologies for some valid PLFs/SLFs, which guarantees that the local Lyapunov functions are continuous at the boundaries, were discussed in [43].

In the aforementioned stability analysis approaches, the stability conditions are in terms of LMIs, which can be solved by some convex programming techniques. In [28, 58, 61, 69], the stability conditions are cast into bilinear matrix inequalities (BMIs). However, the solution to BMIs cannot be simply solved by the convex programming techniques. Considering that some decision variables for each BMI term are kept constant, the BMI-based stability conditions become LMI-based ones. Taking advantage of this property, a genetic-algorithm (GA) based convex programming technique was proposed in [58, 61, 69] to find the solution to the BMIs.

A brief review for the MFSI stability analysis of FMB control systems is given in Chapter 2.

MFSD Stability Analysis

The membership-function-shape-dependent (MFSD) analysis approach, unlike the MFSI analysis approach, is able to bring the information of the membership functions to the stability analysis. Consequently, as more information is considered and included in the stability analysis, more relaxed stability conditions can be obtained as compared to the MFSI analysis approach. In general, the MFSD stability analysis can be conducted in two ways, namely the membership-function-boundary (MFB) and the staircase-membership-function (SMF) analysis approaches. On applying the MFB analysis approach, the boundary information of the membership functions [2, 55, 67, 68, 91, 98, 99], such as the upper bounds of the membership functions and/or their products, is utilized in the stability analysis. By proposing membership-function inequalities and the S-procedure [6, 23], membership-function-dependent and independent slack matrix variables can be introduced to facilitate the stability analysis. In [98, 99], stability analysis under the PDC design that the TS fuzzy model and fuzzy controller share the same premise membership was conducted. In [2, 55, 67, 68], FMB control systems under the non-PDC design were considered that the fuzzy controller does not share the membership functions of the TS fuzzy model.

On performing the MFB analysis, only the boundary information of the membership functions is considered. In order to bring more information into the stability analysis, some sample points of the membership functions (which are approximated by some staircase membership functions) are taken into account in the SMF analysis approach. In this approach, the membership functions of both the fuzzy model and the fuzzy controller are able to be absorbed into the stability conditions. When the step size tends to zero, the staircase membership functions tends to the original ones. Under the MFB analysis approach, if a FMB control system is guaranteed to be stable for some given membership functions, it is also stable for any shapes of membership functions providing the same boundary information. The SMF analysis approach, when the step size of the staircase membership functions is sufficiently small, approximately deals with the specified membership functions (characterized by the staircase membership functions) considered in the stability analysis. Consequently, it has a potential to further relax the stability conditions by considering a specific FMB control system as compared with the MFSI and MFB analysis approaches.

The MFSD stability analysis of FMB control systems is discussed in Chapter 3. In the first part of Chapter 3, the MFB information is employed to facilitate the analysis for the FMB control systems subject to imperfectly matched membership functions. Then, in the second part of Chapter 3, the SMF approach is employed to investigate the system stability. In Chapter 4, a time-varying state-feedback fuzzy controller is proposed. BMI-based stability conditions are derived to guarantee the system stability. In Chapter 5, the stability of the FMB control systems based on a PDLF is investigated. In Chapter 10, the system stability of interval type-2 (IT2) FMB control systems is investigated based on the MFSD stability analysis approach.

1.2.3 Switching Fuzzy Control

In switching FMB control systems, the fuzzy controller is designed based on the TS fuzzy model, and it consists of some switching components. The switching fuzzy controller is good at handling nonlinear plants with parameter uncertainties with known bounds. The boundary information of the parameter uncertainties offers important information for the design of the switching controller. By switching between the lower and upper bounds according to some switching rules derived under the consideration of system stability, the equivalent magnitudes of the parameter uncertainties can be obtained and compensated. However, due to the existence of the switching components in the systems, the control signals are high-frequency switching signals which cause undesired chattering effect [100] in the system output. The chattering effect can be alleviated by replacing the switching function with a saturation function. However, it may cause some finite steady-state error in the system output.

Referring to Fig. 1.2, in general, there are three types of switching FMB control approaches found in the literature. The first type of switching fuzzy controller is developed based on the sliding mode control theory [100] and a TS fuzzy model with unknown membership functions subject to unknown system parameters. A switching fuzzy controller with switching membership functions was then developed [57, 63, 65]. The second type of switching fuzzy controller is realized as a fuzzy combined switching fuzzy controller that softly switched among some controllers. In [52–54], a fuzzy model consisting of a local and a global TS fuzzy model was proposed. When the system is operating globally, a switching/sliding-mode controller is employed to drive the system states towards the origin. Once the system states are near the local operating domain, a local state-feedback fuzzy controller gradually replace the switching/sliding-mode controller. In the local operating region, the local state-feedback fuzzy controller determined by some fuzzy rules will dominate the control process. Consequently, a good transient response can be ensured by the switching/sliding-mode controller and the chattering effect can be eliminated by employing the local state-feedback fuzzy controller. The third type of switching fuzzy controller is designed based on a switching fuzzy model [21, 22, 106, 128]. The full operating domain is divided into some operating sub-domains. A local TS fuzzy model is then constructed for each operating sub-domain. Corresponding to each local TS fuzzy model, a local fuzzy controller is designed. When the switching FMB control system is working in one operating sub-domain, the corresponding local fuzzy controller is employed for the control process.

The switching FMB control approach is covered in Chapter 6 that the regional information of membership functions is employed to facilitate the stability analysis. The fuzzy combined fuzzy controller is discussed in Chapter 7.

1.2.4 Time-Delay Fuzzy Control

A time-delay nonlinear system is a dynamic system that depends on both the current and time-delayed system states. The time delay can be constant or time varying. To investigate the system stability of time-delay FMB control systems, two approaches can be found in the literature, namely the delay-independent and delay-dependent approaches. Delay-independent stability conditions for time-delay fuzzy control systems were derived in [7, 8, 124] based on the Lyapunov-Krasovskii or Lyapunov-Razumikhin approaches. In the delay-independent approach, the stability conditions are not related to the time-delay information. Once the time-delay fuzzy control system is guaranteed to be stable, it is stable for any value of time delay. Hence, the delay-independent stability conditions are particularly useful for nonlinear systems subject to unknown or inestimable value of time delay. In [10–12, 34, 75, 79, 113, 126, 129], delay-dependent stability conditions were derived based on the Lyapunov-Krasovskii approach. During the stability analysis, the time-delay information is considered. To deal with the time-delay information, various inequalities have been proposed. In [7], the Leibniz-Newton formula was employed to approximate the time-delay system states with the current system states. To relax the conservativeness of the stability analysis, other forms of inequalities were proposed in [127, 133]. These inequalities have been employed in [10–12, 34, 75, 79, 113, 126, 129] to investigate the stability of time-delay FMB control systems. It has been shown in [12, 75, 79, 126, 129] that relaxed inequalities have led to relaxed stability analysis results. Furthermore, by introducing some slack matrices, the stability conditions can be further relaxed. Compared with the delay-independent approach, the analysis procedure of the delay-dependent approach is more complicated. However, as the time-delay information is one of the elements to determine the system stability in the delay-dependent approach, less conservative stability conditions may be produced. The delay-dependent stability conditions are good for time-delay fuzzy control systems with known or estimable values of time delay. Consequently, both delay-independent and delay-dependent stability analysis results have their own advantages for different kinds of the time-delay nonlinear systems.

The time-delay FMB control approach is discussed in Chapter 8. Based on the time-delay independent/dependent stability analysis approaches, LMI-based conditions are derived to guarantee the system stability and synthesize the fuzzy controller.

1.2.5 Sampled-Data Fuzzy Control

A sampled-data FMB control system is shown in Fig. 1.7 It consists of a continuous-time nonlinear plant and a sampled-data fuzzy controller connected in a closed loop. To realize the control process, the system state vector $\mathbf{x}(t)$ is first sampled by the sampler at every h_s seconds. The sampled state vector is then fed to the discrete-time fuzzy controller to produce the discrete-time control signal $\mathbf{u}(t_\gamma)$. After going through the zero-order-hold (ZOH) unit, the discrete-time control signal is held constant during the sampling period.

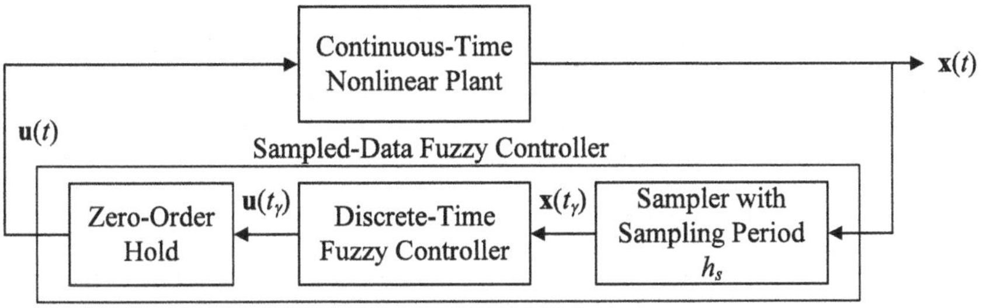

Fig. 1.7 A block diagram of sampled-data FMB control system.

Because of the ZOH and the sampling activity, discontinuities are introduced to the closed-loop control systems. The Lyapunov function for the continuous-time state-feedback FBM systems cannot be directly applied to investigate the system stability. Furthermore, as the system states can be obtained only at the sampling instant, the current system states cannot be obtained for feedback compensation. In [60], the system stability of sampled-data FMB control systems was investigated by casting the sampled-data FMB control system as a time-varying delay FMB control system. By employing the Lyapunov-Krasovskii functional, LMI stability conditions were provided to check for the system stability and facilitate the controller synthesis. In [32, 92], a hybrid system approach, in which the fuzzy controller contains both continuous-time and discrete-time components, was proposed. In [44, 130, 131], an equivalent jump system was proposed to represent the dynamics of the sampled-data FMB control systems at the sampling instant. The system stability is guaranteed if both the sampled-data FMB control system governing the system dynamics during the sampling period, and the jump system governing the system dynamics at the sampling instant, are both stable subject to a common time-varying Lyapunov function. In [9, 72, 73], an intelligent digital redesign approach, which is to approximate the nonlinear plant by a discrete-time fuzzy model, was proposed. Based on the discrete-time fuzzy model, a discrete-time fuzzy controller is then proposed to close the feedback loop. However, the discretization error may become a source to causing system instability.

The stability of sampled-data FMB control systems is investigated in Chapter 9. LMI-based conditions are derived to guarantee the system stability and synthesize the sampled-data controller.

Chapter 2
Stability and Performance Conditions for MFSI State-Feedback FMB Control Systems

2.1 Introduction

Fuzzy-model-based (FMB) control approach [27] offers a systematic and effective way to handle the nonlinear control problems. With the powerful TS fuzzy model [101, 102], a nonlinear system can be generally represented as a weighted sum of some local linear systems. The TS fuzzy model provides a systematic and general framework that effectively separates the linear and nonlinear dynamics of the nonlinear plant. This semi-linear property of the TS fuzzy model allows some linear analysis and control approaches to be applied to facilitate the stability analysis and controller synthesis.

Based on the TS fuzzy model, a fuzzy controller [109] was proposed to close the feedback loop to form a FMB control system. The Lyapunov stability theory provides one of the mathematical tools to investigate the stability of the FMB control systems. It was shown in [15, 109] that the FMB control system is guaranteed to be asymptotically stable if a set of linear matrix inequalities (LMIs) [6] are satisfied. The solution to the LMIs can be found numerically by using convex programming techniques. As the stability analysis in [15, 109] did not consider the membership functions of both the TS fuzzy model and the fuzzy controller, the stability conditions are very conservative. Nevertheless, there is not much restriction on the design of membership functions of the fuzzy controller. As a result, the implementation cost of the fuzzy controller can be reduced by using some simple membership functions. To relax the stability conditions, a parallel compensation distribution (PDC) design approach was proposed [122] in which the TS fuzzy model and the fuzzy controller share the same membership functions. Although the conservativeness of the stability analysis can be reduced, the structural complexity of the fuzzy controller may be increased when the membership functions of the TS fuzzy model are complicated. Further relaxed stability conditions were achieved in [24, 48, 81, 82, 97, 105, 112, 125].

H.-K. Lam and F.H.F. Leung: FMB Control Systems, STUDFUZZ 264, pp. 13–22.
springerlink.com

In this book, the stability conditions in [24, 48, 81, 82, 97, 105, 112, 122, 125] are classified as membership-function-shape-independent (MFSI) ones as the stability analysis does not depend on the shapes of the membership functions. The MFSI stability conditions can be applied to FMB control systems with membership functions of any shape. However, it was revealed in [2, 55, 67, 98, 99] that the shape information of the membership functions play an import role for the relaxation of stability conditions. By bringing the membership function information to the analysis, the stability conditions applied only to the FMB control systems with some specified shapes of membership functions. Yet, thanks to the additional information, the resulting membership-function-shape-dependent (MFSD) stability conditions are more relaxed.

This chapter introduces the fundamental concepts and properties of the TS fuzzy model and fuzzy controller. Some developed LMI-based stability and performance conditions, which guarantee the system stability and performance of the MFSI state-feedback FMB control systems, are then presented.

2.2 TS Fuzzy Model

The TS fuzzy model offers a fixed framework to represent nonlinear systems with a number of linguistic rules. The system dynamics of a nonlinear system can be expressed as a weighted average of linear sub-systems. The linear and nonlinear characteristics of the nonlinear plant are extracted and expressed as the linear sub-systems and nonlinear weights, respectively. Consequently, the TS fuzzy model exhibits a favourable semi-linear property that enables some linear analysis and design methods to be applied for carrying out system analysis.

Let p be the number of fuzzy rules describing the nonlinear plant. The i-th rule is of the following format:

$$\begin{aligned} &\text{Rule } i\text{: IF } f_1(\mathbf{x}(t)) \text{ is } M_1^i \text{ AND } \cdots \text{ AND } f_\Psi(\mathbf{x}(t)) \text{ is } M_\Psi^i \\ &\qquad \text{THEN } \dot{\mathbf{x}}(t) = \mathbf{A}_i\mathbf{x}(t) + \mathbf{B}_i\mathbf{u}(t) \end{aligned} \tag{2.1}$$

where M_α^i is a fuzzy set of rule i corresponding to the function $f_\alpha(\mathbf{x}(t))$, $\alpha = 1, 2, \cdots, \Psi$; $i = 1, 2, \cdots, p$; Ψ is a positive integer; $\mathbf{x}(t) \in \Re^n$ is the system state vector; $\mathbf{A}_i \in \Re^{n\times n}$ and $\mathbf{B}_i \in \Re^{n\times m}$ are known system and input matrices, respectively; $\mathbf{u}(t) \in \Re^m$ is the input vector. The system dynamics is described by the following equation:

$$\dot{\mathbf{x}}(t) = \sum_{i=1}^{p} w_i(\mathbf{x}(t))(\mathbf{A}_i\mathbf{x}(t) + \mathbf{B}_i\mathbf{u}(t)) \tag{2.2}$$

where

$$w_i(\mathbf{x}(t)) \geq 0 \; \forall \, i, \sum_{i=1}^{p} w_i(\mathbf{x}(t)) = 1, \tag{2.3}$$

$$w_i(\mathbf{x}(t)) = \frac{\prod_{l=1}^{\Psi} \mu_{M_l^i}(f_l(\mathbf{x}(t)))}{\sum_{k=1}^{p} \prod_{l=1}^{\Psi} \mu_{M_l^k}(f_l(\mathbf{x}(t)))} \; \forall \, i, \tag{2.4}$$

$w_i(\mathbf{x}(t))$, $i = 1, 2, \cdots, p$, are the normalized grades of membership function, $\mu_{M_l^i}(f_l(\mathbf{x}(t)))$, $l = 1, 2, \cdots, \Psi$, are the membership functions corresponding to the fuzzy set M_l^i.

In general, there are two methods to obtain the TS fuzzy model for a nonlinear plant.

1. The TS fuzzy model can be obtained using the system identification algorithms [49, 101, 102] based on given input-output data pairs.
2. The TS fuzzy model can be obtained by directly deriving from the mathematical equations of the nonlinear plants based on the sector nonlinearity technique [109, 122].

Remark 2.1. For the second method, the grades of the membership can be uncertain if they are in terms of uncertain system parameters. As a result, a nonlinear plant subject to parameter uncertainties can be represented by a TS fuzzy model with uncertain grades of membership.

2.3 State-Feedback Fuzzy Controller

In the fuzzy control literature, the most popular fuzzy controller is the state-feedback fuzzy controller, which is called the fuzzy controller hereafter. The fuzzy controller, which shares a similar structure of the TS fuzzy model, is a weighted sum of some linear state-feedback sub-controllers. The control action is described by some linguistic rules.

Let c be the number of fuzzy rules describing the fuzzy controller, the j-th rule is of the following format:

$$\begin{aligned} &\text{Rule } j\text{: IF } g_1(\mathbf{x}(t)) \text{ is } N_1^j \text{ AND } \cdots \text{ AND } g_\Omega(\mathbf{x}(t)) \text{ is } N_\Omega^j \\ &\qquad \text{THEN } \mathbf{u}(t) = \mathbf{G}_j\mathbf{x}(t) \end{aligned} \tag{2.5}$$

where N_β^j is a fuzzy set of rule j corresponding to the function $g_\beta(\mathbf{x}(t))$, $\beta = 1, 2, \cdots, \Omega$; $j = 1, 2, \cdots, c$; Ω is a positive integer; $\mathbf{G}_j \in \Re^{m \times n}$, $j = 1, 2, \cdots, c$, are the constant feedback gains to be determined. The fuzzy controller is defined as follows,

$$\mathbf{u}(t) = \sum_{j=1}^{c} m_j(\mathbf{x}(t))\mathbf{G}_j\mathbf{x}(t) \tag{2.6}$$

where

$$m_j(\mathbf{x}(t)) \geq 0 \; \forall \; j, \sum_{j=1}^{c} m_j(\mathbf{x}(t)) = 1, \tag{2.7}$$

$$m_j(\mathbf{x}(t)) = \frac{\prod_{l=1}^{\Omega} \mu_{N_l^j}(g_l(\mathbf{x}(t)))}{\sum_{k=1}^{c}\prod_{l=1}^{\Omega} \mu_{N_l^k}(g_l(\mathbf{x}(t)))} \; \forall \; j, \tag{2.8}$$

$m_j(\mathbf{x}(t))$, $j = 1, 2, \cdots, c$, are the normalized grades of membership, $\mu_{N_l^j}(g_l(\mathbf{x}(t)))$, $l = 1, 2, \cdots, \Omega$, are the membership functions corresponding to the fuzzy set N_l^j.

2.4 FMB Control Systems

An FMB control system consists of nonlinear plant represented by the TS fuzzy model (2.2) and the fuzzy controller (2.6) connected in a closed loop. Throughout this book, from (2.3) and (2.7), the following property will be used during the system analysis.

$$\sum_{i=1}^{p} w_i(\mathbf{x}(t)) = \sum_{j=1}^{c} m_j(\mathbf{x}(t)) = \sum_{i=1}^{p}\sum_{j=1}^{c} w_i(\mathbf{x}(t))m_j(\mathbf{x}(t)) = 1 \tag{2.9}$$

From (2.2), (2.6) and (2.9), the FMB control system is obtained as follows.

$$\begin{aligned}\dot{\mathbf{x}}(t) &= \sum_{i=1}^{p} w_i(\mathbf{x}(t))\big(\mathbf{A}_i\mathbf{x}(t) + \mathbf{B}_i \sum_{j=1}^{c} m_j(\mathbf{x}(t))\mathbf{G}_j\mathbf{x}(t)\big) \\ &= \sum_{i=1}^{p}\sum_{j=1}^{c} w_i(\mathbf{x}(t))m_j(\mathbf{x}(t))(\mathbf{A}_i + \mathbf{B}_i\mathbf{G}_j)\mathbf{x}(t) \end{aligned} \tag{2.10}$$

2.5 LMI-Based MFSI Stability Conditions

It was reported in [15, 122] that the FMB control system of (2.10) is asymptotically stable if the LMI-based MFSI stability conditions in the following theorems are satisfied.

Theorem 2.1. *[15, 122]: The FMB control system (2.10), formed by the nonlinear plant represented by the fuzzy model (2.2) and the fuzzy controller*

(2.6) connected in a closed loop, is asymptotically stable if there exist matrices $\mathbf{N}_j \in \Re^{m \times n}$, $j = 1, 2, \cdots, c$ *and* $\mathbf{X} = \mathbf{X}^T \in \Re^{n \times n}$ *such that the following LMIs hold:*

$$\mathbf{X} > 0;$$

$$\mathbf{X}\mathbf{A}_i^T + \mathbf{A}_i\mathbf{X} + \mathbf{N}_j^T\mathbf{B}_i^T + \mathbf{B}_i\mathbf{N}_j < 0 \,\forall\, i,\, j;$$

and the feedback gains are designed as $\mathbf{G}_j = \mathbf{N}_j\mathbf{X}^{-1}$ *for all* j.

Remark 2.2. Theorem 2.1 offers a great deal of design flexibility to the fuzzy controller as the membership functions can be freely designed. This favourable property leads to a simple controller structure which may lower the computational demand and implementation cost. Furthermore, the fuzzy controller designed based on Theorem 2.1 exhibits an inherent robustness property. As the stability conditions are not related to the membership functions of both the TS fuzzy model and the fuzzy controller, the fuzzy controller can stabilize the nonlinear plant subject to parameter uncertainties which are embedded in the membership functions of the TS fuzzy model. However, one drawback of Theorem 2.1 is that it often leads to conservative stability result. Hence, it is suggested to design the fuzzy controller using Theorem 2.1 as a first trial to take advantage of the design flexibility and robustness property. If the design fails, other relaxed stability conditions can then be applied.

In the following, the stability conditions based on the design criterion that the TS fuzzy model and fuzzy controller share the same premise membership functions, i.e., $c = p$ and $m_i(\mathbf{x}(t)) = w_i(\mathbf{x}(t))$, $i = 1, 2, \cdots, p$, are presented. This design criterion, generally known as the PDC design, leads to relaxed stability conditions as the membership functions are taken into account in the stability analysis. However, the sharing of premise membership functions will lead to a complicated controller structure when the membership functions of the TS fuzzy model is complicated.

Remark 2.3. Throughout this book, the design criterion that the TS fuzzy model and fuzzy controller share the same premises, i.e. $c = p$ and $m_i(\mathbf{x}(t)) = w_i(\mathbf{x}(t))$, $i = 1, 2, \cdots, p$, is referred to as the PDC design. Otherwise, it is referred to as the non-PDC design.

In [24, 48, 81, 82, 97, 105, 112, 122, 125], the stability analysis based on the PDC design approach was proposed. Different levels of relaxed stability conditions were derived based on Polya's theorem. Some published stability analysis results are given below.

Theorem 2.2. *[105, 122]: The FMB control system (2.10), formed by the nonlinear plant represented by the fuzzy model (2.2) and the fuzzy controller (2.6) under the PDC design technique, i.e., with* $c = p$ *and* $m_i(\mathbf{x}(t)) = w_i(\mathbf{x}(t))$ *for all* i*, connected in a closed loop, is asymptotically stable if there exist matrices* $\mathbf{N}_j \in \Re^{m \times n}$, $j = 1, 2, \cdots, p$, *and* $\mathbf{X} = \mathbf{X}^T \in \Re^{n \times n}$ *such that the following LMIs hold:*

$$\mathbf{X} > 0;$$

$$\mathbf{X}\mathbf{A}_i^T + \mathbf{A}_i\mathbf{X} + \mathbf{N}_i^T\mathbf{B}_i^T + \mathbf{B}_i\mathbf{N}_i < 0 \;\forall\, i;$$

$$\mathbf{X}\mathbf{A}_i^T + \mathbf{A}_i\mathbf{X} + \mathbf{N}_j^T\mathbf{B}_i^T + \mathbf{B}_i\mathbf{N}_j + \mathbf{X}\mathbf{A}_j^T + \mathbf{A}_j\mathbf{X} + \mathbf{N}_i^T\mathbf{B}_j^T + \mathbf{B}_j\mathbf{N}_i \le 0 \;\forall\, j, i < j;$$

and the feedback gains are designed as $\mathbf{G}_j = \mathbf{N}_j\mathbf{X}^{-1}$ *for all* j.

Theorem 2.3. *[82]: The FMB control system (2.10), formed by the nonlinear plant represented by the fuzzy model (2.2) and the fuzzy controller (2.6) under the PDC design technique, i.e., with* $c = p$ *and* $m_i(\mathbf{x}(t)) = w_i(\mathbf{x}(t))$ *for all* i *connected in a closed loop, is asymptotically stable if there exist matrices* $\mathbf{N}_j \in \Re^{m\times n}$, $\mathbf{X} = \mathbf{X}^T \in \Re^{n\times n}$ *and* $\mathbf{X}_{ij} = \mathbf{X}_{ji}^T \in \Re^{n\times n}$, $i, j = 1, 2, \cdots, p$ *such that the following LMIs hold:*

$$\mathbf{X} > 0;$$

$$\mathbf{X}\mathbf{A}_i^T + \mathbf{A}_i\mathbf{X} + \mathbf{N}_i^T\mathbf{B}_i^T + \mathbf{B}_i\mathbf{N}_i < \mathbf{X}_{ii} \;\forall\, i;$$

$$\begin{aligned}&\mathbf{X}\mathbf{A}_i^T + \mathbf{A}_i\mathbf{X} + \mathbf{N}_j^T\mathbf{B}_i^T + \mathbf{B}_i\mathbf{N}_j + \mathbf{X}\mathbf{A}_j^T + \mathbf{A}_j\mathbf{X}\\ &+ \mathbf{N}_i^T\mathbf{B}_j^T + \mathbf{B}_j\mathbf{N}_i \le \mathbf{X}_{ij} + \mathbf{X}_{ij}^T \;\forall\, j, i < j;\end{aligned}$$

$$\tilde{\mathbf{X}} = \begin{bmatrix} \mathbf{X}_{11} & \mathbf{X}_{12} & \cdots & \mathbf{X}_{1p} \\ \mathbf{X}_{21} & \mathbf{X}_{22} & \cdots & \mathbf{X}_{2p} \\ \vdots & \vdots & \vdots & \vdots \\ \mathbf{X}_{p1} & \mathbf{X}_{p2} & \cdots & \mathbf{X}_{pp} \end{bmatrix} < 0;$$

and the feedback gains are designed as $\mathbf{G}_j = \mathbf{N}_j\mathbf{X}^{-1}$ *for all* j.

Theorem 2.4. *[24]: The FMB control system (2.10), formed by the nonlinear plant represented by the fuzzy model (2.2) and the fuzzy controller (2.6) under the PDC design technique, i.e., with* $c = p$ *and* $m_i(\mathbf{x}(t)) = w_i(\mathbf{x}(t))$ *for all* i *connected in a closed loop, is asymptotically stable if there exist matrices* $\mathbf{N}_j \in \Re^{m\times n}$, $j = 1, 2, \cdots, p$, $\mathbf{X} = \mathbf{X}^T \in \Re^{n\times n}$, $\mathbf{Y}_{iii} = \mathbf{Y}_{iii}^T \in \Re^{n\times n}$, $i = 1, 2, \cdots, p$, $\mathbf{Y}_{iij} = \mathbf{Y}_{jii}^T \in \Re^{n\times n}$, $\mathbf{Y}_{iji} = \mathbf{Y}_{iji}^T \in \Re^{n\times n}$, $i, j = 1, 2, \cdots, p$; $i \neq j$, $\mathbf{Y}_{ijk} = \mathbf{Y}_{kji}^T \in \Re^{n\times n}$, $\mathbf{Y}_{ikj} = \mathbf{Y}_{jki}^T \in \Re^{n\times n}$ *and* $\mathbf{Y}_{jik} = \mathbf{Y}_{kij}^T \in \Re^{n\times n}$, $i = 1, 2, \cdots, p\text{-}2$; $j = i{+}1, i{+}2, \cdots, p\text{-}1$; $k = j{+}1, j{+}2, \cdots, p$ *such that the following LMIs hold:*

$$\mathbf{X} > 0;$$

$$\mathbf{X}\mathbf{A}_i^T + \mathbf{A}_i\mathbf{X} + \mathbf{N}_i^T\mathbf{B}_i^T + \mathbf{B}_i\mathbf{N}_i < \mathbf{Y}_{iii} \;\forall\, i;$$

$$\begin{aligned}&2\mathbf{X}\mathbf{A}_i^T + \mathbf{X}\mathbf{A}_j^T + 2\mathbf{A}_i\mathbf{X} + \mathbf{A}_j\mathbf{X} + (\mathbf{N}_i + \mathbf{N}_j)^T\mathbf{B}_i^T + \mathbf{N}_i^T\mathbf{B}_j^T\\ &+ \mathbf{B}_i(\mathbf{N}_i + \mathbf{N}_j) + \mathbf{B}_j\mathbf{N}_i \le \mathbf{Y}_{iij} + \mathbf{Y}_{iji} + \mathbf{Y}_{iij}^T \;\forall\, i,\ j;\ j \neq i;\end{aligned}$$

$$2\mathbf{X}(\mathbf{A}_i + \mathbf{A}_j + \mathbf{A}_k)^T + (\mathbf{N}_j + \mathbf{N}_k)^T \mathbf{B}_i^T + (\mathbf{N}_i + \mathbf{N}_k)^T \mathbf{B}_j^T + (\mathbf{N}_i + \mathbf{N}_j)^T \mathbf{B}_k^T$$
$$+ 2(\mathbf{A}_i + \mathbf{A}_j + \mathbf{A}_k)\mathbf{X} + \mathbf{B}_i(\mathbf{N}_j + \mathbf{N}_k) + \mathbf{B}_j(\mathbf{N}_i + \mathbf{N}_k) + \mathbf{B}_k(\mathbf{N}_i + \mathbf{N}_j)$$
$$\leq \mathbf{Y}_{ijk} + \mathbf{Y}_{ikj} + \mathbf{Y}_{jik} + \mathbf{Y}_{ijk}^T + \mathbf{Y}_{ikj}^T + \mathbf{Y}_{jik}^T, i = 1, 2, \cdots, p-2;$$
$$j = i+1, 2, \cdots, p-1; i, k = j+1, 2, \cdots, p;$$

$$\tilde{\mathbf{Y}}_i = \begin{bmatrix} \mathbf{Y}_{1i1} & \mathbf{Y}_{1i2} & \cdots & \mathbf{Y}_{1ip} \\ \mathbf{Y}_{2i1} & \mathbf{Y}_{2i2} & \cdots & \mathbf{Y}_{2ip} \\ \vdots & \vdots & \vdots & \vdots \\ \mathbf{Y}_{pi1} & \mathbf{Y}_{pi2} & \cdots & \mathbf{Y}_{pip} \end{bmatrix} < 0 \; \forall \; i;$$

where the feedback gains are designed as $\mathbf{G}_j = \mathbf{N}_j \mathbf{X}^{-1}$ *for all* j.

Remark 2.4. It was shown in [24] that the stability conditions in Theorem 2.2 and Theorem 2.3 are particular cases of those in Theorem 2.4. The analysis approach in [24] can be generalized by the Polya's theorem [97]. It was reported that the stability conditions in [97] covers all stability conditions in [24, 48, 81, 82, 105, 112, 122, 125].

Denote $I_q = \{\mathbf{i} = (i_1, i_2, \cdots, i_q) \in N^q | 1 \leq i_j \leq p \; \forall \; j = 1, 2, \cdots, q\}$, $I_q^+ = \{\mathbf{i} \in I_q | i_k \leq i_{k+1}, k = 1, 2, \cdots, q-1\}$ as a subset of I_q, $\sum_{\mathbf{i} \in I_q} \mathbf{w_i}(\mathbf{x}(t)) = \sum_{i_1=1}^p \sum_{i_2=1}^p \cdots \sum_{i_q=1}^p w_{i_1}(\mathbf{x}(t)) w_{i_2}(\mathbf{x}(t)) \cdots w_{i_q}(\mathbf{x}(t))$, and the set of permutations as $P(\mathbf{i}) \subset I_q$ where $\mathbf{i} \in I_q$. The stability conditions for the FMB control system (2.10) with two-dimensional fuzzy summations based on Polya's theorem are given in the theorem below.

Theorem 2.5. *[97]: The FMB control system (2.10), formed by the nonlinear plant represented by the fuzzy model (2.2) and the fuzzy controller (2.6) under the PDC design technique, i.e., with* $c = p$ *and* $m_i(\mathbf{x}(t)) = w_i(\mathbf{x}(t))$ *for all* i *connected in a closed loop, is asymptotically stable if the following LMIs given in the following* h *steps are satisfied, where* $h = 0, 1, 2, \cdots$, $h_{max} = floor\left(\frac{d-1}{2}\right)$ *and* $d \geq 2$. *The dimension of the multi-indices in the iteration step* h *is denoted by* $d_h = d - 2h$.

1. *(Initialization) Choose the degree of the fuzzy summation as* $d \geq 2$ *and set* $\mathbf{Q_i}^{[0]} = \mathbf{Q}_{i_1 i_2 \cdots i_d}$ *for* $\mathbf{i} \in I_d$ *and* $h = 0$. *Define matrices* $\mathbf{Q_j}^{[0]} = \mathbf{Q_j}^{[0]^T} \in \Re^{n \times n}$ *for* $\mathbf{j} \in P(\mathbf{i})$, $\mathbf{i} \in I_{d_o}^+$.
2. *(Recursive procedure) In the iterative step* h, *when* $h < h_{max}$, *the following inequality is included as the LMI condition:*

$$\sum_{\mathbf{j} \in P(\mathbf{i})} \mathbf{Q_j}^{[h]} < \frac{1}{2} \sum_{\mathbf{j} \in P(\mathbf{i})} (\mathbf{X_j}^{[h]} + \mathbf{X_j}^{[h]^T}) \; \forall \; \mathbf{i} \in I_{d_h}^+,$$

and set

$$\mathbf{Q_k}^{[h+1]} = \begin{bmatrix} \mathbf{X}^{[h]}_{(\mathbf{k},1,1)} & \mathbf{X}^{[h]}_{(\mathbf{k},1,2)} & \cdots & \mathbf{X}^{[h]}_{(\mathbf{k},1,p)} \\ \mathbf{X}^{[h]}_{(\mathbf{k},2,1)} & \mathbf{X}^{[h]}_{(\mathbf{k},2,2)} & \cdots & \mathbf{X}^{[h]}_{(\mathbf{k},2,p)} \\ \vdots & \vdots & \vdots & \vdots \\ \mathbf{X}^{[h]}_{(\mathbf{k},p,1)} & \mathbf{X}^{[h]}_{(\mathbf{k},p,2)} & \cdots & \mathbf{X}^{[h]}_{(\mathbf{k},p,p)} \end{bmatrix} \quad \forall\, \mathbf{k} \in I^{+}_{d_h-2}$$

where $\mathbf{X}^{[h]}_{(\mathbf{k},i_{d-1},i_d)} = \mathbf{X}^{[h]}_{(\mathbf{k},i_d,i_{d-1})}{}^{T} \in \Re^{n\times n}$ *when* $i_{d-1} = i_d \;\forall\, \mathbf{k} \in I^{+}_{d_h-2}$, $i_{d-1} = 1, 2, \cdots, p$. *It should be noted that* $d_{h+1} = d_h - 2$.

3. *Set* $h = h + 1$. *If* $h < h_{max}$, *go to step 2, otherwise, go to next step.*
4. *(Termination) When* $d_{h_{max}} = 1$, *the stability conditions in Theorem 2.2 are included as the LMI conditions in this theorem. When* $d_{h_{max}} = 2$, *the stability conditions in Theorem 2.3 are included as the LMI conditions in this theorem.*

Remark 2.5. The stability conditions in Theorem 2.3 and Theorem 2.4 are special cases of Theorem 2.5 with $d = 2$ and $d = 3$, respectively.

Example 2.1. In this simulation example, a 3-rule fuzzy model in the form of (2.2) is considered and a 3-rule fuzzy controller in the form of (2.6) is employed to close the feedback loop. The membership functions of the TS fuzzy model and the fuzzy controller are assumed to be the same and can take any shapes satisfying the membership function properties (2.3) and (2.7).

The 3-rule TS fuzzy model is chosen as the one in [24] with $\mathbf{A}_1 = \begin{bmatrix} 1.59 & -7.29 \\ 0.01 & 0 \end{bmatrix}$, $\mathbf{A}_2 = \begin{bmatrix} 0.02 & -4.64 \\ 0.35 & 0.21 \end{bmatrix}$, $\mathbf{A}_3 = \begin{bmatrix} -a & -4.33 \\ 0 & -0.05 \end{bmatrix}$, $\mathbf{B}_1 = \begin{bmatrix} 1 \\ 0 \end{bmatrix}$, $\mathbf{B}_2 = \begin{bmatrix} 8 \\ 0 \end{bmatrix}$, $\mathbf{B}_3 = \begin{bmatrix} -b+6 \\ -1 \end{bmatrix}$ where a and b are constant parameters.

The stability conditions in Theorem 2.1 to Theorem 2.5 ($d = 4$ for Theorem 2.5) are employed to check for the stability region characterized by the system parameters $2 \le a \le 12$ and $2 \le b \le 12$ at the interval of 1. With the help of the MATLAB LMI toolbox, the stability regions given by different theorems are shown in Fig. 2.1 indicated by different symbols. An empty point means that no feasible solution is found. It should be noted that there is no feasible solution found based on the stability conditions in Theorem 2.1 and Theorem 2.2. It can also be seen from Fig. 2.1 that the stability conditions in [97] offer the most relaxed result demonstrated by the largest size of the stability region.

It is shown in this example that the more slack matrices introduced to the stability conditions, the more relaxed result indicated by a larger stability region can be obtained. However, the introduction of more slack matrices to the stability conditions will increase the computational demand on searching for the solution to the stability conditions.

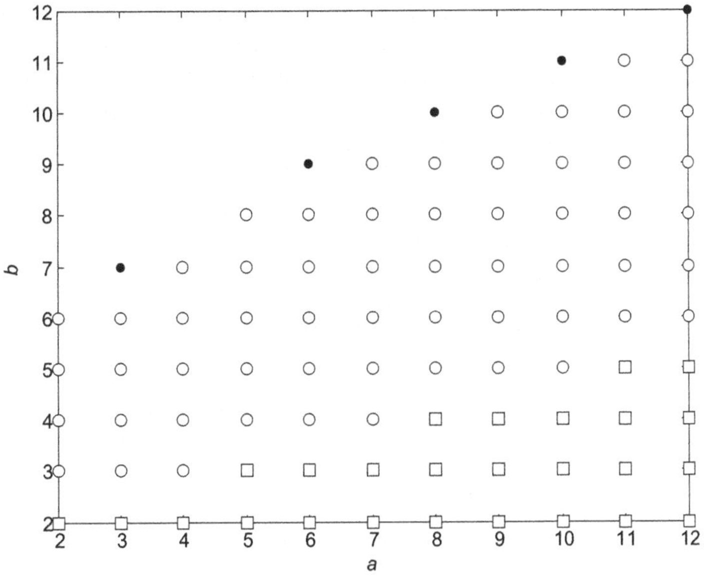

Fig. 2.1 Stability regions given by Theorem 2.3 ('□'), Theorem 2.4 ('□' and '○') and Theorem 2.5 ('□', '○' and '•').

2.6 LMI-Based Performance Conditions

On tackling control problems, apart from the system stability, the system performance is an essential issue to be considered. The stability conditions introduced in Theorem 2.1 to Theorem 2.4 govern the system stability only but tell nothing about the system performance such as the transient responses and control requirements. In this section, some LMI-based performance conditions are introduced, which have to be applied simultaneously with the stability conditions to realize the final design of the fuzzy controller.

Two LMI-based conditions are introduced in the following to constrain the system states, $\mathbf{x}(t)$ and control input, $\mathbf{u}(t)$. The system state constraint is to suppress the energy of the system states such that it satisfies $\mathbf{x}(t)\mathbf{X}^{-1}\mathbf{x}(t) \leq \mu_x^2$ where μ_x is a non-zero scalar. By satisfying this constraint, the energy the system states will never be beyond the level of μ_x. Similarly, the constraint of $\mathbf{u}(t)^T\mathbf{u}(t) \leq \mu_u^2$, where μ_u is a non-zero scalar, is considered to limit the control energy. By satisfaction of this constraint, the energy of the control input will never be beyond the level of μ_u. Imposing these constraints on the stability conditions will limit the feasible domain of the solution such that the feasible solution will satisfy both the stability and performance conditions.

2.6.1 System State Constraint

Theorem 2.6. *[110]: The system state constraint* $\mathbf{x}(t)\mathbf{X}^{-1}\mathbf{x}(t) \leq \mu_x^2$*, where* μ_x *is a pre-defined non-zero positive scalar, is achieved at all time* $t \geq 0$ *if there exists a matrix* $\mathbf{X} = \mathbf{X}^T \in \Re^{n\times n}$*, such that the following LMIs are satisfied:*

$$\mathbf{X} > 0;$$

$$\begin{bmatrix} \mu_x^2 & \mathbf{0}^T \\ \mathbf{0} & \mathbf{X} \end{bmatrix} \geq 0;$$

2.6.2 Control Input Constraint

Theorem 2.7. *[110]: The control input constraint* $\mathbf{u}(t)^T\mathbf{u}(t) \leq \mu_u^2$*, where* μ_u *is a predefined non-zero positive scalar, is achieved at all time* $t \geq 0$ *if there exists matrices* $\mathbf{X} = \mathbf{X}^T \in \Re^{n\times n}$ *and* $\mathbf{N}_j \in \Re^{m\times n}$*,* $j = 1, 2, \ldots, c$*, such that the following LMIs are satisfied:*

$$\mathbf{X} > 0;$$

$$\begin{bmatrix} \mathbf{I} & \mathbf{0}^T \\ \mathbf{0} & \mathbf{X} \end{bmatrix} \geq 0;$$

$$\begin{bmatrix} \mathbf{X} & \mathbf{N}_j^T \\ \mathbf{N}_j & \mu_u^2\mathbf{I} \end{bmatrix} \geq 0;$$

Remark 2.6. The LMI-based system state and control input constraints can be applied simultaneously.

2.7 Conclusion

The fundamental concepts and properties of the TS fuzzy models and fuzzy controllers have been presented. An FMB control system which is formed by a nonlinear plant represented by the TS fuzzy model and a fuzzy controller connected in a closed loop has been introduced. Some published LMI-based stability and performance conditions have been introduced that help design a stable and well-performed MFSI state-feedback FMB control system.

Chapter 3
Stability Analysis of FMB Control Systems under MFSD Approach

3.1 Introduction

Stability analysis on FMB control systems has been investigated extensively in the past decades. Flourish stability analysis results have been obtained to facilitate the design of stable FMB control systems. Some published LMI-based MFSI stability conditions are presented in Chapter 2 under the PDC design technique.

The PDC design technique leads to an FMB control system with perfectly matched premise membership functions of which both the TS fuzzy model and the fuzzy controller share the same premise membership functions. The perfectly matched premise membership functions are able to produce less conservative stability analysis result by grouping the cross terms of the membership functions. However, in general, there are two main drawbacks coming with the PDC design technique. First, as the fuzzy controller shares the premise membership functions of the fuzzy model, the design flexibility of the controller membership functions vanishes. Furthermore, if the membership functions of the TS fuzzy model are complex, the implementation cost of the fuzzy controller will increase. Second, the membership functions of the TS fuzzy model must be uncertainty free. Thus, the inherent robustness property of the fuzzy controller vanishes. This drawback makes the non-PDC design technique (of which the TS fuzzy model and fuzzy controller do not share the same premise membership functions) attractive. The non-PDC design technique can enhance the design flexibility and robustness property of the fuzzy controller. However, under the non-PDC design technique, the imperfectly matched membership functions lead to conservative stability analysis results. Yet, it is possible to widen the applicability of the FMB control approach by integrating the advantages of the PDC and non-PDC design techniques.

In this chapter, the membership-function-shape-dependent (MFSD) analysis approach is employed to facilitate the stability analysis and controller synthesis. The shape information of the membership functions is taken into

H.-K. Lam and F.H.F. Leung: FMB Control Systems, STUDFUZZ 264, pp. 23–58.
springerlink.com

consideration in the stability analysis for relaxing the stability conditions. Under the MFSD analysis, two approaches, namely membership-function-boundary (MFB) and staircase-membership-function (SMF) approaches, are considered. The MFB approach considers the lower and upper bounds of the membership functions in the stability analysis. Through the S-procedure [6], some slack matrices are introduced to the stability conditions. In order to bring more information to the stability analysis, the SMF approach approximates the membership functions with the staircase membership functions. Under this approach, the SMFs are considered in the stability anslysis and brought to the stability conditions. Consequently, as the SMFs carries much more information of the nonlinearities of the plant as compared with the MFB approach, the SMF-based stability conditions are more dedicated to the nonlinear plant, which considers the specified SMFs rather than any shapes. The MFSD relaxed stability conditions offer an effective tool to achieve a stable FMB control systems with imperfectly matched membership functions.

3.2 Membership-Function-Boundary Approach

Consider the FMB control system of (2.10), which is formed by a nonlinear plant represented by a TS fuzzy model (2.2) and a fuzzy controller of (2.6). In this section, the system stability of the FMB control systems subject to imperfectly matched membership functions is investigated using the Lyapunov stability theory. MFB information is considered for the relaxation of stability conditions in terms of LMIs. LMI-based performance conditions are also derived to govern the system performance.

3.2.1 Stability Analysis

Consider the following quadratic Lyapunov function for investigating the system stability of the FMB control system of (2.10).

$$V(t) = \mathbf{x}(t)^T \mathbf{P} \mathbf{x}(t) \tag{3.1}$$

where $0 < \mathbf{P} = \mathbf{P}^T \in \Re^{n \times n}$.

In the following, for brevity, $w_i(\mathbf{x}(t))$ and $m_j(\mathbf{x}(t))$ are denoted as w_i and m_j, respectively. The property of the membership functions (2.9) is utilized in the following stability analysis. From (2.10) and (3.1), denoting $\mathbf{X} = \mathbf{P}^{-1}$ and $\mathbf{z}(t) = \mathbf{X}^{-1}\mathbf{x}(t)$ and defining the feedback gains of the fuzzy controller as $\mathbf{G}_j = \mathbf{N}_j \mathbf{X}^{-1}$, where $\mathbf{N}_j \in \Re^{m \times n}$, $j = 1, 2, \cdots, c$, we have,

$$
\begin{aligned}
\dot{V}(t) &= \dot{\mathbf{x}}(t)^T \mathbf{P}\mathbf{x}(t) + \mathbf{x}(t)^T \mathbf{P}\dot{\mathbf{x}}(t) \\
&= \sum_{i=1}^{p}\sum_{j=1}^{c} w_i m_j \mathbf{x}(t)^T \big((\mathbf{A}_i + \mathbf{B}_i\mathbf{G}_j)^T\mathbf{P} + \mathbf{P}(\mathbf{A}_i + \mathbf{B}_i\mathbf{G}_j)\big)\mathbf{x}(t) \\
&= \sum_{i=1}^{p}\sum_{j=1}^{c} w_i m_j \mathbf{z}(t)^T \big(\mathbf{X}(\mathbf{A}_i + \mathbf{B}_i\mathbf{G}_j)^T + (\mathbf{A}_i + \mathbf{B}_i\mathbf{G}_j)\mathbf{X}\big)\mathbf{z}(t) \\
&= \sum_{i=1}^{p}\sum_{j=1}^{c} w_i m_j \mathbf{z}(t)^T \mathbf{Q}_{ij}\mathbf{z}(t)
\end{aligned} \tag{3.2}
$$

where $\mathbf{Q}_{ij} = \mathbf{A}_i\mathbf{X} + \mathbf{X}\mathbf{A}_i^T + \mathbf{B}_i\mathbf{N}_j + \mathbf{N}_j^T\mathbf{B}_i^T$.

Remark 3.1. Based on the Lyapunov stability theory [100], it can be concluded that the FMB control system (2.10) is asymptotically stable if $V(t) > 0$ and $\dot{V}(t) < 0$ (excluding $\mathbf{x}(t) = \mathbf{0}$) can be achieved. Throughout the stability analysis in this book, the primary objective is to obtain some conditions (stability conditions) such that $V(t) > 0$ and $\dot{V}(t) < 0$ are achieved.

Remark 3.2. Under the MFSI stability analysis approach, it was reported in [15, 122] that the FMB control system (2.10) is guaranteed to be asymptotically stable if the stability conditions in Theorem 2.1 are satisfied. Referring to Theorem 2.1, on satisfying the stability conditions, $V(t) > 0$ and $\dot{V}(t) < 0$ are achieved. As it is only required in Theorem 2.1 that $\mathbf{Q}_{ij} < 0$, the membership functions are not considered in the stability analysis and/or stability conditions. It is anticipated that the stability conditions in Theorem 2.1 are very conservative, motivating the investigation of the MFSD stability analysis approach considered in this chapter.

Under the proposed MFB approach in this section, it is required the fuzzy model and fuzzy controller share the same number of fuzzy rules, i.e. $c = p$. To relax the stability conditions, some slack matrix variables are introduced by considering the following equations given by the property of the membership functions.

$$
\begin{aligned}
\sum_{i=1}^{p}\sum_{j=1}^{p} w_i(w_j - m_j)\mathbf{\Lambda}_i &= \sum_{i=1}^{p} w_i\Big(\sum_{j=1}^{p} w_j - \sum_{j=1}^{p} m_j\Big)\mathbf{\Lambda}_i \\
&= \sum_{i=1}^{p} w_i(1-1)\mathbf{\Lambda}_i \\
&= \mathbf{0},
\end{aligned} \tag{3.3}
$$

$$
\sum_{i=1}^{p}\sum_{j=1}^{p} w_i \rho_j w_j (\mathbf{V}_{ij} - \mathbf{V}_{ij}) = \mathbf{0} \tag{3.4}
$$

where $0 < \mathbf{\Lambda}_i = \mathbf{\Lambda}_i^T \in \Re^{n \times n}$, $\mathbf{V}_{ij} = \mathbf{V}_{ij}^T \in \Re^{n \times n}$, $i, j = 1, 2, \cdots, p$, are arbitrary matrices. From (3.2), (3.3) and (3.4), we have,

$$\begin{aligned}
\dot{V}(t) &= \sum_{i=1}^{p}\sum_{j=1}^{p} w_i m_j \mathbf{z}(t)^T \mathbf{Q}_{ij} \mathbf{z}(t) + \sum_{i=1}^{p}\sum_{j=1}^{p} w_i (w_j - m_j) \mathbf{z}(t)^T \mathbf{\Lambda}_i \mathbf{z}(t) \\
&= \sum_{i=1}^{p}\sum_{j=1}^{p} w_i m_j \mathbf{z}(t)^T \mathbf{Q}_{ij} \mathbf{z}(t) \\
&\quad + \sum_{i=1}^{p}\sum_{j=1}^{p} w_i (w_j - m_j + \rho_j w_j - \rho_j w_j) \mathbf{z}(t)^T \mathbf{\Lambda}_i \mathbf{z}(t) \\
&= \sum_{i=1}^{p}\sum_{j=1}^{p} w_i m_j \mathbf{z}(t)^T \mathbf{Q}_{ij} \mathbf{z}(t) + \sum_{i=1}^{p}\sum_{j=1}^{p} w_i (w_j - \rho_j w_j) \mathbf{z}(t)^T \mathbf{\Lambda}_i \mathbf{z}(t) \\
&\quad - \sum_{i=1}^{p}\sum_{j=1}^{p} w_i (m_j - \rho_j w_j) \mathbf{z}(t)^T \mathbf{\Lambda}_i \mathbf{z}(t) \\
&\quad + \sum_{i=1}^{p}\sum_{j=1}^{c} w_i \rho_j w_j \mathbf{z}(t)^T (\mathbf{V}_{ij} - \mathbf{V}_{ij}) \mathbf{z}(t)
\end{aligned} \tag{3.5}$$

where the scalars $\rho_j > 0$ and δ_j, $j = 1, 2, \cdots, p$, are designed such that $m_j - \rho_j w_j + \delta_j \geq 0$ for all j and $\mathbf{x}(t)$ (w_j and m_j are function of $\mathbf{x}(t)$) [67]. These additional matrices and conditions are introduced to further reduce the conservativeness of the stability analysis.

From (3.5), we have,

$$
\begin{aligned}
\dot{V}(t) &= \sum_{i=1}^{p}\sum_{j=1}^{p}\mathbf{z}(t)^T w_i(m_j + \rho_j w_j - \rho_j w_j)\mathbf{z}(t)\mathbf{Q}_{ij}\mathbf{z}(t) \\
&\quad + \sum_{i=1}^{p}\sum_{j=1}^{p} w_i(w_j - \rho_j w_j)\mathbf{z}(t)^T\mathbf{\Lambda}_i\mathbf{z}(t) \\
&\quad - \sum_{i=1}^{p}\sum_{j=1}^{p} w_i(m_j - \rho_j w_j)\mathbf{z}(t)^T\mathbf{\Lambda}_i\mathbf{z}(t) \\
&\quad + \sum_{i=1}^{p}\sum_{j=1}^{p} w_i\rho_j w_j\mathbf{z}(t)^T\mathbf{V}_{ij}\mathbf{z}(t) - \sum_{i=1}^{p}\sum_{j=1}^{p} w_i\rho_j w_j\mathbf{z}(t)^T\mathbf{V}_{ij}\mathbf{z}(t) \\
&= \sum_{i=1}^{p}\sum_{j=1}^{p} w_i w_j\mathbf{z}(t)^T\rho_j(\mathbf{Q}_{ij} - \mathbf{\Lambda}_i - \mathbf{V}_{ij})\mathbf{z}(t) \\
&\quad + \sum_{i=1}^{p}\sum_{j=1}^{p} w_i(m_j - \rho_j w_j)\mathbf{z}(t)^T(\mathbf{Q}_{ij} - \mathbf{\Lambda}_i)\mathbf{z}(t) \\
&\quad + \sum_{i=1}^{p}\sum_{j=1}^{p} w_i w_j\mathbf{z}(t)^T(\mathbf{\Lambda}_i + \rho_j\mathbf{V}_{ij})\mathbf{z}(t) \\
&\quad + \sum_{i=1}^{p}\sum_{j=1}^{p} w_i(\delta_j - \delta_j)\mathbf{z}(t)^T(\mathbf{Q}_{ij} - \mathbf{\Lambda}_i)\mathbf{z}(t) \\
&= \sum_{i=1}^{p}\sum_{j=1}^{p} w_i w_j\mathbf{z}(t)^T\big(\rho_j(\mathbf{Q}_{ij} - \mathbf{\Lambda}_i - \mathbf{V}_{ij}) - \sum_{k=1}^{p}\delta_k(\mathbf{Q}_{ik} - \mathbf{\Lambda}_i)\big)\mathbf{z}(t) \\
&\quad + \sum_{i=1}^{p}\sum_{j=1}^{p} w_i(m_j - \rho_j w_j + \delta_j)\mathbf{z}(t)^T(\mathbf{Q}_{ij} - \mathbf{\Lambda}_i)\mathbf{z}(t) \\
&\quad + \sum_{i=1}^{p}\sum_{j=1}^{p} w_i w_j\mathbf{z}(t)^T(\mathbf{\Lambda}_i + \rho_j\mathbf{V}_{ij})\mathbf{z}(t).
\end{aligned} \tag{3.6}
$$

Consider the following LMIs and $m_j - \rho_j w_j + \delta_j \geq 0$ for all j and $\mathbf{x}(t)$,

$$
\mathbf{Q}_{ij} - \mathbf{\Lambda}_i < 0 \ \forall\, i, j \tag{3.7}
$$

From (3.6) and (3.7), we have,

$$\begin{aligned}
\dot{V}(t) \leq & \sum_{i=1}^{p}\sum_{j=1}^{p} w_i w_j \mathbf{z}(t)^T \Big(\rho_j(\mathbf{Q}_{ij} - \mathbf{\Lambda}_i - \mathbf{V}_{ij}) - \sum_{k=1}^{p} \delta_k(\mathbf{Q}_{ik} - \mathbf{\Lambda}_i)\Big)\mathbf{z}(t) \\
& + \sum_{i=1}^{p}\sum_{j=1}^{p} w_i w_j \mathbf{z}(t)^T (\mathbf{\Lambda}_i + \rho_j \mathbf{V}_{ij})\mathbf{z}(t) \\
= & \sum_{i=1}^{p} w_i^2 \mathbf{z}(t)^T \Big(\rho_i(\mathbf{Q}_{ii} - \mathbf{\Lambda}_i - \mathbf{V}_{ii}) - \sum_{k=1}^{p} \delta_k(\mathbf{Q}_{ik} - \mathbf{\Lambda}_i)\Big)\mathbf{z}(t) \\
& + \sum_{j=1}^{p}\sum_{i<j} w_i w_j \mathbf{z}(t)^T \Big(\rho_j(\mathbf{Q}_{ij} - \mathbf{\Lambda}_i - \mathbf{V}_{ij}) - \sum_{k=1}^{p} \delta_k(\mathbf{Q}_{ik} - \mathbf{\Lambda}_i) \\
& + \rho_i(\mathbf{Q}_{ji} - \mathbf{\Lambda}_j - \mathbf{V}_{ji}) - \sum_{k=1}^{p} \delta_k(\mathbf{Q}_{jk} - \mathbf{\Lambda}_j)\Big)\mathbf{z}(t) \\
& + \sum_{i=1}^{p} w_i^2 \mathbf{z}(t)^T (\mathbf{\Lambda}_i + \rho_i \mathbf{V}_{ii})\mathbf{z}(t) \\
& + \sum_{j=1}^{p}\sum_{i<j} w_i w_j \mathbf{z}(t)^T (\mathbf{\Lambda}_i + \rho_j \mathbf{V}_{ij} + \mathbf{\Lambda}_j + \rho_i \mathbf{V}_{ji})\mathbf{z}(t). \qquad (3.8)
\end{aligned}$$

Introducing matrices $\mathbf{R}_{ij} = \mathbf{R}_{ji}^T \in \Re^{n \times n}$ and $\mathbf{S}_{ij} = \mathbf{S}_{ji}^T \in \Re^{n \times n}$, $i, j = 1, 2, \cdots, p$, we consider

$$\mathbf{R}_{ii} > \rho_i(\mathbf{Q}_{ii} - \mathbf{\Lambda}_i - \mathbf{V}_{ii}) - \sum_{k=1}^{p} \delta_k(\mathbf{Q}_{ik} - \mathbf{\Lambda}_i) \;\forall\, i; \qquad (3.9)$$

$$\begin{aligned}
\mathbf{R}_{ij} + \mathbf{R}_{ij}^T \geq & \rho_j(\mathbf{Q}_{ij} - \mathbf{\Lambda}_i - \mathbf{V}_{ij}) - \sum_{k=1}^{p} \delta_k(\mathbf{Q}_{ik} - \mathbf{\Lambda}_i) \\
& + \rho_i(\mathbf{Q}_{ji} - \mathbf{\Lambda}_j - \mathbf{V}_{ji}) - \sum_{k=1}^{p} \delta_k(\mathbf{Q}_{jk} - \mathbf{\Lambda}_j) \;\forall\, j; i < j; \qquad (3.10)
\end{aligned}$$

$$\mathbf{S}_{ii} > \mathbf{\Lambda}_i + \rho_i \mathbf{V}_{ii} \;\forall\, i; \qquad (3.11)$$

$$\mathbf{S}_{ij} + \mathbf{S}_{ij}^T \geq \mathbf{\Lambda}_i + \rho_j \mathbf{V}_{ij} + \mathbf{\Lambda}_j + \rho_i \mathbf{V}_{ji} \;\forall\, j; i < j. \qquad (3.12)$$

From (3.8) to (3.12), (3.8) can be written in the folowing compact form.

$$\dot{V}(t) \leq \mathbf{Z}(t)^T (\mathbf{R} + \mathbf{S})\mathbf{Z}(t) \qquad (3.13)$$

where $\mathbf{Z}(t) = \begin{bmatrix} w_1\mathbf{z}(t) \\ w_2\mathbf{z}(t) \\ \vdots \\ w_p\mathbf{z}(t) \end{bmatrix}$, $\mathbf{R} = \begin{bmatrix} \mathbf{R}_{11} & \mathbf{R}_{12} & \cdots & \mathbf{R}_{1p} \\ \mathbf{R}_{21} & \mathbf{R}_{22} & \cdots & \mathbf{R}_{2p} \\ \vdots & \vdots & \vdots & \vdots \\ \mathbf{R}_{p1} & \mathbf{R}_{p2} & \cdots & \mathbf{R}_{pp} \end{bmatrix}$ and $\mathbf{S} = \begin{bmatrix} \mathbf{S}_{11} & \mathbf{S}_{12} & \cdots & \mathbf{S}_{1p} \\ \mathbf{S}_{21} & \mathbf{S}_{22} & \cdots & \mathbf{S}_{2p} \\ \vdots & \vdots & \vdots & \vdots \\ \mathbf{S}_{p1} & \mathbf{S}_{p2} & \cdots & \mathbf{S}_{pp} \end{bmatrix}$.
From (3.1) and (3.13), based on the the Lyapunov stability theory, $V(t) > 0$ and $\dot{V}(t) < 0$ for $\mathbf{z}(t) \neq \mathbf{0}$ ($\mathbf{x}(t) \neq \mathbf{0}$) imply the asymptotic stability of the FMB control system (2.10), i.e., $\mathbf{x}(t) \to 0$ when time $t \to \infty$, can be achieved if the stability conditions summarized in the following theorem are satisfied.

Theorem 3.1. *The FMB control system (2.10), formed by the nonlinear plant represented by the fuzzy model (2.2) and the fuzzy controller (2.6) connected in a closed loop, is asymptotically stable if there exist predefined constant scalars* ρ_j *and* δ_j, $j = 1, 2, \cdots, p$, *satisfying* $m_j(\mathbf{x}(t)) - \rho_j w_j(\mathbf{x}(t)) + \delta_j \geq 0$ *for all* j *and* $\mathbf{x}(t)$ *and there exist matrices* $\mathbf{N}_j \in \Re^{m\times n}$, $\mathbf{R}_{ij} = \mathbf{R}_{ji}^T \in \Re^{n\times n}$, $\mathbf{S}_{ij} = \mathbf{S}_{ji}^T \in \Re^{n\times n}$, $\mathbf{V}_{ij} = \mathbf{V}_{ij}^T \in \Re^{n\times n}$, $\mathbf{X} = \mathbf{X}^T \in \Re^{n\times n}$ *and* $\mathbf{\Lambda}_j = \mathbf{\Lambda}_j^T \in \Re^{n\times n}$ *such that the following LMIs are satisfied.*

$$\mathbf{X} > 0;$$

LMIs in (3.7); (3.9); (3.10); (3.11); (3.12);

$$\mathbf{R} < 0;$$
$$\mathbf{S} < 0;$$

and the feedback gains are designed as $\mathbf{G}_j = \mathbf{N}_j\mathbf{X}^{-1}$ *for all* j.

Remark 3.3. Consider the inequalities $m_j(\mathbf{x}(t)) - \rho_j w_j(\mathbf{x}(t)) + \delta_j \geq 0$ for all j and $\mathbf{x}(t)$. It can be seen that the parameters ρ_j and δ_j carry the information of the membership functions. The parameter δ_j is the upper bound of the difference of the scaled membership functions by ρ_j of the TS fuzzy model and the membership functions of the fuzzy controller. Through the parameters ρ_j and δ_j, the boundary information of the membership functions is brought to the stability conditions. As a result, the stability conditions in Theorem 3.1 are applicable to FMB control systems with membership functions satisfying the boundary conditions but not any requirement for the shapes. Various MFB information can be employed to further relax the stability conditions [55, 91, 98, 99].

3.2.2 LMI-Based Performance Conditions

Other than the system stability, system performance is an important issue to be considered in the FMB control systems. To measure the system performance quantitatively, a commonly used scalar performance index [1] J defined as follows can be employed.

$$J = \int_0^{\infty} \begin{bmatrix} \mathbf{x}(t) \\ \mathbf{u}(t) \end{bmatrix}^T \begin{bmatrix} \mathbf{J}_1 & \mathbf{0} \\ \mathbf{0} & \mathbf{J}_2 \end{bmatrix} \begin{bmatrix} \mathbf{x}(t) \\ \mathbf{u}(t) \end{bmatrix} dt \tag{3.14}$$

where $\mathbf{J}_1 \geq 0$ and $\mathbf{J}_2 \geq 0$ are predefined weighting matrices governing the contribution of the system states and control signal to the performance index. Based on the fuzzy controller (2.6) and the performance index (3.14), we have,

$$J = \int_0^{\infty} \begin{bmatrix} \mathbf{x}(t) \\ \mathbf{x}(t) \end{bmatrix}^T \begin{bmatrix} \mathbf{I} & \mathbf{0} \\ \mathbf{0} & \sum_{j=1}^{p} m_j \mathbf{G}_j \end{bmatrix}^T \begin{bmatrix} \mathbf{J}_1 & \mathbf{0} \\ \mathbf{0} & \mathbf{J}_2 \end{bmatrix} \begin{bmatrix} \mathbf{I} & \mathbf{0} \\ \mathbf{0} & \sum_{k=1}^{p} m_k \mathbf{G}_k \end{bmatrix} \begin{bmatrix} \mathbf{x}(t) \\ \mathbf{x}(t) \end{bmatrix} dt. \tag{3.15}$$

Let the performance index J satisfy the following condition.

$$J < \eta \int_0^{\infty} \begin{bmatrix} \mathbf{x}(t) \\ \mathbf{x}(t) \end{bmatrix}^T \begin{bmatrix} \mathbf{X}^{-1} & \mathbf{0} \\ \mathbf{0} & \mathbf{X}^{-1} \end{bmatrix} \begin{bmatrix} \mathbf{x}(t) \\ \mathbf{x}(t) \end{bmatrix} dt \tag{3.16}$$

where η is a non-zero positive scalar to attenuate the scalar performance index J to a prescribed level.

In the following, the LMI-based performance conditions are derived such that the inequality (3.16) is satisfied. It can be seen that the term on the right hand side can be regarded as the upper bound of the cost function J. By satisfying the inequality (3.16) with a smaller value of η, it implies that a better performance can be achieved as a lower value of J can be achieved.

Combining (3.15) and (3.16), recalling the feedback gains of the fuzzy controller as $\mathbf{G}_j = \mathbf{N}_j \mathbf{X}^{-1}$, we have,

$$\int_0^{\infty} \begin{bmatrix} \mathbf{x}(t) \\ \mathbf{x}(t) \end{bmatrix}^T \begin{bmatrix} \mathbf{X}^{-1} & \mathbf{0} \\ \mathbf{0} & \mathbf{X}^{-1} \end{bmatrix} \mathbf{W} \begin{bmatrix} \mathbf{X}^{-1} & \mathbf{0} \\ \mathbf{0} & \mathbf{X}^{-1} \end{bmatrix} \begin{bmatrix} \mathbf{x}(t) \\ \mathbf{x}(t) \end{bmatrix} < 0 \tag{3.17}$$

where

$$\mathbf{W} = \begin{bmatrix} \mathbf{X} & \mathbf{0} \\ \mathbf{0} & \sum_{j=1}^{p} m_j \mathbf{N}_j \end{bmatrix}^T \begin{bmatrix} \mathbf{J}_1 & \mathbf{0} \\ \mathbf{0} & \mathbf{J}_2 \end{bmatrix} \begin{bmatrix} \mathbf{X} & \mathbf{0} \\ \mathbf{0} & \sum_{k=1}^{p} m_j \mathbf{N}_k \end{bmatrix} - \eta \begin{bmatrix} \mathbf{X} & \mathbf{0} \\ \mathbf{0} & \mathbf{X} \end{bmatrix}. \tag{3.18}$$

To make sure that the inequality (3.17) is satisfied, it is required that $\mathbf{W} < 0$. It should be noted that the inequality $\mathbf{W} < 0$ cannot be formulated as an LMI problem. By Schur complement [120], $\mathbf{W} < 0$ is equivalent to the follow inequality.

$$\overline{\mathbf{W}} = \sum_{j=1}^{p} m_j \overline{\mathbf{W}}_j < 0 \tag{3.19}$$

where $\overline{\mathbf{W}}_j = \begin{bmatrix} -\eta\mathbf{X} & \mathbf{0} & \mathbf{X} & \mathbf{0} \\ \mathbf{0} & -\eta\mathbf{X} & \mathbf{0} & \mathbf{N}_j^T \\ \mathbf{X} & \mathbf{0} & -\mathbf{J}_1^{-1} & \mathbf{0} \\ \mathbf{0} & \mathbf{N}_j & \mathbf{0} & -\mathbf{J}_2^{-1} \end{bmatrix}$, $j = 1, 2, \cdots, p$. The LMIs of $\overline{\mathbf{W}}_j < 0$ for all j (guaranteeing $\mathbf{W} < 0$) are regarded as the LMI-based performance conditions which have to be applied with the stability conditions in Theorem 3.1 to realize the system performance with pre-defined weighting matrices $\mathbf{J}_1$ and $\mathbf{J}_2$.

Theorem 3.2. *The system performance of a stable FMB control system (2.10), formed by the nonlinear plant represented by the fuzzy model (2.2) and the fuzzy controller (2.6) connected in a closed loop, satisfies the performance index J defined in (3.16) characterized by pre-defined weighting matrices $\mathbf{J}_1 \geq 0$ and $\mathbf{J}_2 \geq 0$ that is attenuated to the level $\eta > 0$ if any stability conditions reported in this book and the LMIs $\overline{\mathbf{W}}_j < 0$ for all j are satisfied.*

Remark 3.4. It can be seen from (3.19) that the information of the membership functions is not considered to achieve the performance conditions in Theorem 3.2. By following the MFSD analysis approach, the performance conditions can be relaxed by considering the information of the membership functions.

Remark 3.5. There are different approaches reported in some published work to considering the system performance. For example, a decay-rate design was given in [105] that the decaying rate of the Lyapunov function can be controlled. A faster decaying rate will offer a better performed FMB control system in terms of faster transient response. Some LMI-based constraints [105] introduced in Section 2.6 can be imposed on the feasible set of the solutions to the stability conditions to restrain the system state energy and/or control power. A guaranteed-cost approach [108] can also be applied that, similar to the decay-rate design, the Lyapunov function is bounded by a cost function related to the system states and control signal. These approaches can also be applied to achieve the system performance.

Example 3.1. Consider the same FMB control system in Example 2.1. Under the MFSD approach, we have to know the membership functions to obtain the parameters ρ_j and δ_j. The membership functions for the TS fuzzy model and fuzzy controller, which are shown graphically in Fig. 3.1, are defined as follows

$$w_1(x_1(t)) = \mu_{M_1^1}(x_1(t)) = \begin{cases} 1 & \text{for } x_1(t) < -10 \\ \frac{-x_1(t)+2}{12} & \text{for } -10 \leq x_1(t) \leq 2 \\ 0 & \text{for } x_1(t) > 2 \end{cases},$$

$$w_2(x_1(t)) = \mu_{M_1^2}(x_1(t)) = 1 - w_1(x_1(t)) - w_3(x_1(t)),$$

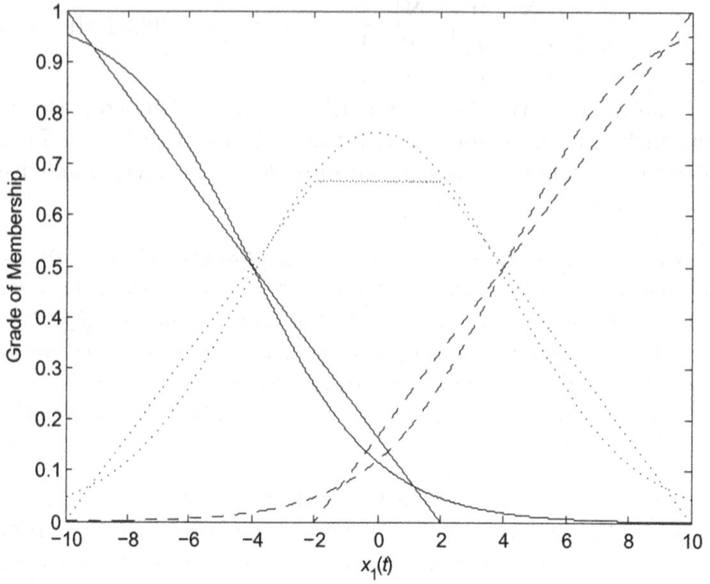

Fig. 3.1 Membership functions of the fuzzy model: $w_1(x_1(t))$ (left triangle in solid line), $w_2(x_1(t))$ (trapezoid in dotted line) and $w_3(x_1(t))$ (right triangle in dash line). Membership functions of the fuzzy controller: $m_1(x_1(t))$ (left z shape in solid line), $m_2(x_1(t))$ (bell shape in dotted line) and $m_3(x_1(t))$ (right s shape in dash line).

$$w_3(x_1(t)) = \mu_{M_1^3}(x_1(t)) = \begin{cases} 0 & \text{for } x_1(t) < -2 \\ \frac{x_1(t)+2}{12} & \text{for } -2 \le x_1(t) \le 10 \\ 1 & \text{for } x_1(t) > 10 \end{cases},$$

$m_1(x_1(t)) = \mu_{N_1^1}(x_1(t)) = 1 - \frac{1}{1+e^{-(x_1(t)+4)}}$, $m_2(x_1(t)) = \mu_{N_1^2}(x_1(t)) = 1 - m_1(x_1(t)) - m_3(x_1(t))$, $m_3(x_1(t)) = \mu_{N_1^3}(x_1(t)) = \frac{1}{1+e^{-(x_1(t)-4)}}$.

Based on the chosen membership functions, it can be seen that the inequalities of $m_j(x_1(t)) - \rho_j w_j(x_1(t)) + \delta_j \ge 0$ for all j and $x_1(t)$ are satisfied for $\rho_1 = \rho_2 = \rho_3 = 1$, $\delta_1 = 6.9182 \times 10^{-2}$, $\delta_2 = 7.4195 \times 10^{-2}$ and $\delta_3 = 6.9182 \times 10^{-2}$. By using the MATLAB LMI toolbox, the stability region for the system parameters of the TS fuzzy model in the ranges $2 \le a \le 9$ and $-8 \le b \le -4.5$ is shown in Fig. 3.2, indicated by 'o'.

It should be noted that the stability conditions in Theorem 2.2 to Theorem 2.5 under the PDC design technique, which require that both the TS fuzzy model and the fuzzy controller share the same premise membership functions, cannot be applied in this example.

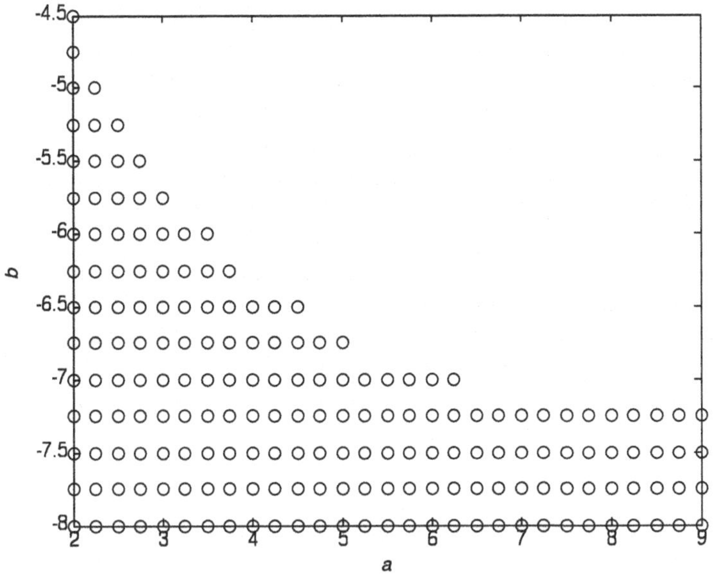

Fig. 3.2 Stability region indicated by 'o' given by the stability conditions in Theorem 3.1.

Example 3.2. An inverted pendulum on a cart [84] is employed to illustrate the design flexibility of the proposed fuzzy controller. In this example, membership functions of the fuzzy controller are designed different from those of the TS fuzzy model. Simple membership functions with lower structural complexity are employed to lower the implementation cost. Under this design, the published stability conditions in Theorem 2.2 to Theorem 2.5 for perfect premise matching cannot be applied to guarantee the system stability. The proposed LMI-based stability conditions in Theorem 3.1 and the performance conditions in Theorem 3.2 show an effective approach to help design a stable and well-performed FMB control system under imperfect premise matching in this simulation example.

Step I) The dynamic equations of the inverted pendulum on a cart [84] is given by,

$$\dot{x}_1(t) = x_2(t) \tag{3.20}$$

$$\dot{x}_2(t) = \frac{h_1(\mathbf{x}(t))}{(M_c + m_p)(J_0 + m_p l^2) - m_p^2 l^2 \cos^2(x_1(t))} \tag{3.21}$$

$$\dot{x}_3(t) = x_4(t) \tag{3.22}$$

$$\dot{x}_4(t) = \frac{h_2(\mathbf{x}(t))}{(M_c + m_p)(J_0 + m_p l^2) - m_p^2 l^2 \cos^2(x_1(t))} \tag{3.23}$$

where $x_1(t)$ and $x_2(t)$ denote the angular displacement (rad) and the angular velocity (rad/s) of the pendulum from vertical, respectively, $x_3(t)$ and $x_4(t)$ denote the displacement (m) and the velocity (m/s) of the cart, respectively, $h_1(\mathbf{x}(t)) = -F_1(M_c+m_p)x_2(t) - m_p^2l^2x_2^2(t)\sin(x_1(t))\cos(x_1(t)) + F_0m_plx_4(t)\cos(x_1(t)) + (M_c+m_p)m_pgl\sin(x_1(t)) - m_pl\cos(x_1(t))u(t)$, $h_2(\mathbf{x}(t)) = F_1m_plx_2(t)\cos(x_1(t)) + (J_0+m_pl^2)m_plx_2(t)^2\sin(x_1(t)) - F_0(J_0+m_pl^2)x_4(t) - m_p^2gl^2\sin(x_1(t))\cos(x_1(t)) + (J_0+m_pl^2)u(t)$, $g = 9.8\ m/s^2$ is the acceleration due to gravity, $m_p = 0.22\ kg$ is the mass of the pendulum, $M_c = 1.3282kg$ is the mass of the cart, $l = 0.304m$ is the length from the center of mass of the pendulum to the shaft axis, $J_0 = m_pl^2/3\ kgm^2$ is the moment of inertia of the pendulum around the center of mass, $F_0 = 22.915N/m/s$ and $F_1 = 0.007056\ N/rad/s$ are the friction factors of the cart and the pendulum respectively, and $u(t)$ is the force (N) applied to the cart. The nonlinear plant can be represented by a fuzzy model with two fuzzy rules [84]. The i-th rule is given by,

$$\begin{aligned}&\text{Rule } i\text{: IF } f_1(x_1(t)) \text{ is } M_1^i\\&\qquad\text{THEN } \dot{\mathbf{x}}(t) = \mathbf{A}_i\mathbf{x}(t) + \mathbf{B}_iu(t), i = 1, 2.\end{aligned} \tag{3.24}$$

The system dynamics is described by,

$$\dot{\mathbf{x}}(t) = \sum_{i=1}^{2} w_i(\mathbf{A}_i\mathbf{x}(t) + \mathbf{B}_iu(t)). \tag{3.25}$$

where

$$\mathbf{x}(t) = \begin{bmatrix} x_1(t) & x_2(t) & x_3(t) & x_4(t) \end{bmatrix}^T;$$

$$a_1 = (M_c + m_p)(J_0 + m_pl^2) - m_p^2l^2;$$

$$a_2 = (M_c + m_p)(J_0 + m_pl^2) - m_p^2l^2\cos^2(\frac{\pi}{3});$$

$$\mathbf{A}_1 = \begin{bmatrix} 0 & 1 & 0 & 0 \\ (M_c+m_p)m_pgl/a_1 & -F_1(M_c+m_p)/a_1 & 0 & F_0m_pl/a_1 \\ 0 & 0 & 0 & 1 \\ -m_p^2gl^2/a_1 & F_1m_pl/a_1 & 0 & -F_0(J_0+m_pl^2)/a_1 \end{bmatrix};$$

$$\mathbf{A}_2 = \begin{bmatrix} 0 & 1 & 0 & 0 \\ \frac{3\sqrt{3}}{2\pi}(M_c+m_p)m_pgl/a_2 & -F_1(M_c+m_p)/a_2 & 0 & F_0m_pl\cos(\frac{\pi}{3})/a_2 \\ 0 & 0 & 0 & 1 \\ -\frac{3\sqrt{3}}{2\pi}m_p^2gl^2\cos(\frac{\pi}{3})/a_1 & F_1m_pl\cos(\frac{\pi}{3})/a_2 & 0 & -F_0(J_0+m_pl^2)/a_1 \end{bmatrix};$$

$$\mathbf{B}_1 = \begin{bmatrix} 0 \\ -m_pl/a_1 \\ 0 \\ (J_0+m_pl^2)/a_1 \end{bmatrix};$$

$$\mathbf{B}_2 = \begin{bmatrix} 0 \\ -m_p l \cos(\frac{\pi}{3})/a_2 \\ 0 \\ (J_0 + m_p l^2)/a_2 \end{bmatrix}.$$

The membership functions are defined as $w_1(x_1(t)) = \mu_{M_1^1}(x_1(t)) = \left(1 - \frac{1}{1+e^{-7(x_1(t)-\pi/6)}}\right)\frac{1}{1+e^{-7(x_1(t)+\pi/6)}}$ and $w_2(x_1(t)) = \mu_{M_1^2}(x_1(t)) = 1 - w_1(x_1(t))$ that are shown graphically in Fig. 3.3.

Step II) A 2-rule fuzzy controller is employed to close the feedback loop to achieve the control objective, i.e. $\mathbf{x}(t) \to 0$ as $t \to \infty$. The j-th rule of the fuzzy controller is given by,

$$\begin{aligned} &\text{Rule } j\text{: IF } g_1(x_1(t)) \text{ is } N_1^j \\ &\qquad \text{THEN } u(t) = \mathbf{G}_j\mathbf{x}(t), j = 1, 2. \end{aligned} \tag{3.26}$$

The fuzzy controller is defined by,

$$u(t) = \sum_{j=1}^{2} m_i \mathbf{G}_j \mathbf{x}(t). \tag{3.27}$$

The membership functions are designed as $m_1(x_1(t)) = \mu_{N_1^1}(x_1(t)) = 0.925e^{-\frac{x_1(t)}{2\times1.5^2}}$ and $m_2(x_1(t)) = \mu_{N_1^2}(x_1(t)) = 1 - m_1(x_1(t))$, which are shown graphically in Fig. 3.3.

Step III) The inequalities of $m_j(x_1(t)) - \rho_j w_j(x_1(t)) + \delta_j \geq 0$ for all j and $x_1(t)$ are satisfied for $\rho_1 = \rho_2 = 0.82$; $\delta_1 = 6.6581 \times 10^{-2}$ and $\delta_2 = 4.0061 \times 10^{-2}$. By solving the stability conditions in Theorem 3.1 and the LMI-based performance conditions in Theorem 3.2, i.e., $\overline{\mathbf{W}}_j < 0$ for all j using the MATLAB LMI toolbox, the feedback gains of three fuzzy controllers, referred to as fuzzy controllers 1 to 3, are obtained and listed in Table 3.1.

The fuzzy controllers 1 to 3 are employed to stabilize the inverted pendulum described in (3.20) to (3.23). The system state responses and control signals under the initial system state condition $\left[\frac{5\pi}{12}\ 0\ 0\ 0\right]$ are shown in Fig. 3.4 and Fig. 3.5, respectively. It can be seen from this figure that fuzzy controllers 1 to 3 are able to stabilize the inverted pendulum. The fuzzy controller 1 offers the fastest transient response at the cost of a large control signal. With the LMI-based performance conditions, $\mathbf{J}_2$ is employed to constrain the control signals of fuzzy controllers 2 and 3 during the design. Referring to Fig. 3.5, the magnitudes of the control signals offered by the fuzzy controllers 2 and 3 are smaller than that of the fuzzy controller 1. Furthermore, comparing fuzzy controllers 2 and 3, $\mathbf{J}_1 = \begin{bmatrix} 1\,0\,0\,0 \\ 0\,1\,0\,0 \\ 0\,0\,5\,0 \\ 0\,0\,0\,1 \end{bmatrix}$ of fuzzy controller 3 has a heavier

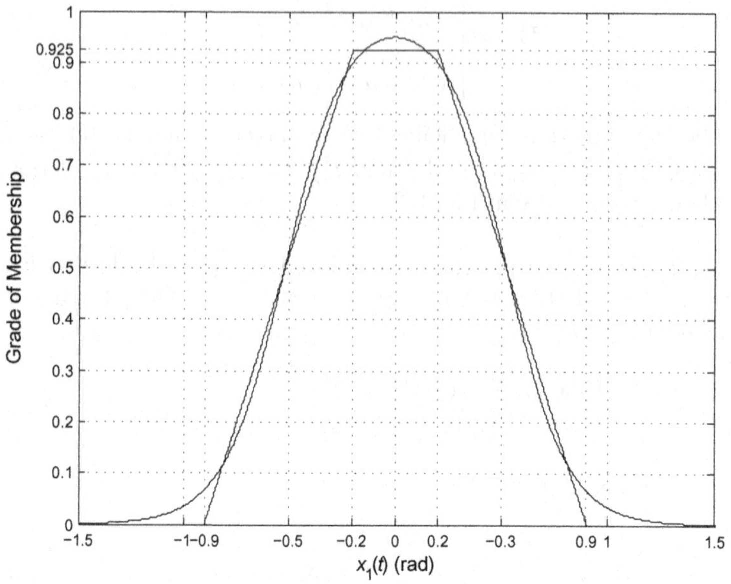

(a) $\mu_{M_1^1}(x_1(t))$ (Gaussian) and $\mu_{N_1^1}(x_1(t))$ (trapezoid).

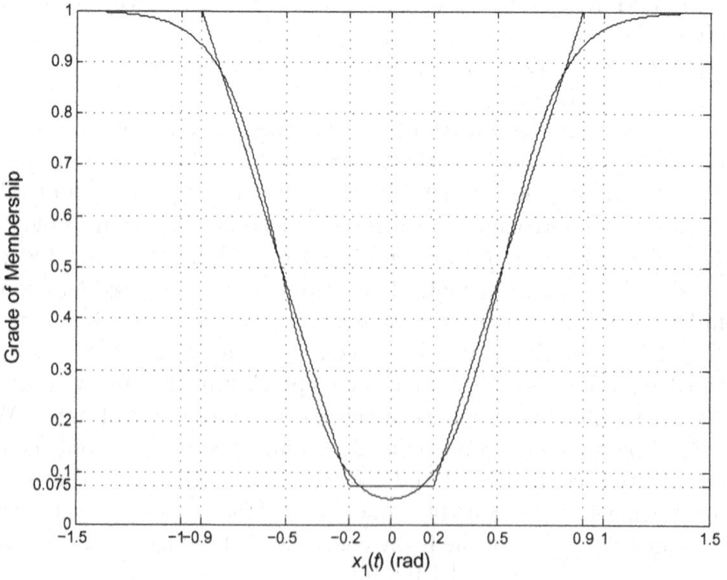

(b) $\mu_{M_1^2}(x_1(t))$ (Gaussian) and $\mu_{N_1^2}(x_1(t))$ (trapezoid).

Fig. 3.3 Membership functions of the TS fuzzy model and fuzzy controller in Example 3.2

Table 3.1 Feedback gains for fuzzy controllers 1 to 3 in Example 3.2.

Feedback Gains	η, $\mathbf{J}_1$ and $\mathbf{J}_2$
FC[†] 1 $\mathbf{G}_1 = \begin{bmatrix} 1495.8418 & 102.2374 & 11.3099 & 87.3309 \end{bmatrix}$ $\mathbf{G}_2 = \begin{bmatrix} 1993.1825 & 134.8019 & 14.9267 & 107.8680 \end{bmatrix}$	NA[††]
FC[†] 2 $\mathbf{G}_1 = \begin{bmatrix} 330.4341 & 31.4571 & 0.1883 & 36.0413 \end{bmatrix}$ $\mathbf{G}_2 = \begin{bmatrix} 615.4812 & 46.3514 & 0.1960 & 36.6065 \end{bmatrix}$	$\eta = 0.1$ $\mathbf{J}_2 = 1$ $\mathbf{J}_1 = \begin{bmatrix} 1 & 0 & 0 & 0 \\ 0 & 1 & 0 & 0 \\ 0 & 0 & 1 & 0 \\ 0 & 0 & 0 & 1 \end{bmatrix}$
FC[†] 3 $\mathbf{G}_1 = \begin{bmatrix} 426.3628 & 39.1925 & 0.7414 & 43.2931 \end{bmatrix}$ $\mathbf{G}_2 = \begin{bmatrix} 781.1192 & 59.2399 & 0.8995 & 47.7560 \end{bmatrix}$	$\eta = 0.1$ $\mathbf{J}_2 = 1$ $\mathbf{J}_1 = \begin{bmatrix} 1 & 0 & 0 & 0 \\ 0 & 1 & 0 & 0 \\ 0 & 0 & 5 & 0 \\ 0 & 0 & 0 & 1 \end{bmatrix}$

[†] FC stands for Fuzzy Controller.
[††] LMI performance conditions are not considered.

weight on $x_3(t)$ in the performance index to suppress its magnitude. Consequently, the system state response of $x_3(t)$ with the fuzzy controller 3 offers better system performance than that with the fuzzy controller 2 in terms of transient response and settling time.

For comparison purposes, the fuzzy controller (3.27) is designed based on the decay-rate design approach [105] such that the performance condition of $\dot{V}(t) \leq -2\alpha V(t)$, where $\alpha > 0$, is achieved. This performance condition is incorporated into Theorem 3.1 and replaces the proposed LMI-based performance conditions in Theorem 3.2. Choosing the value of α arbitrarily as 0.01, 0.1 and 0.5, and with the MATLAB LMI toolbox, we obtain 3 decay-rate fuzzy controllers 1 to 3, respectively with the feedback gains shown in Table 3.2.

The system responses of the control signals offered by the decay-rate fuzzy controllers 1 to 3 are shown in Fig. 3.6 and Fig. 3.7, respectively. All decay-rate fuzzy controllers are able to stabilize the inverted pendulum. A larger value of α leads to better system performance with larger values of feedback gains. Comparing to the simulation results in Fig. 3.4 and Fig. 3.5, the decay-rate fuzzy controllers 1 to 3 can only improve the system responses in terms of a faster rising time and settling time at the cost of a large magnitude of control signal. However, unlike the proposed LMI-based performance conditions, the decay-rate performance condition is for tackling the overall system performance without specifically considering the system states or control signals. With the LMI-based performance conditions in Theorem 3.2, a more flexible way for realizing the system performance by specifying the weighting to each system state and control signal using $\mathbf{J}_1$ and $\mathbf{J}_2$ can be offered.

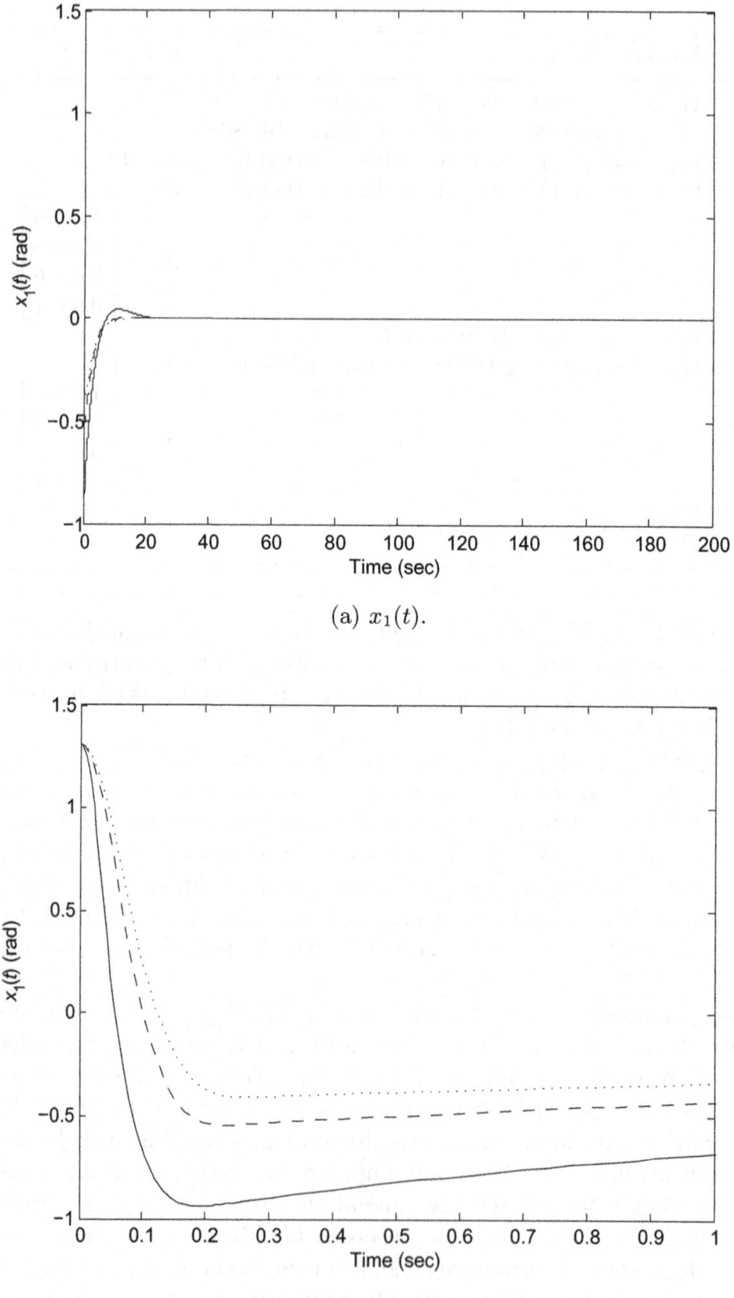

(a) $x_1(t)$.

(b) $x_1(t)$ for $0s \leq t \leq 1s$.

Fig. 3.4 System state responses and control signals of the inverted pendulum with fuzzy controllers 1 (solid lines), 2 (dotted lines) and 3 (dash lines).

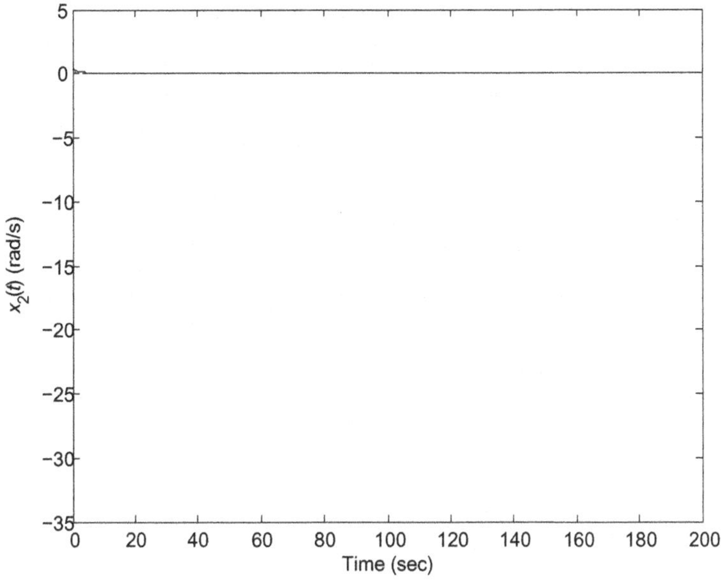

(c) $x_2(t)$.

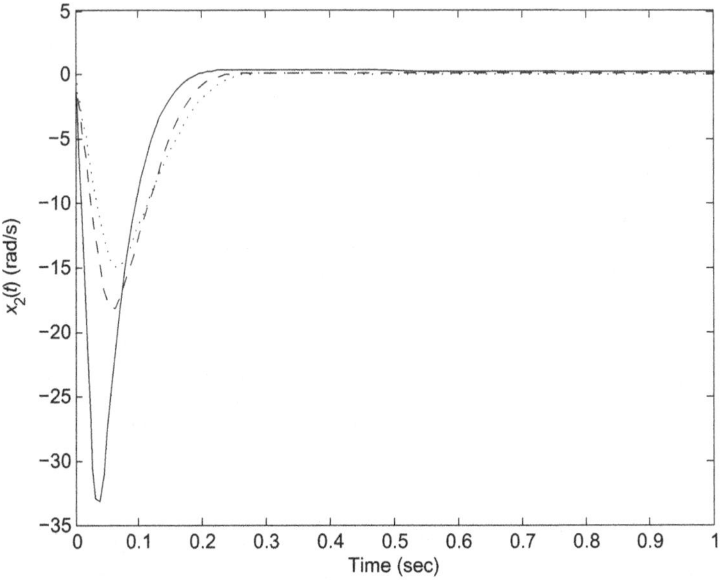

(d) $x_2(t)$ for $0s \leq t \leq 1s$.

Fig. 3.4 (*continued*)

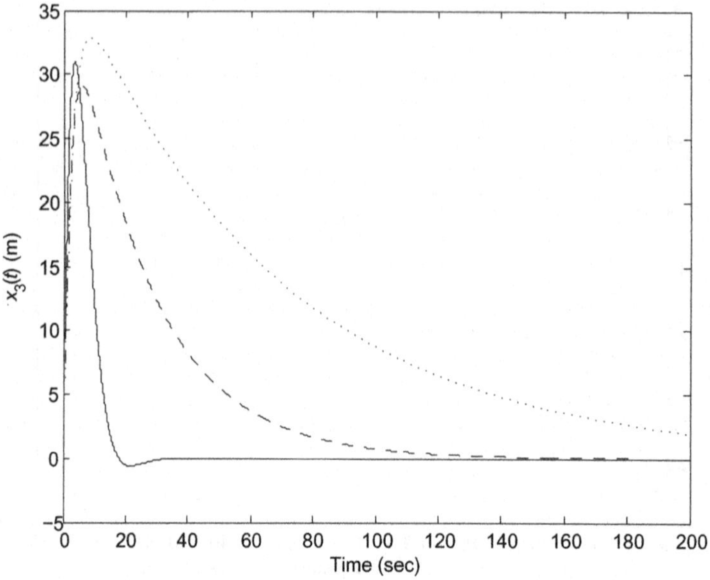

(e) $x_3(t)$.

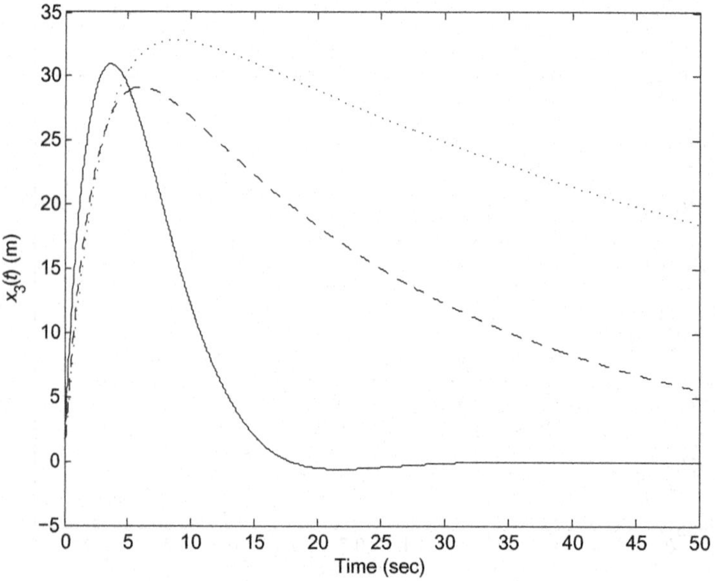

(f) $x_3(t)$ for $0s \leq t \leq 50s$.

Fig. 3.4 (*continued*)

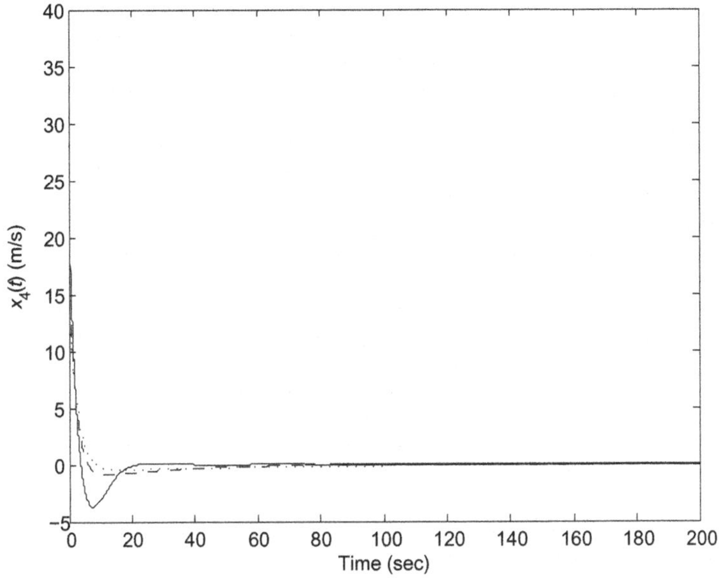

(g) $x_4(t)$.

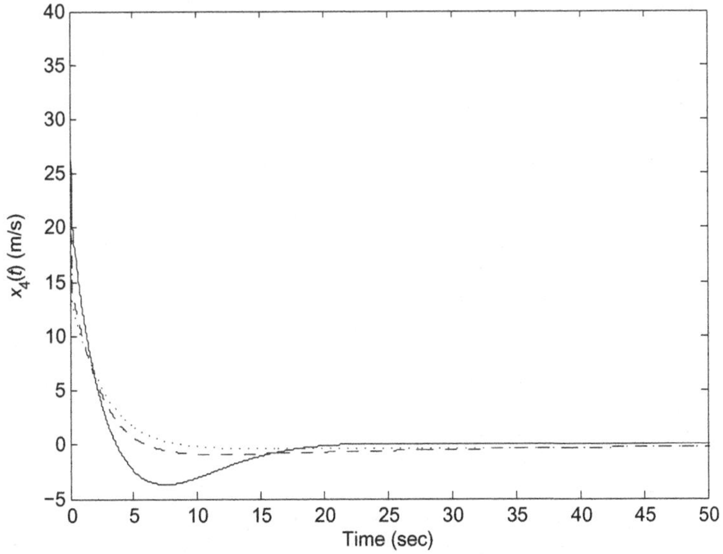

(h) $x_4(t)$ for $0s \leq t \leq 50s$.

Fig. 3.4 (*continued*)

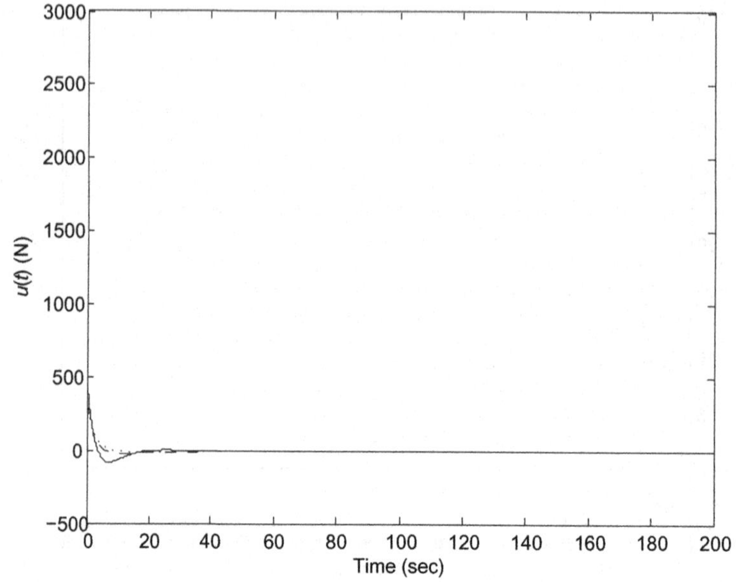

(a) $u(t)$.

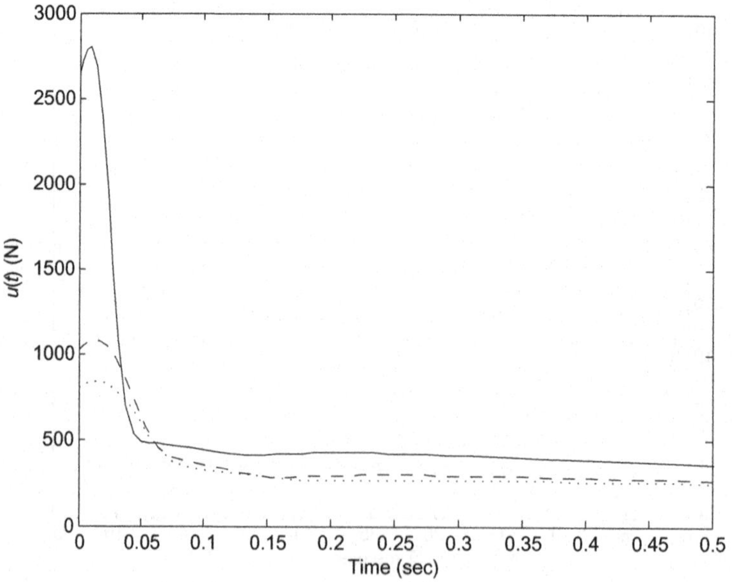

(b) $u(t)$ for $0s \leq t \leq 0.5s$.

Fig. 3.5 Control signals of the inverted pendulum with fuzzy controllers 1 (solid lines), 2 (dotted lines) and 3 (dash lines).

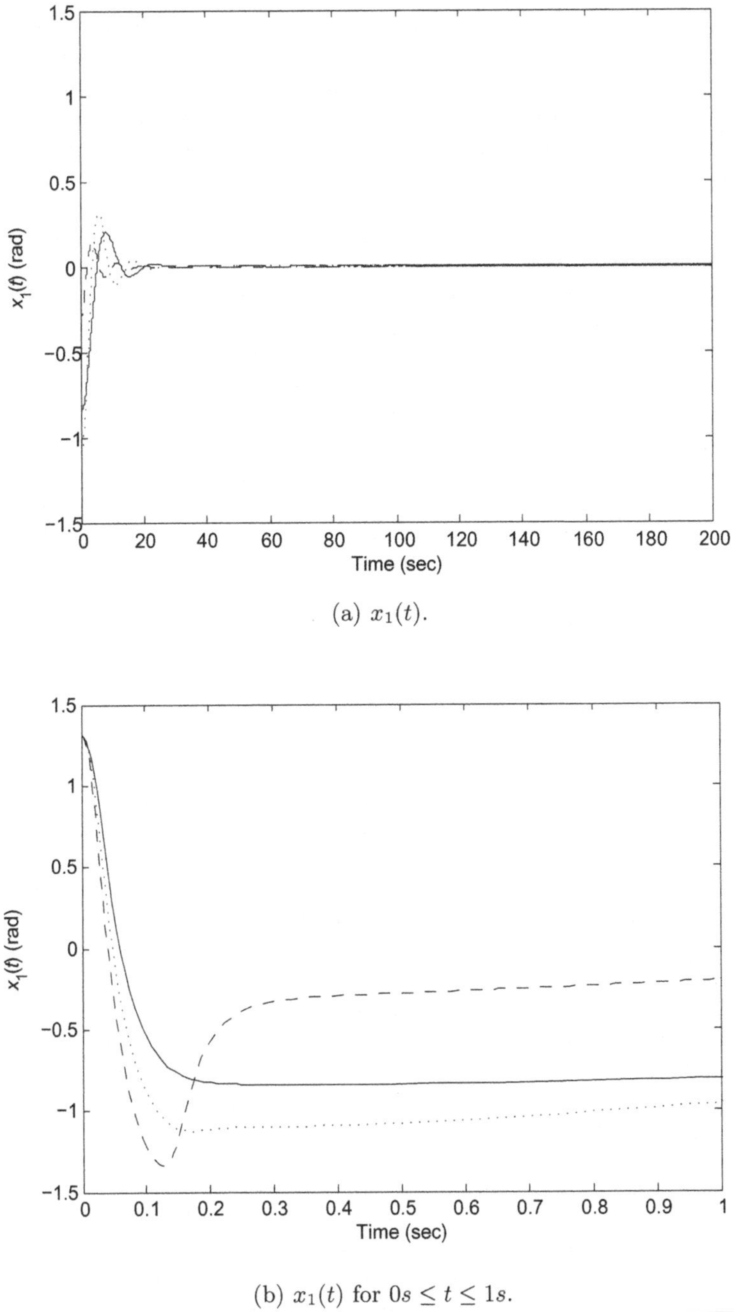

(a) $x_1(t)$.

(b) $x_1(t)$ for $0s \leq t \leq 1s$.

Fig. 3.6 System state responses and control signals of the inverted pendulum with decay-rate fuzzy controllers 1 (solid lines), 2 (dotted lines) and 3 (dash lines).

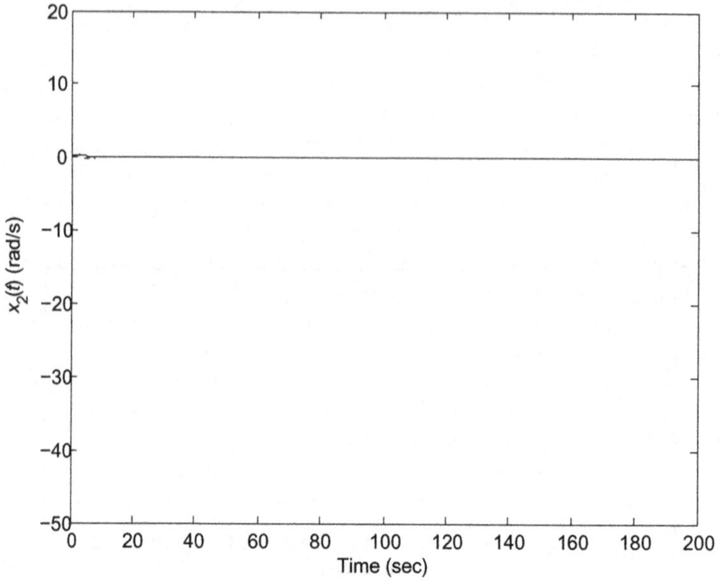

(c) $x_2(t)$.

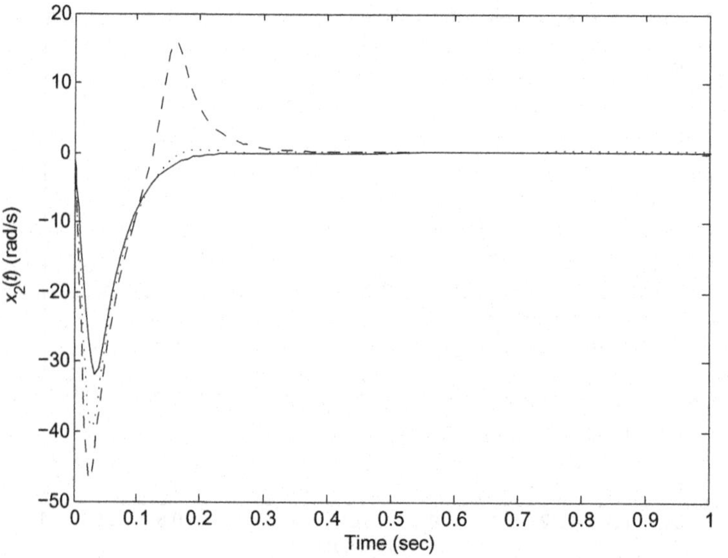

(d) $x_2(t)$ for $0s \leq t \leq 1s$.

Fig. 3.6 *(continued)*

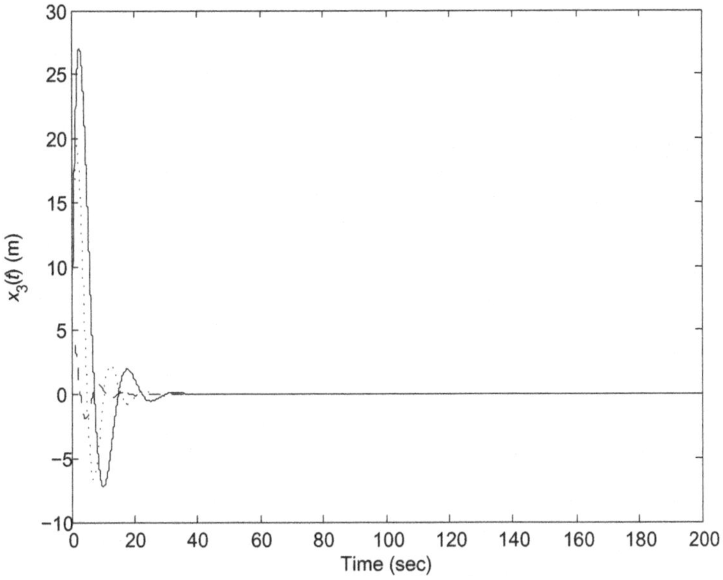

(e) $x_3(t)$.

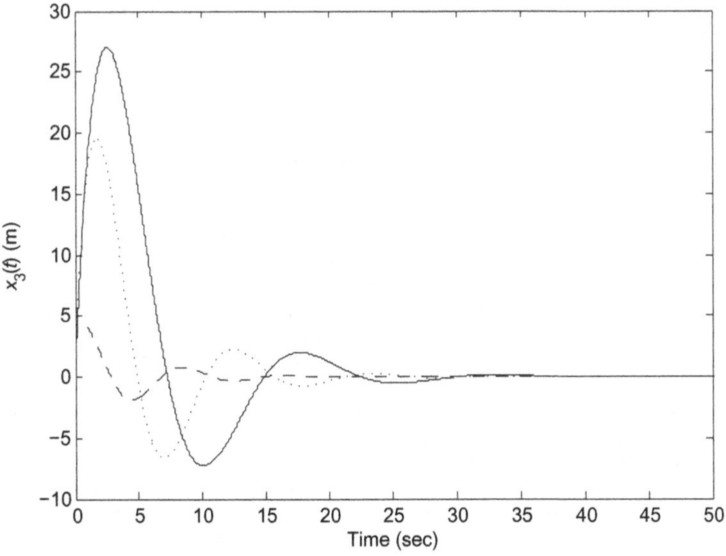

(f) $x_3(t)$ for $0s \leq t \leq 50s$.

Fig. 3.6 (*continued*)

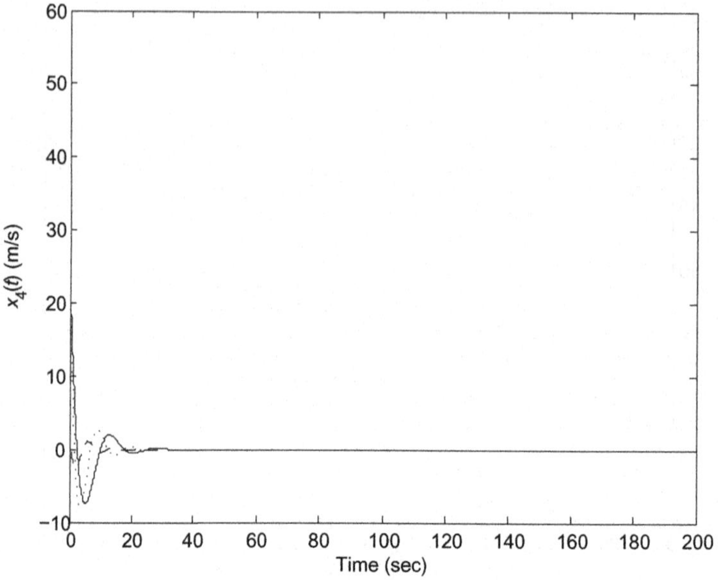

(g) $x_4(t)$.

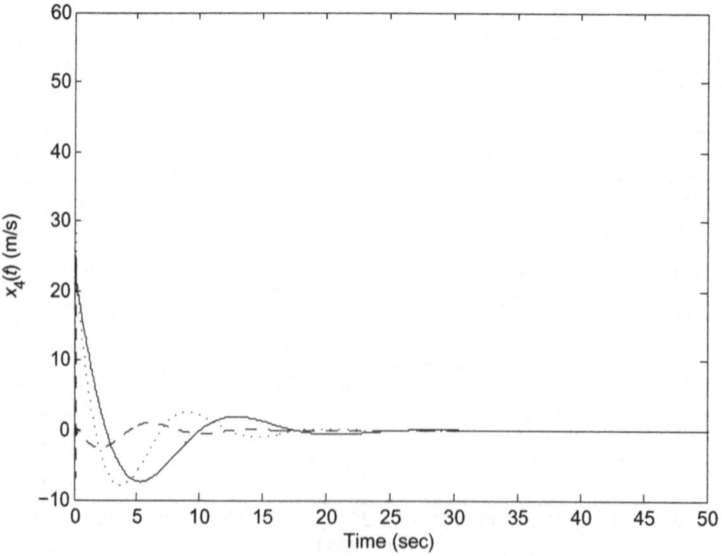

(h) $x_4(t)$ for $0s \leq t \leq 50s$.

Fig. 3.6 (*continued*)

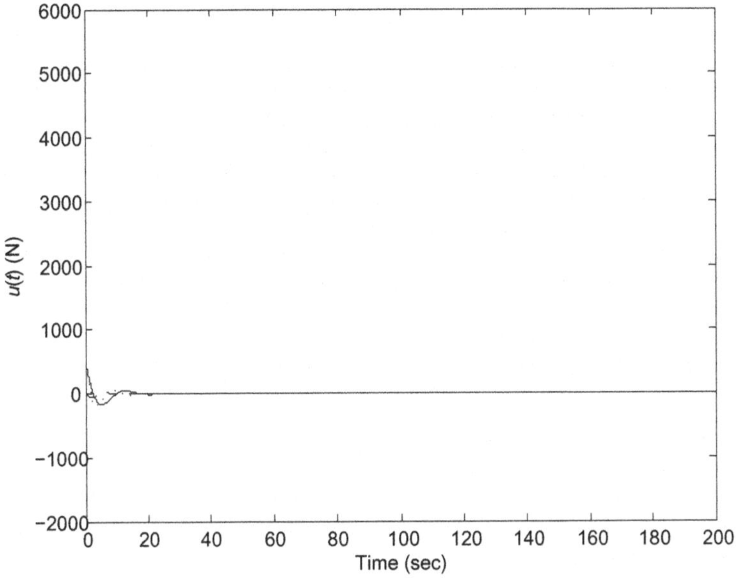

(a) $u(t)$.

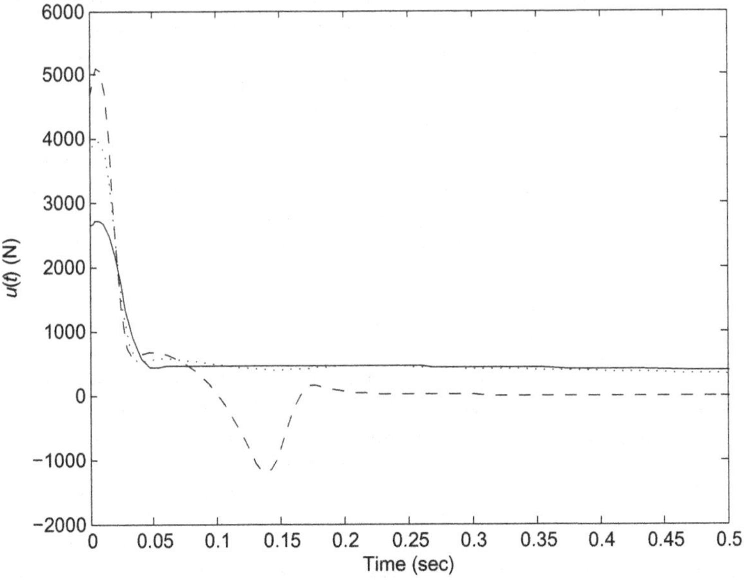

(b) $u(t)$ for $0s \leq t \leq 0.5s$.

Fig. 3.7 Control signals of the inverted pendulum with decay-rate fuzzy controllers 1 (solid lines), 2 (dotted lines) and 3 (dash lines).

Table 3.2 Feedback gains for decay-rate fuzzy controllers 1 to 3 in Example 3.2.

Feedback Gains		α
DRFC† 1	$\mathbf{G}_1 = [1985.1149 \quad 121.5018 \quad 38.6393 \quad 91.9798]$	$\alpha = 0.01$
	$\mathbf{G}_2 = [2007.1015 \quad 122.7442 \quad 39.0031 \quad 92.6289]$	
DRFC† 2	$\mathbf{G}_1 = [2828.6308 \quad 179.1381 \quad 102.4027 \quad 138.0461]$	$\alpha = 0.1$
	$\mathbf{G}_2 = [2863.6978 \quad 181.2361 \quad 103.5343 \quad 139.3103]$	
DRFC† 3	$\mathbf{G}_1 = [3494.2023 \quad 224.6229 \quad 220.5881 \quad 179.6569]$	$\alpha = 0.5$
	$\mathbf{G}_2 = [3537.8243 \quad 227.3355 \quad 223.1618 \quad 181.4586]$	

† DRFC stands for Decay-Rate Fuzzy Controller.

3.3 Staircase-Membership-Function Approach

The Lyapunov stability theory is one of the mathematical tools to investigate the system stability of FMB control systems. It was shown in [15, 122] that the FMB control system is guaranteed to be asymptotically stable if the solution to a set of LMIs can be found numerically by using some convex programming techniques. Considering the quadratic Lyapunov function candidate (3.1) and its time derivative (3.2), the system stability of the FMB control system is guaranteed by the satisfaction of $\sum_{i=1}^{p}\sum_{j=1}^{p} w_i m_j \mathbf{Q}_{ij} < 0$. It was reported in [15, 122] that the sufficient stability conditions are $\mathbf{Q}_{ij} < 0$ for all i and j.

When the PDC design technique is applied, the membership functions are designed such that $m_i = w_i$ for all i. The stability conditions become $\sum_{i=1}^{p}\sum_{j=1}^{p} w_i w_j \mathbf{Q}_{ij} = \frac{1}{2}\sum_{i=1}^{p}\sum_{j=1}^{p} w_i w_j (\mathbf{Q}_{ij} + \mathbf{Q}_{ji}) < 0$ by grouping the common cross terms of $w_i w_j$. The sufficient stability conditions become $\mathbf{Q}_{ij} + \mathbf{Q}_{ji} < 0$. Comparatively, these sufficient stability conditions are more relaxed than the previous ones as the property of the membership functions is considered. However, it can be seen that if either set of sufficient stability conditions, i.e., $\mathbf{Q}_{ij} < 0$ or $\mathbf{Q}_{ij} + \mathbf{Q}_{ji} < 0$ for all i and j, is satisfied, it is satisfied for all grades of membership w_i and w_j. In this regard, it can be seen that the published stability conditions in Theorem 2.1 to Theorem 2.5 do not include the membership functions in the stability conditions (i.e. only $\mathbf{Q}_{ij}$ appear in the stability conditions). However, when the membership functions are brought into the stability conditions, we have the sufficient and necessary condition of $\sum_{i=1}^{p}\sum_{j=1}^{p} w_i m_j \mathbf{Q}_{ij} < 0$ for $\dot{V}(t) < 0$ excluding $\mathbf{x}(t) = \mathbf{0}$ (under the quadratic Lyapunov function candidate (3.1)) that is required to be satisfied for every single value of w_i and m_j. As both w_i and m_j are governed by continuous membership functions, consequently, the number of stability conditions to ensure $\sum_{i=1}^{p}\sum_{j=1}^{p} w_i m_j \mathbf{Q}_{ij} < 0$ is infinity. It is thus impossible to investigate $\sum_{i=1}^{p}\sum_{j=1}^{p} w_i m_j \mathbf{Q}_{ij} < 0$ directly, and researchers thus turn to the MFSI stability analysis approaches presented in Chapter 2 to circumvent the difficulty.

It was revealed in [2, 55, 67, 68, 91, 98, 99] that the information of membership functions play an import role for the relaxation of stability conditions.

However, bringing the membership function into the stability conditions is difficult mainly because of the continuity of the membership functions. In this section, in order to consider the membership functions in the stability analysis for the relaxation of stability conditions, SMFs [68] are proposed to approximate the continuous membership functions of the TS fuzzy model and fuzzy controller. *It is worth mentioning that the SMFs are for the stability analysis only and not necessarily implemented physically.* As the SMFs have finite numbers of levels, they effectively circumvent the difficulty by approximating the infinite number of LMIs into a finite one. To make the stability analysis possible using the SMFs, some slack matrix variables are introduced using the property of the membership functions. Stability conditions in terms of LMIs are derived to achieve a stable FMB control system based on the Lyapunov stability theory. The SMF-based FMB control approach offers the following advantages over some published work based the MFSI and MFSD stability analysis.

1. The SMFs approximating the original ones of the TS fuzzy model and fuzzy controller that carry the information of the system nonlinearities are allowed to be brought to the stability conditions. Consequently, the stability conditions are applied to a dedicated nonlinear plant (characterized by the SMFs) but not a family of FMB control systems with any shapes of membership functions as considered in the work of MFSI analysis.
2. It does not require that the TS fuzzy model and fuzzy controller share the same membership functions. Consequently, it offers larger design flexibility for choosing the membership functions of the fuzzy controller. By employing some simple membership functions and/or less number of fuzzy control rules, the implementation cost of the fuzzy controller can be reduced.
3. Unlike the MFSI analysis, the SMF-based approach does not require the membership functions of the TS fuzzy controller to be exactly known. Consequently, by embedding the parameter uncertainties (with know bounds) to the membership functions of the TS fuzzy model, the SMF-based controller is robust to parameter uncertainties to a certain extent.

3.3.1 Stability Analysis Using Staircase Membership Functions

In the previous section, we consider the MFB stability analysis that requires the fuzzy model and fuzzy controller share the same number of fuzzy rules, i.e. $c = p$. Under the SMF approach, this limitation can be eliminated. Define the staircase membership function $\overline{w}_i(\mathbf{x}(t))$ and $\overline{m}_j(\mathbf{x}(t))$ approximating the original membership functions w_i and m_j, respectively. For brevity, $\overline{w}_i(\mathbf{x}(t))$ and $\overline{m}_j(\mathbf{x}(t))$ are denoted by $\overline{w}_i$ and $\overline{m}_j$, respectively. Examples of the continuous and staircase membership functions are shown in Fig. 3.8. It can be seen that the continuous membership function is approximated by a staircase membership function with finite number of levels. The approximation error will become smaller when the step size of the SMFs is reduced.

Remark 3.6. The SMFs are designed to satisfy the properties of the membership functions in (2.3) and (2.7), respectively, i.e.,

$$\sum_{i=1}^{p}\overline{w}_i = \sum_{j=1}^{c}\overline{m}_j = \sum_{i=1}^{p}\sum_{j=1}^{c}\overline{w}_i\overline{m}_j = 1. \tag{3.28}$$

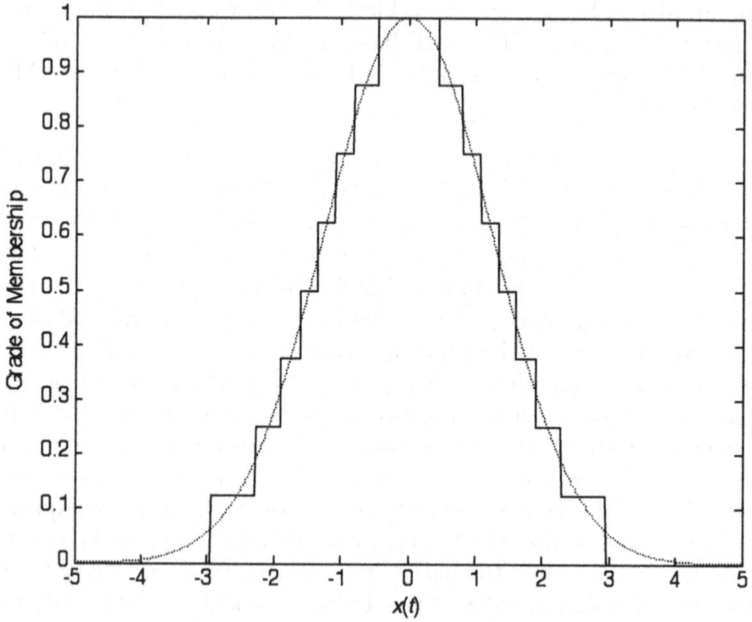

Fig. 3.8 Continuous (dotted line) and staircase (solid line) membership functions.

To investigate the system stability of the FMB control systems (2.10), we consider the quadratic Lyapunov function (3.1). From (2.10) and (3.1), we have $\dot{V}(t)$ in (3.2) and it follows that

$$\begin{aligned}\dot{V}(t) &= \sum_{i=1}^{p}\sum_{j=1}^{c} w_i m_j \mathbf{z}(t)^T \mathbf{Q}_{ij}\mathbf{z}(t)\\ &= \sum_{i=1}^{p}\sum_{j=1}^{c}\overline{w}_i\overline{m}_j\mathbf{z}(t)^T\mathbf{Q}_{ij}\mathbf{z}(t) + \sum_{i=1}^{p}\sum_{j=1}^{c}(w_i m_j - \overline{w}_i\overline{m}_j)\mathbf{z}(t)^T\mathbf{Q}_{ij}\mathbf{z}(t).\end{aligned} \tag{3.29}$$

Referring to (3.29), there are two terms on the right hand side. A possible way to achieve $V(t) > 0$, $\dot{V}(t) < 0$ (excluding $\mathbf{z}(t) = \mathbf{0}$) implying the asymptotically stability of the FMB control system (2.10) is to ensure that these two terms are negative definite. The first term $\sum_{i=1}^{p}\sum_{j=1}^{c}\overline{w}_i\overline{m}_j\mathbf{Q}_{ij}$ can be determined if it is negative definite by checking all the combinations of $\overline{w}_i\overline{m}_j$ using the nice property of the SMFs $\overline{w}_i$ and $\overline{m}_j$ that have finite numbers of levels. However, the second term containing the original membership functions will make it impossible to determine if $\sum_{i=1}^{p}\sum_{j=1}^{c}(w_i m_j - \overline{w}_i\overline{m}_j)\mathbf{Q}_{ij}$ is negative definite for every signal value of $w_i m_j$.

In order to proceed further, some slack matrices and inequalities are introduced in the following. Based on the property of the membership functions in (2.9) and (3.28), introducing the slack matrix variables $\mathbf{M} = \mathbf{M} \in \Re^{n\times n}$ to facilitate the stability analysis, we have

$$\sum_{i=1}^{p}\sum_{j=1}^{c}(w_i m_j - \overline{w}_i\overline{m}_j)\mathbf{M} = \mathbf{0}. \tag{3.30}$$

Furthermore, considering the slack matrices $0 \leq \mathbf{W}_{ij} = \mathbf{W}_{ij}^T \in \Re^{n\times n}$, it is obvious that the following inequality holds.

$$\sum_{i=1}^{p}\sum_{j=1}^{c} w_i m_j \mathbf{W}_{ij} \geq 0 \tag{3.31}$$

We further consider the following inequalities.

$$w_i m_j - \overline{w}_i\overline{m}_j - \gamma_{ij} \geq 0 \;\forall\; i,\; j,\; \mathbf{x}(t) \tag{3.32}$$

where γ_{ij} are constant scalars to be determined.

From (3.29) to (3.32), we have

$$\begin{aligned}
\dot{V}(t) \leq & \sum_{i=1}^{p}\sum_{j=1}^{c}\overline{w}_i\overline{m}_j\mathbf{z}(t)^T\mathbf{Q}_{ij}\mathbf{z}(t) + \sum_{i=1}^{p}\sum_{j=1}^{c}(w_im_j - \overline{w}_i\overline{m}_j)\mathbf{z}(t)^T\mathbf{Q}_{ij}\mathbf{z}(t) \\
& + \sum_{i=1}^{p}\sum_{j=1}^{c}(w_im_j - \overline{w}_i\overline{m}_j)\mathbf{z}(t)^T\mathbf{M}\mathbf{z}(t) + \sum_{i=1}^{p}\sum_{j=1}^{c}w_im_j\mathbf{z}(t)^T\mathbf{W}_{ij}\mathbf{z}(t) \\
= & \sum_{i=1}^{p}\sum_{j=1}^{c}\overline{w}_i\overline{m}_j\mathbf{z}(t)^T\mathbf{Q}_{ij}\mathbf{z}(t) \\
& + \sum_{i=1}^{p}\sum_{j=1}^{c}(w_im_j - \overline{w}_i\overline{m}_j)\mathbf{z}(t)^T(\mathbf{Q}_{ij} + \mathbf{M})\mathbf{z}(t) \\
& + \sum_{i=1}^{p}\sum_{j=1}^{c}(w_im_j - \overline{w}_i\overline{m}_j + \overline{w}_i\overline{m}_j)\mathbf{z}(t)^T\mathbf{W}_{ij}\mathbf{z}(t) \\
= & \sum_{i=1}^{p}\sum_{j=1}^{c}\overline{w}_i\overline{m}_j\mathbf{z}(t)^T(\mathbf{Q}_{ij} + \mathbf{W}_{ij})\mathbf{z}(t) \\
& + \sum_{i=1}^{p}\sum_{j=1}^{c}(w_im_j - \overline{w}_i\overline{m}_j)\mathbf{z}(t)^T(\mathbf{Q}_{ij} + \mathbf{W}_{ij} + \mathbf{M})\mathbf{z}(t) \\
& + \sum_{i=1}^{p}\sum_{j=1}^{c}(\gamma_{ij} - \gamma_{ij})\mathbf{z}(t)^T(\mathbf{Q}_{ij} + \mathbf{W}_{ij} + \mathbf{M})\mathbf{z}(t) \\
= & \sum_{i=1}^{p}\sum_{j=1}^{c}\mathbf{z}(t)^T\big((\overline{w}_i\overline{m}_j + \gamma_{ij})(\mathbf{Q}_{ij} + \mathbf{W}_{ij}) + \gamma_{ij}\mathbf{M}\big)\mathbf{z}(t) \\
& + \sum_{i=1}^{p}\sum_{j=1}^{c}(w_im_j - \overline{w}_i\overline{m}_j - \gamma_{ij})\mathbf{z}(t)^T(\mathbf{Q}_{ij} + \mathbf{W}_{ij} + \mathbf{M})\mathbf{z}(t). \qquad (3.33)
\end{aligned}$$

From (3.1) and (3.33), based on the the Lyapunov stability theory, $V(t) > 0$ and $\dot{V}(t) < 0$ for $\mathbf{z}(t) \neq \mathbf{0}$ ($\mathbf{x}(t) \neq \mathbf{0}$) implying the asymptotic stability of the FMB control system (2.10), i.e., $\mathbf{x}(t) \to 0$ when time $t \to \infty$ can be achieved if the stability conditions summarized in the following theorem are satisfied.

Theorem 3.3. *The FMB control system (2.10), formed by the nonlinear plant represented by the fuzzy model (2.2) and the fuzzy controller (2.6) connected in a closed loop, is asymptotically stable if there exist predefined constant scalars* γ_{ij} *satisfying* $w_i(\mathbf{x}(t))m_j(\mathbf{x}(t)) - \overline{w}_i(\mathbf{x}(t))\overline{m}_j(\mathbf{x}(t)) - \gamma_{ij} \geq 0$ *for all* i, j *and* $\mathbf{x}(t)$ *and there exist matrices* $\mathbf{M} = \mathbf{M}^T \in \Re^{n \times n}$, $\mathbf{N}_j \in \Re^{m \times n}$, $\mathbf{W}_{ij} = \mathbf{W}_{ij}^T \in \Re^{n \times n}$ *and* $\mathbf{X} = \mathbf{X}^T \in \Re^{n \times n}$ *such that the following LMIs are satisfied.*

$$\mathbf{X} > 0;$$

$$\mathbf{W}_{ij} \geq 0 \; \forall \, i, j;$$

$$\sum_{i=1}^{p}\sum_{j=1}^{c}\left((\overline{w}_i\overline{m}_j+\gamma_{ij})(\mathbf{Q}_{ij}+\mathbf{W}_{ij})+\gamma_{ij}\mathbf{M}\right)<0\,\forall\, i,\ j \text{ and valid values of } \overline{w}_i\overline{m}_j;$$

$$\mathbf{Q}_{ij}+\mathbf{W}_{ij}+\mathbf{M}<0\,\forall\, i,\ j;$$

and the feedback gains are designed as $\mathbf{G}_j=\mathbf{N}_j\mathbf{X}^{-1}$ *for all* j.

Remark 3.7. The LMIs $\sum_{i=1}^{p}\sum_{j=1}^{c}\left((\overline{w}_i\overline{m}_j+\gamma_{ij})(\mathbf{Q}_{ij}+\mathbf{W}_{ij})+\gamma_{ij}\mathbf{M}\right)<0$ in Theorem 3.3 include the SMFs of $\overline{w}_i$ and $\overline{m}_j$ (which are the approximations of original membership functions w_i and m_j, respectively). Through the SMFs, the information of the membership functions of both the fuzzy model and fuzzy controller can be taken into the stability analysis for the relaxation of the stability conditions. With the slack variable matrices $\mathbf{M}$ and $\mathbf{W}_{ij}$, the information of the membership functions can be transferred to the last LMIs in Theorem 3.3, i.e., $\mathbf{Q}_{ij}+\mathbf{W}_{ij}+\mathbf{M}<0$, to make the stability conditions to be satisfied more easily.

Remark 3.8. It can be shown that the solution $\mathbf{X}$ and $\mathbf{N}_j$ in Theorem 2.1 [15, 122] is also the solution of the stability conditions in Theorem 3.3. It was reported in [15, 122] that the FMB control system is asymptotically stable if there exist $\mathbf{X}$ and $\mathbf{N}_j$ such that $\mathbf{X}>0$ and $\mathbf{Q}_{ij}=\mathbf{X}\mathbf{A}_i^T+\mathbf{A}_i\mathbf{X}+\mathbf{N}_j^T\mathbf{B}_i^T+\mathbf{B}_i\mathbf{N}_j<0$ for all i and j (Theorem 2.1) under the case that the fuzzy model and fuzzy controller do not share the same premise membership functions. From Theorem 3.3, choosing $\mathbf{W}_{ij}=\mathbf{0}$ for all i and j and $\mathbf{M}=-\varepsilon\mathbf{I}<0$ where $\varepsilon>0$ is a scalar, the LMIs in Theorem 3.3 become $\sum_{i=1}^{p}\sum_{j=1}^{c}\overline{w}_i\overline{m}_j\mathbf{Q}_{ij}+\sum_{i=1}^{p}\sum_{j=1}^{c}\gamma_{ij}(\mathbf{Q}_{ij}-\varepsilon\mathbf{I})<0$ and $\mathbf{Q}_{ij}-\varepsilon\mathbf{I}<0$ for all i and j. As $\mathbf{Q}_{ij}<0$ for all i and j, the second LMI is satisfied and there must exist a sufficiently small value of γ_{ij} such that the first LMI is satisfied. As the SMFs $\overline{w}_i$ and $\overline{m}_j$ can be chosen arbitrarily, they can be chosen such that $w_im_j-\overline{w}_i\overline{m}_j\geq\gamma_{ij}$ are satisfied for sufficiently small values of γ_{ij}. Hence, it can be seen that the solutions of the stability conditions in [15, 122] are particular cases of Theorem 3.3.

Remark 3.9. Consider the MFB stability conditions in [2, 55, 67] and Theorem 3.1, which require that $c=p$. It can be shown that the solutions in [2, 55, 67] and Theorem 3.1 are also covered by the solution of Theorem 3.3. If there exists a solution to the stability conditions in [2, 55, 67] and Theorem 3.1, it implies that $\sum_{i=1}^{p}\sum_{j=1}^{c}w_im_j\mathbf{Q}_{ij}<0$. In this case, considering the stability conditions in Theorem 3.3, it is obvious that $\sum_{i=1}^{p}\sum_{j=1}^{c}\overline{w}_i\overline{m}_j\mathbf{Q}_{ij}<0$ as $\overline{w}_i$ and $\overline{m}_j$ can be regarded as the sampled points of w_i and m_j, respectively. Choosing $\mathbf{W}_{ij}=\mathbf{0}$ for all i and j, and $\mathbf{M}=-\varepsilon\mathbf{I}<0$ where $\varepsilon>0$ is a scalar, the LMI in Theorem 1 becomes $\sum_{i=1}^{p}\sum_{j=1}^{c}\overline{w}_i\overline{m}_j\mathbf{Q}_{ij}+\sum_{i=1}^{p}\sum_{j=1}^{c}\gamma_{ij}(\mathbf{Q}_{ij}-\varepsilon\mathbf{I})<0$, which is satisfied for sufficiently small values of γ_{ij} subject to $\sum_{i=1}^{p}\sum_{j=1}^{c}\overline{w}_i\overline{m}_j\mathbf{Q}_{ij}<0$. As the SMFs $\overline{w}_i$ and $\overline{m}_j$ can be chosen arbitrarily, they can be chosen such that $w_im_j-\overline{w}_i\overline{m}_j\geq\gamma_{ij}$ are satisfied for sufficiently small values of γ_{ij}. Similarly, the LMIs $\mathbf{Q}_{ij}+\mathbf{W}_{ij}+\mathbf{M}<0$ in Theorem 3.3 becomes $\mathbf{Q}_{ij}-\varepsilon\mathbf{I}<0$ for all i and j, which is satisfied by choosing a sufficiently large positive value of ε. Hence, it can be seen that the solutions of the stability conditions in [2, 55, 67] and Theorem 3.1 are particular cases of Theorem 3.3.

In the following, we consider the PDC design [122], i.e., $c = p$ and $m_i = w_i$ for all i, for the SMF-based stability analysis. To investigate the system stability of the FMB control system (2.10) under the PDC design approach, we proceed from (3.33) with $0 < \mathbf{W}_{ij} = \mathbf{W}_{ji}^T$ and $\gamma_{ij} = \gamma_{ji}$ for all i and j. Rewrite (3.33) as follows.

$$\begin{aligned}
\dot{V}(t) \leq & \sum_{i=1}^{p}\sum_{j=1}^{p} \mathbf{z}(t)^T \left((\overline{w}_i\overline{w}_j + \gamma_{ij})(\mathbf{Q}_{ij} + \mathbf{W}_{ij}) + \gamma_{ij}\mathbf{M}\right)\mathbf{z}(t) \\
& + \sum_{i=1}^{p}\sum_{j=1}^{p} (w_i w_j - \overline{w}_i\overline{w}_j - \gamma_{ij})\mathbf{z}(t)^T (\mathbf{Q}_{ij} + \mathbf{W}_{ij} + \mathbf{M})\mathbf{z}(t) \\
= & \sum_{i=1}^{p}\sum_{j=1}^{p} \mathbf{z}(t)^T \left((\overline{w}_i\overline{w}_j + \gamma_{ij})(\mathbf{Q}_{ij} + \mathbf{W}_{ij}) + \gamma_{ij}\mathbf{M}\right)\mathbf{z}(t) \\
& + \frac{1}{2}\sum_{i=1}^{p}\sum_{j=1}^{p} (w_i w_j - \overline{w}_i\overline{w}_j - \gamma_{ij})\mathbf{z}(t)^T (\mathbf{Q}_{ij} + \mathbf{Q}_{ji} \\
& + \mathbf{W}_{ij} + \mathbf{W}_{ji} + 2\mathbf{M})\mathbf{z}(t)
\end{aligned} \tag{3.34}$$

It is required that the inequality $\sum_{i=1}^{p}\sum_{j=1}^{p} w_i w_j \mathbf{W}_{ij} \geq 0$ holds that can be written as $\begin{bmatrix} w_1\mathbf{I} \\ w_2\mathbf{I} \\ \vdots \\ w_p\mathbf{I} \end{bmatrix}^T \mathbf{W} \begin{bmatrix} w_1\mathbf{I} \\ w_2\mathbf{I} \\ \vdots \\ w_p\mathbf{I} \end{bmatrix} \geq 0$ where $\mathbf{W} = \begin{bmatrix} \mathbf{W}_{11} & \mathbf{W}_{12} & \cdots & \mathbf{W}_{1p} \\ \mathbf{W}_{21} & \mathbf{W}_{22} & \cdots & \mathbf{W}_{2p} \\ \vdots & \vdots & \vdots & \vdots \\ \mathbf{W}_{p1} & \mathbf{W}_{p2} & \cdots & \mathbf{W}_{pp} \end{bmatrix}$. It can be seen that $\mathbf{W} \geq 0$ implies $\sum_{i=1}^{p}\sum_{j=1}^{p} w_i w_j \mathbf{W}_{ij} \geq 0$. Consequently, the stability conditions under the PDC design can be summarized in the following theorem.

Theorem 3.4. *The FMB control system (2.10), formed by the nonlinear plant represented by the fuzzy model (2.2) and the fuzzy controller (2.6) under the PDC design technique (i.e., $c = p$, $m_i(\mathbf{x}(t)) = w_i(\mathbf{x}(t))$ for all i) connected in a closed loop, is asymptotically stable if there exist predefined constant scalars $\gamma_{ij} = \gamma_{ji}$ satisfying $w_i(\mathbf{x}(t))w_j(\mathbf{x}(t)) - \overline{w}_i(\mathbf{x}(t))\overline{w}_j(\mathbf{x}(t)) - \gamma_{ij} \geq 0$ for all i, j and $\mathbf{x}(t)$, and there exist matrices $\mathbf{M} = \mathbf{M}^T \in \Re^{n \times n}$, $\mathbf{N}_j \in \Re^{m \times n}$, $\mathbf{W}_{ij} = \mathbf{W}_{ji}^T \in \Re^{n \times n}$ and $\mathbf{X} = \mathbf{X}^T \in \Re^{n \times n}$ such that the following LMIs are satisfied.*

$$\mathbf{X} > 0;$$

$$\mathbf{W} = \begin{bmatrix} \mathbf{W}_{11} & \mathbf{W}_{12} & \cdots & \mathbf{W}_{1p} \\ \mathbf{W}_{21} & \mathbf{W}_{22} & \cdots & \mathbf{W}_{2p} \\ \vdots & \vdots & \vdots & \vdots \\ \mathbf{W}_{p1} & \mathbf{W}_{p2} & \cdots & \mathbf{W}_{pp} \end{bmatrix} \geq 0;$$

$$\sum_{i=1}^{p}\sum_{j=1}^{c}\big((\overline{w}_i\overline{w}_j + \gamma_{ij})(\mathbf{Q}_{ij} + \mathbf{W}_{ij}) + \gamma_{ij}\mathbf{M}\big) < 0 \,\forall\, i,\ j \text{ and valid values of } \overline{w}_i\overline{w}_j;$$

$$\mathbf{Q}_{ij} + \mathbf{Q}_{ji} + \mathbf{W}_{ij} + \mathbf{W}_{ji} + 2\mathbf{M} < 0 \,\forall\, i,\ j;$$

and the feedback gains are designed as $\mathbf{G}_j = \mathbf{N}_j\mathbf{X}^{-1}$ *for all* j.

Remark 3.10. It can be shown that the solution of some published stability conditions under PDC design with MFSI analysis (Theorem 2.2 to Theorem 2.5) and MFSD analysis [24, 48, 81, 82, 91, 97–99, 105, 112, 122, 125] are also covered by Theorem 3.4. If there exists a solution for the stability conditions in [24, 48, 81, 82, 91, 97–99, 105, 112, 122, 125], it implies that $\sum_{i=1}^{p}\sum_{j=1}^{c} w_i w_j \mathbf{Q}_{ij} < 0$. As $\overline{w}_i$ can be regarded as the sampled points of w_i, it is obvious that $\sum_{i=1}^{p}\sum_{j=1}^{c} \overline{w}_i\overline{w}_j\mathbf{Q}_{ij} < 0$. We choose $\mathbf{W}_{ij} = \mathbf{0}$ for all i and j, and $\mathbf{M} = -\varepsilon\mathbf{I} < 0$ where $\varepsilon > 0$ is a scalar, and $\overline{w}_i$ such that $w_i w_j - \overline{w}_i\overline{w}_j \geq \gamma_{ij}$ with a sufficiently small value of γ_{ij}. It can be seen that the LMIs in Theorem 3.4 become $\sum_{i=1}^{p}\sum_{j=1}^{c} \overline{w}_i\overline{w}_j\mathbf{Q}_{ij} + \sum_{i=1}^{p}\sum_{j=1}^{c} \gamma_{ij}(\mathbf{Q}_{ij} - \varepsilon\mathbf{I}) < 0$ and $\mathbf{Q}_{ij} + \mathbf{Q}_{ji} - 2\varepsilon\mathbf{I} < 0$ which are satisfied for sufficiently small values of γ_{ij} and sufficiently large value of ε subject to $\sum_{i=1}^{p}\sum_{j=1}^{c} \overline{w}_i\overline{w}_j\mathbf{Q}_{ij} < 0$. Hence, it can be seen that the solutions of the stability conditions with MFSI analysis (Theorem 2.2 to Theorem 2.5) and MFSD analysis [24, 48, 81, 82, 91, 97–99, 105, 112, 122, 125] are particular cases of Theorem 3.4.

Remark 3.11. The LMI-based performance conditions given in Theorem 3.2 can be applied with the stability conditions in Theorem 3.3 and Theorem 3.4 to realize the performance design.

Example 3.3. Consider the example FMB control system in Example 2.1 with the system parameters $2 \leq a \leq 9$ and $2 \leq b \leq 22$. The membership functions in Example 3.1 as shown in Fig. 3.1 are used in this example. The SMFs are chosen as $\overline{w}_i(x_1(t)) = w_i(h\delta)$ and $\overline{m}_j(x_1(t)) = m_j(h\delta)$ for $(h - 0.5)\delta < x_1(t) \leq (h+0.5)\delta$ where $i, j = 1, 2, 3$ and $h = -\infty, \cdots, -10, -9, ...10, \cdots, \infty$. As the grades of membership keep constant for $x_1(t) > 10$ or $x_1(t) < -10$, we have $w_1(x_1(t)) = w_1(10)$ for $x_1(t) > 10$ and $w_1(x_1(t)) = w_1(-10)$ for $x_1(t) < -10$, and we only need to consider $h = -10, -9, \cdots, 10$.

The system stability of the FMB control system is examined by using the stability conditions in Theorem 3.3 with the MATLAB LMI toolbox. For demonstration purposes, we choose $\delta = 0.1$ and $\delta = 0.05$ to demonstrate the effect of the value of δ to the stability conditions. Table 3.3 lists the values of γ_{ij} which satisfies the inequalities of $w_i(x_1(t))m_j(x_1(t)) - \overline{w}_i(x_1(t))\overline{m}_j(x_1(t)) - \gamma_{ij} \geq 0$ for all i and j.

Table 3.3 Values of δ and γ_{ij} for Example 3.3.

$\delta = 0.1$	$\gamma_{11} = -0.006327$, $\gamma_{12} = -0.003277$, $\gamma_{13} = -0.001006$,
	$\gamma_{21} = -0.003955$, $\gamma_{22} = -0.005526$, $\gamma_{23} = -0.003955$,
	$\gamma_{31} = -0.001060$, $\gamma_{32} = -0.003279$, $\gamma_{33} = -0.006326$.
$\delta = 0.05$	$\gamma_{11} = -0.003164$, $\gamma_{12} = -0.001642$, $\gamma_{13} = -0.000545$,
	$\gamma_{21} = -0.001981$, $\gamma_{22} = -0.002763$, $\gamma_{23} = -0.001966$,
	$\gamma_{31} = -0.000545$, $\gamma_{32} = -0.001642$, $\gamma_{33} = -0.003164$.

The stability regions are shown in Fig. 3.9 indicated by '×' and '∘', respectively. It can be seen from Fig. 3.9 that a smaller value of δ is able to produce a larger stability region as the SMFs are able to better approximate their corresponding original membership functions with smaller approximation error. For comparison purposes, the non-PDC stability conditions in Theorem 2.1 are also applied to the FMB control systems. However, there is no stability region found.

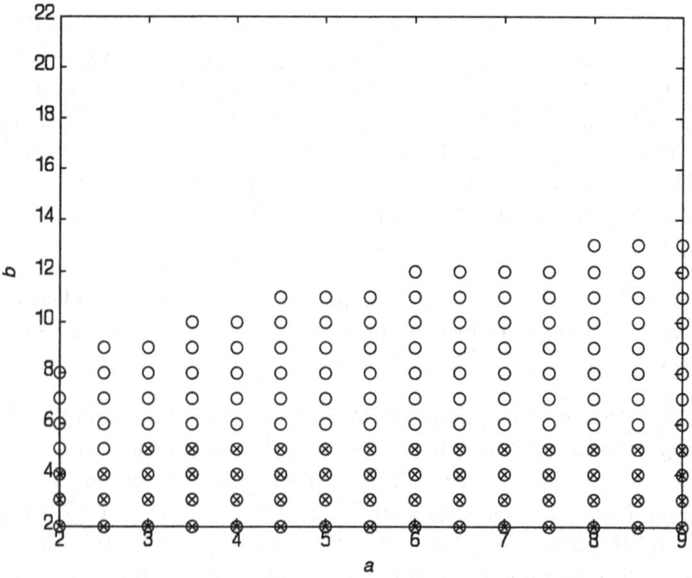

Fig. 3.9 Stability regions given by the stability conditions in Theorem 3.3 with $\delta = 0.1$ ('×') and $\delta = 0.05$ ('∘') for Example 3.3.

Example 3.4. We continue from Example 3.3 and consider the situation that $c = p$ and $m_i = w_i$ for the TS fuzzy model and the fuzzy controller (both of them share the same premise membership functions). The SMFs are chosen

Table 3.4 Values of δ and γ_{ij} for Example 3.4.

$\delta = 0.1$	$\gamma_{11} = -0.008177$, $\gamma_{12} = \gamma_{21} = -0.004115$,
	$\gamma_{13} = \gamma_{31} = -0.001337$, $\gamma_{22} = -0.005538$,
	$\gamma_{23} = \gamma_{32} = -0.004115$, $\gamma_{33} = -0.008247$.
$\delta = 0.05$	$\gamma_{11} = -0.004145$, $\gamma_{12} = \gamma_{21} = -0.002070$,
	$\gamma_{13} = \gamma_{31} = -0.000681$, $\gamma_{22} = -0.002773$,
	$\gamma_{23} = \gamma_{32} = -0.002018$, $\gamma_{33} = -0.004162$.

as $\overline{w}_i(x_1(t)) = w_i(h\delta)$ for $(h - 0.5)\delta < x_1(t) \leq (h + 0.5)\delta$ where $i = 1, 2, 3$ and $h = -\infty, \cdots, -10, -9, ...10, \cdots, \infty$.

Similarly, choosing $\delta = 0.1$ and $\delta = 0.05$, we have the values of γ_{ij} listed in Table 3.4 which satisfies the inequalities of $w_i(x_1(t))w_j(x_1(t)) - \overline{w}_i(x_1(t))\overline{w}_j(x_1(t)) - \gamma_{ij} \geq 0$ for all i and j.

The stability regions given by the stability conditions in Theorem 3.4 are shown in Fig. 3.10, indicated by '$\times$' and '$\circ$', respectively. For comparison purposes, the stability conditions in Theorem 2.2 to Theorem 2.5, where Theorem 2.5 is applied with the dimension of fuzzy summation $d = 4$, are employed and the corresponding stability regions are shown in Fig. 2.1. It can be seen from these figures that the proposed stability conditions in Theorem 3.4 are able to produce a larger stability region.

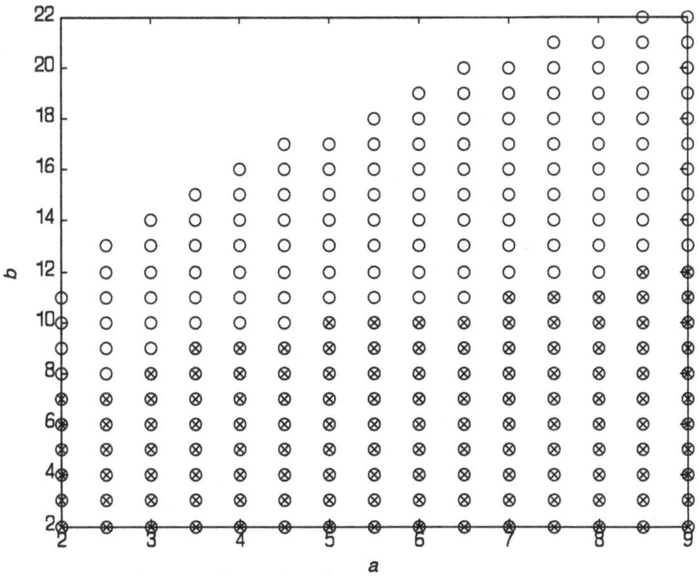

Fig. 3.10 Stability regions given by the stability conditions in Theorem 3.4 with $\delta = 0.1$ ('$\times$') and $\delta = 0.05$ ('$\circ$') for Example 3.4.

3.4 Conclusion

The stability analysis of FMB control systems with imperfectly matched premise membership functions have been investigated under the MFSD approaches. Two approaches, namely the MFB and the SMF approaches, have been proposed. In the MFB approach, the boundary information of the membership functions has been utilized to conduct stability analysis. Based on the Lyapunov stability theory, LMI stability and performance conditions have been derived to guarantee the system stability and performance. In the SMF approach, staircase membership functions have been employed to approximate the continuous membership functions of the TS fuzzy model and fuzzy controller. Through SMFs, the nonlinearity of membership functions can be brought into the stability conditions. Consequently, the stability conditions are more dedicated to the FMB control systems as compared with the MFSI-based ones. Unlike the MFSI analysis approach, the shape information of the membership functions is able to be brought into the stability conditions in the MFSD analysis approach for relaxing the stability conditions. The SMF approach has also been extended to the PDC design (with perfectly matched premise membership functions). The SMF approach has also been extended to the PDC design (with perfectly matched premise membership functions.) Simulation and comparison examples have been presented to illustrate the effectiveness of the proposed MFSD-based stability conditions.

Chapter 4
BMI Stability Conditions for FMB Control Systems

4.1 Introduction

In general, two classes of FMB control systems, namely with imperfectly and perfectly matched premise membership functions, have been investigated in the past decades. The first class of FMB control systems with imperfectly matched premise membership functions has the TS fuzzy model and the fuzzy controller not sharing the same premise membership functions. This class of FMB control systems has been extensively investigated recently. Based on the Lyapunov stability theory, LMI-based stability conditions were derived to guarantee the system stability and facilitate the synthesis of the fuzzy controller. With the imperfectly matched premise membership functions [2, 55, 67, 122], the fuzzy controller exhibits two favourable features. One, the premise membership functions of the fuzzy controller can be freely chosen to enhance the design flexibility. Some simple and commonly-used membership functions can be employed to lower the structural complexity, computational demand and implementation cost of the fuzzy controller. Two, the fuzzy controller displays an inherent robustness property to handle parameter uncertainties of the nonlinear plant. As the stability analysis and the fuzzy controller do not involve the membership functions of the TS fuzzy plant model, the fuzzy controller designed with imperfectly matched premise membership functions is able to stabilize nonlinear plant with its fuzzy model subject to uncertain grades of membership owing to the presence of plant parameter uncertainties. However, it will lead to conservative stability conditions as the membership functions of the fuzzy model are not considered in the stability analysis. This problem is partially alleviated by the second class of FMB control systems design. This class of FMB control systems is designed with perfectly matched premise membership functions in which the fuzzy controller shares the same premise membership functions as those of the TS fuzzy model for relaxing the stability conditions [24, 48, 81, 82, 97, 105, 112, 122, 125]. However, as the grades of

H.-K. Lam and F.H.F. Leung: FMB Control Systems, STUDFUZZ 264, pp. 59–84.
springerlink.com

membership function are required to be known, the TS fuzzy model in the form of (2.2) must be uncertainty free. Hence, with the perfectly matched premise membership functions, the stability conditions are relaxed by sacrificing the inherent robustness property of the fuzzy controller. Both fuzzy controllers, designed with the imperfectly and perfectly matched premise membership functions, have their own advantages to various applications.

In this chapter, the stability of FMB control systems with imperfectly matched premise membership functions is investigated. To facilitate the stability analysis, the information of the membership functions of both the TS fuzzy model and fuzzy controller are employed. In order to further relax the stability conditions, a fuzzy controller with time-varying state-feedback gains [19, 35, 36] is proposed. Based on the Lyapunov-based approach and the knowledge on the membership functions of the fuzzy model, stability conditions in bilinear matrix inequalities (BMI) with imperfectly matched premise membership functions are derived to guarantee the system stability. As the stability conditions are in terms of BMIs, convex programming techniques cannot be applied directly to find the solution. Taking advantage of the powerful global searching ability of the genetic algorithm (GA) [90], a GA-based convex programming technique is proposed to search for the solution of the BMI-based stability conditions. Some simulation examples are given to illustrate the effectiveness of the proposed BMI-based fuzzy control scheme.

4.2 Fuzzy Controller with Time-Varying Feedback Gains

A p-rule fuzzy controller with time-varying feedback gains is employed to control the nonlinear plant represented by the TS fuzzy model (2.2). The j-th rule of the fuzzy controller is of the following format:

$$\begin{aligned}&\text{Rule } j\text{: IF } g_1(\mathbf{x}(t)) \text{ is } N_1^j \text{ AND } \cdots \text{ AND } g_\Omega(\mathbf{x}(t)) \text{ is } N_\Omega^j\\&\qquad\text{THEN } \mathbf{u}(t) = \mathbf{F}_j(\mathbf{x}(t))\mathbf{x}(t)\end{aligned} \tag{4.1}$$

where $\mathbf{F}_j(\mathbf{x}(t)) \in \Re^{m\times n}$, $j = 1, 2, \cdots, c$, are time-varying feedback gains to be determined and the rest variables are defined in Section 2.3.

The fuzzy controller with time-varying feedback gains is defined as follows,

$$\mathbf{u}(t) = \sum_{j=1}^{p} m_j(\mathbf{x}(t))\mathbf{F}_j(\mathbf{x}(t))\mathbf{x}(t) \tag{4.2}$$

where the membership functions exhibit the property in (2.7).

The feedback gains are defined as $\mathbf{F}_j(\mathbf{x}(t)) = \frac{\mathbf{G}_j}{\sum_{k=1}^{p} m_k(\mathbf{x}(t))a_k}$ for all j, where $\mathbf{G}_j \in \Re^{m\times n}$ are some constant feedback gains to be determined. The time-varying feedback gains $\mathbf{F}_j(\mathbf{x}(t))$ are used to enhance the nonlinearity compensation of the nonlinear feedback plant dynamics. From (4.2), we have,

$$\mathbf{u}(t) = \frac{\sum_{j=1}^{p} m_j(\mathbf{x}(t))\mathbf{G}_j\mathbf{x}(t)}{\sum_{k=1}^{p} m_k(\mathbf{x}(t))a_k} \tag{4.3}$$

where a_k, $k = 1, 2, \cdots, p$, are scalars to be designed that satisfy the inequality of $\sum_{k=1}^{p} m_k(\mathbf{x}(t))a_k > 0$.

Remark 4.1. The proposed fuzzy controller of (4.3) is equivalent to that in [36] when $m_i(\mathbf{x}(t)) = w_i(\mathbf{x}(t))$ for all i. It is reduced to the traditional fuzzy controller (2.6) [122] when the constant scalars $a_k = 1$ for all k.

4.3 BMI-Based Stability Analysis

Consider the nonlinear plant represented by the TS fuzzy model (2.2) and the fuzzy controller (4.3) connected in a closed loop. The FMB control system is obtained as follows.

$$\begin{aligned}\dot{\mathbf{x}}(t) &= \sum_{i=1}^{p} w_i(\mathbf{x}(t))\left(\mathbf{A}_i\mathbf{x}(t) + \mathbf{B}_i\frac{\sum_{j=1}^{p} m_j(\mathbf{x}(t))\mathbf{G}_j\mathbf{x}(t)}{\sum_{k=1}^{p} m_k(\mathbf{x}(t))a_k}\right) \\ &= \frac{1}{\sum_{k=1}^{p} m_k(\mathbf{x}(t))a_k}\sum_{i=1}^{p}\sum_{j=1}^{c} w_i(\mathbf{x}(t))m_j(\mathbf{x}(t))(a_k\mathbf{A}_i + \mathbf{B}_i\mathbf{G}_j)\mathbf{x}(t) \end{aligned} \tag{4.4}$$

In the following, for brevity, $w_i(\mathbf{x}(t))$ and $m_j(\mathbf{x}(t))$ are denoted as w_i and m_j. To investigate the stability of the FMB control system (4.4), with the membership function property (2.9), we consider the quadratic Lyapunov function (3.1). From (3.1) and (4.4), we have

$$\begin{aligned}\dot{V}(t) &= \dot{\mathbf{x}}(t)^T\mathbf{P}\mathbf{x}(t) + \mathbf{x}(t)^T\mathbf{P}\dot{\mathbf{x}}(t) \\ &= \left(\frac{1}{\sum_{k=1}^{p} m_k a_k}\sum_{i=1}^{p}\sum_{j=1}^{p} w_i m_j(a_j\mathbf{A}_i + \mathbf{B}_i\mathbf{G}_j)\mathbf{x}(t)\right)^T\mathbf{P}\mathbf{x}(t) \\ &\quad + \mathbf{x}(t)^T\mathbf{P}\left(\frac{1}{\sum_{k=1}^{p} m_k a_k}\sum_{i=1}^{p}\sum_{j=1}^{p} w_i m_j(a_j\mathbf{A}_i + \mathbf{B}_i\mathbf{G}_j)\mathbf{x}(t)\right) \\ &= \frac{1}{\sum_{k=1}^{p} m_k a_k}\sum_{i=1}^{p}\sum_{j=1}^{p} w_i m_j\mathbf{z}(t)^T\overline{\mathbf{Q}}_{ij}\mathbf{z}(t) \end{aligned} \tag{4.5}$$

where $\overline{\mathbf{Q}}_{ij} = a_j\mathbf{A}_i\mathbf{X} + a_j\mathbf{X}\mathbf{A}_i^T + \mathbf{B}_i\mathbf{N}_j + \mathbf{N}_j^T\mathbf{B}_i^T$.

Remark 4.2. Compared with $\dot{V}(t)$ in (3.2), $\overline{\mathbf{Q}}_{ij}$ defined in this chapter has the variables a_j that effectively increases the dimension of the feasible solution and thus possibly leads to more relaxed stability analysis result. However, the terms $a_j\mathbf{A}_i\mathbf{X}$ and $a_j\mathbf{X}\mathbf{A}_i^T$ are not in the form of LMIs but BMIs. Convex programming techniques cannot be applied to obtain numerically the solution to the BMIs. In the following, a GA-based convex programming technique is proposed to search for the solution to the BMIs.

In order to proceed with the stability analysis, a number of slack matrix variables are introduced to (4.5) based on the S-procedure [6]. Introducing slack matrices $\mathbf{\Lambda}_i = \mathbf{\Lambda}_i^T \in \Re^{n\times n}$, $\mathbf{V}_i = \mathbf{V}_i^T \in \Re^{n\times n}$, $0 \leq \mathbf{J}_{ij} = \mathbf{J}_{ij}^T \in \Re^{n\times n}$, $0 \leq \mathbf{K}_{ij} = \mathbf{K}_{ij}^T \in \Re^{n\times n}$, $0 \leq \mathbf{M}_{ij} = \mathbf{M}_{ij}^T \in \Re^{n\times n}$ and $0 \leq \mathbf{W}_{ij} = \mathbf{W}_{ij}^T \in \Re^{n\times n}$; $i, j = 1, 2, \cdots, p$, we consider the following equations and inequalities.

$$\sum_{i=1}^{p}\sum_{j=1}^{p} w_i(w_j - m_j)\mathbf{\Lambda}_i = \mathbf{0} \tag{4.6}$$

$$\sum_{i=1}^{p}\sum_{j=1}^{p} m_i(w_j - m_j)\mathbf{V}_i = \mathbf{0} \tag{4.7}$$

$$\sum_{i=1}^{p}\sum_{j=1}^{p} w_i(w_j - m_j + \sigma_j)\mathbf{J}_{ij} \geq 0 \tag{4.8}$$

$$\sum_{i=1}^{p}\sum_{j=1}^{p} m_i(\gamma_j - w_j + m_j)\mathbf{K}_{ij} \geq 0 \tag{4.9}$$

$$\sum_{i=1}^{p}\sum_{j=1}^{p}(w_i w_j - w_i m_j - m_i w_j + m_i m_j + \delta_{ij})\mathbf{M}_{ij} \geq 0 \tag{4.10}$$

$$\sum_{i=1}^{p}\sum_{j=1}^{p}(\rho_{ij} - w_i w_j + w_i m_j + m_i w_j - m_i m_j)\mathbf{W}_{ij} \geq 0 \tag{4.11}$$

where σ_j, γ_j, δ_{ij} and ρ_{ij} are scalars to be determined such that the following inequalities hold for all i, j and $\mathbf{x}(t)$.

$$w_j - m_j + \sigma_j \geq 0 \;\forall\, j \tag{4.12}$$

$$\gamma_j - w_j + m_j \geq 0 \;\forall\, j \tag{4.13}$$

$$w_i w_j - w_i m_j - m_i w_j + m_i m_j + \delta_{ij} \geq 0 \,\forall\, i,\ j \tag{4.14}$$

$$\rho_{ij} - w_i w_j + w_i m_j + m_i w_j - m_i m_j \geq 0 \,\forall\, i,\ j \tag{4.15}$$

Remark 4.3. $\mathbf{\Lambda}_i$ and $\mathbf{V}_i$ in (4.6) and (4.7), respectively, are referred to as MFSI slack matrices as they do not depend on the membership functions. As a result, (4.6) and (4.7) are satisfied for any shapes of membership function. These two equations serve an important purpose that they introduce free matrix variables transferring and compensating some unstable components between LMI terms to make the stability conditions to be satisfied more easily. In order words, the free matrix variables increase the dimension of the feasible solution. Conversely, $\mathbf{J}_{ij}$, $\mathbf{K}_{ij}$, $\mathbf{M}_{ij}$ and $\mathbf{M}_{ij}$ in the inequalities (4.8) to (4.11), respectively, are refereed to as MFSD slack matrices as they depend on the shapes of the membership functions. It is required that the membership functions satisfy the inequalities (4.12) to (4.15). The scalars of σ_j, γ_j, δ_{ij} and ρ_{ij} provide the boundary information of the membership functions and their multiplications. The slack matrix variables and the membership function boundary information offer useful information to facilitate the stability analysis for relaxing the stability conditions.

Remark 4.4. It was reported that further relaxed stability conditions can be achieved by considering some polynomial inequality [99] of membership functions. The higher the degree of the polynomial inequality, the more the number of slack matrices can be introduced. More slack variables can be introduced to the stability analysis based on the Polya's theorem [91] that the boundary information of the higher order of multiplications of membership functions is considered. However, it should be noted that a larger number of slack matrices will lead to higher computational demand on searching for the solution of the stability conditions.

From (4.5) to (4.11), recalling that $\sum_{k=1}^{p} m_k a_k > 0$ and defining $\mathbf{X} = \mathbf{P}^{-1}$, $\mathbf{z}(t) = \mathbf{X}^{-1}\mathbf{x}(t)$ and $\mathbf{G}_j = \mathbf{N}_j \mathbf{X}^{-1}$ where $\mathbf{N}_j$, $j = 1, 2, \cdots, p$; it follows that

$$\dot{V}(t) \leq \frac{1}{\sum_{k=1}^{p} m_k a_k}\mathbf{z}(t)^T\Big(\sum_{i=1}^{p}\sum_{j=1}^{p} w_i m_j \overline{\mathbf{Q}}_{ij} + \sum_{j=1}^{p} w_i(w_j - m_j)\mathbf{\Lambda}_i$$
$$+ \sum_{j=1}^{p} m_i(w_j - m_j)\mathbf{V}_i + \sum_{j=1}^{p} w_i(w_j - m_j + \sigma_j)\mathbf{J}_{ij}$$
$$+ \sum_{i=1}^{p}\sum_{j=1}^{p} m_i(\gamma_j - w_j + m_j)\mathbf{K}_{ij}$$
$$+ \sum_{i=1}^{p}\sum_{j=1}^{p}(w_i w_j - w_i m_j - m_i w_j + m_i m_j + \delta_{ij})\mathbf{M}_{ij}$$
$$+ \sum_{i=1}^{p}\sum_{j=1}^{p}(\rho_{ij} - w_i w_j + w_i m_j + m_i w_j - m_i m_j)\mathbf{W}_{ij}\Big)\mathbf{z}(t) \quad (4.16)$$

Rearranging the terms in (4.16), it can be written as follows.

$$\dot{V}(t) \leq= \frac{1}{\sum_{k=1}^{p} m_k a_k}\mathbf{z}(t)^T\Big(\sum_{i=1}^{p}\sum_{j=1}^{p} w_i m_j\big(\overline{\mathbf{Q}}_{ij} - \mathbf{\Lambda}_i + \mathbf{V}_j - \mathbf{J}_{ij} - \mathbf{K}_{ji} - \mathbf{M}_{ij}$$
$$- \mathbf{M}_{ji} + \mathbf{W}_{ij} + \mathbf{W}_{ji}\big) + \sum_{i=1}^{p}\sum_{j=1}^{p} w_i w_j\big(\mathbf{M}_{ij} - \mathbf{W}_{ij} + \mathbf{\Lambda}_i + \mathbf{J}_{ij}$$
$$+ \sum_{k=1}^{p}\sigma_k \mathbf{J}_{ik} + \sum_{k=1}^{p}\sum_{l=1}^{p}\delta_{kl}\mathbf{M}_{kl} + \sum_{k=1}^{p}\sum_{l=1}^{p}\rho_{kl}\mathbf{W}_{kl}\big)$$
$$+ \sum_{i=1}^{p}\sum_{j=1}^{p} m_i m_j\big(\mathbf{M}_{ij} - \mathbf{W}_{ij} - \mathbf{V}_i + \mathbf{K}_{ij} + \sum_{k=1}^{p}\gamma_k \mathbf{K}_{ik}\big)\Big)\mathbf{z}(t). \quad (4.17)$$

Introducing matrices $\mathbf{R}_{ij} = \mathbf{R}_{ji}^T \in \Re^{n\times n}$, $\mathbf{S}_{ij} = \mathbf{S}_{ji}^T \in \Re^{n\times n}$ and $\mathbf{T}_{ij} = \mathbf{T}_{ji}^T \in \Re^{n\times n}$, $i, j = 1, 2, \cdots, p$; we consider

$$\mathbf{R}_{ii} > \mathbf{M}_{ii} - \mathbf{W}_{ii} - \mathbf{V}_i + \mathbf{K}_{ii} + \sum_{k=1}^{p}\gamma_k \mathbf{K}_{ik} \ \forall\, i; \quad (4.18)$$

$$\mathbf{R}_{ij} + \mathbf{R}_{ij}^T \geq \mathbf{M}_{ij} - \mathbf{W}_{ij} - \mathbf{V}_i + \mathbf{K}_{ij} + \sum_{k=1}^{p}\gamma_k \mathbf{K}_{ik}$$
$$+ \mathbf{M}_{ji} - \mathbf{W}_{ji} - \mathbf{V}_j + \mathbf{K}_{ji} + \sum_{k=1}^{p}\gamma_k \mathbf{K}_{jk} \ \forall\, j; i < j; \quad (4.19)$$

$$\mathbf{S}_{ij} + \mathbf{S}_{ij}^T \geq \overline{\mathbf{Q}}_{ij} - \mathbf{\Lambda}_i + \mathbf{V}_j - \mathbf{J}_{ij} - \mathbf{K}_{ji} - \mathbf{M}_{ij} - \mathbf{M}_{ji} + \mathbf{W}_{ij} + \mathbf{W}_{ji} \ \forall\, i,\ j; \tag{4.20}$$

$$\mathbf{T}_{ii} > \mathbf{M}_{ii} - \mathbf{W}_{ii} + \mathbf{\Lambda}_i + \mathbf{J}_{ii} + \sum_{k=1}^{p} \sigma_k \mathbf{J}_{ik} + \sum_{k=1}^{p}\sum_{l=1}^{p} \delta_{kl}\mathbf{M}_{kl} + \sum_{k=1}^{p}\sum_{l=1}^{p} \rho_{kl}\mathbf{W}_{kl} \ \forall\, i; \tag{4.21}$$

$$\mathbf{T}_{ij} + \mathbf{T}_{ij}^T \geq \mathbf{M}_{ij} - \mathbf{W}_{ij} + \mathbf{\Lambda}_i + \mathbf{J}_{ij} + \mathbf{M}_{ji} - \mathbf{W}_{ji} + \mathbf{\Lambda}_j + \mathbf{J}_{ji} + \sum_{k=1}^{p} \sigma_k \mathbf{J}_{ik} + \sum_{k=1}^{p} \sigma_k \mathbf{J}_{jk} + 2\sum_{k=1}^{p}\sum_{l=1}^{p} \delta_{kl}\mathbf{M}_{kl} + 2\sum_{k=1}^{p}\sum_{l=1}^{p} \rho_{kl}\mathbf{W}_{kl} \ \forall\, j; i < j. \tag{4.22}$$

From (4.17) to (4.22), we have

$$\begin{aligned}\dot{V}(t) &\leq \frac{1}{\sum_{k=1}^{p} m_k a_k} \mathbf{z}(t)^T \Big(\sum_{i=1}^{p}\sum_{j=1}^{p} w_i m_j (\mathbf{S}_{ij} + \mathbf{S}_{ij}^T) \\ &\quad + \sum_{i=1}^{p} w_i^2 \mathbf{T}_{ii} + \sum_{j=1}^{p}\sum_{i<j} w_i w_j (\mathbf{T}_{ij} + \mathbf{T}_{ij}^T) \\ &\quad + \sum_{i=1}^{p} m_i^2 \mathbf{R}_{ii}^T + \sum_{j=1}^{p}\sum_{i<j} m_i m_j (\mathbf{R}_{ij} + \mathbf{R}_{ij}^T) \Big) \mathbf{z}(t) \\ &= \frac{1}{\sum_{k=1}^{p} m_k a_k} \begin{bmatrix} \mathbf{r}(t) \\ \mathbf{s}(t) \end{bmatrix}^T \begin{bmatrix} \mathbf{R} & \mathbf{S}^T \\ \mathbf{S} & \mathbf{T} \end{bmatrix} \begin{bmatrix} \mathbf{r}(t) \\ \mathbf{s}(t) \end{bmatrix}. \end{aligned} \tag{4.23}$$

where $\mathbf{r}(t) = \begin{bmatrix} m_1\mathbf{z}(t) \\ m_2\mathbf{z}(t) \\ \vdots \\ m_p\mathbf{z}(t) \end{bmatrix}$, $\mathbf{s}(t) = \begin{bmatrix} w_1\mathbf{z}(t) \\ w_2\mathbf{z}(t) \\ \vdots \\ w_p\mathbf{z}(t) \end{bmatrix}$, $\mathbf{R} = \begin{bmatrix} \mathbf{R}_{11} & \mathbf{R}_{12} & \cdots & \mathbf{R}_{1p} \\ \mathbf{R}_{21} & \mathbf{R}_{22} & \cdots & \mathbf{R}_{2p} \\ \vdots & \vdots & \vdots & \vdots \\ \mathbf{R}_{p1} & \mathbf{R}_{p2} & \cdots & \mathbf{R}_{pp} \end{bmatrix}$, $\mathbf{S} = \begin{bmatrix} \mathbf{S}_{11} & \mathbf{S}_{12} & \cdots & \mathbf{S}_{1p} \\ \mathbf{S}_{21} & \mathbf{S}_{22} & \cdots & \mathbf{S}_{2p} \\ \vdots & \vdots & \vdots & \vdots \\ \mathbf{S}_{p1} & \mathbf{S}_{p2} & \cdots & \mathbf{S}_{pp} \end{bmatrix}$ and $\mathbf{T} = \begin{bmatrix} \mathbf{T}_{11} & \mathbf{T}_{12} & \cdots & \mathbf{T}_{1p} \\ \mathbf{T}_{21} & \mathbf{T}_{22} & \cdots & \mathbf{T}_{2p} \\ \vdots & \vdots & \vdots & \vdots \\ \mathbf{T}_{p1} & \mathbf{T}_{p2} & \cdots & \mathbf{T}_{pp} \end{bmatrix}$.

From (3.1) and (3.13), based on the the Lyapunov stability theory, $V(t) > 0$ and $\dot{V}(t) < 0$ for $\mathbf{z}(t) \neq \mathbf{0}$ ($\mathbf{x}(t) \neq \mathbf{0}$) implying the asymptotic stability of the FMB control system (4.4), i.e., $\mathbf{x}(t) \to 0$ when time $t \to \infty$, can be achieved if the stability conditions summarized in the following theorem are satisfied.

Theorem 4.1. *The FMB control system (4.4), formed by the nonlinear plant represented by the fuzzy model (2.2) and the fuzzy controller (4.2) connected in a closed loop, is asymptotically stable if there exist predefined constant scalars* σ_j, γ_j, δ_{ij} *and* ρ_{ij}, $i, j = 1, 2, \cdots, p$, *satisfying the inequalities (4.8) to (4.11) for all* i, j *and* $\mathbf{x}(t)$, *and there exist matrices* $\mathbf{J}_{ij} = \mathbf{J}_{ij}^T \in \Re^{n\times n}$, $\mathbf{K}_{ij} = \mathbf{K}_{ij}^T \in \Re^{n\times n}$, $\mathbf{M}_{ij} = \mathbf{M}_{ij}^T \in \Re^{n\times n}$, $\mathbf{N}_j \in \Re^{m\times n}$, $\mathbf{R}_{ij} = \mathbf{R}_{ji}^T \in \Re^{n\times n}$, $\mathbf{S}_{ij} = \mathbf{S}_{ji}^T \in \Re^{n\times n}$, $\mathbf{T}_{ij} = \mathbf{T}_{ji}^T \in \Re^{n\times n}$, $\mathbf{V}_i = \mathbf{V}_i^T \in \Re^{n\times n}$, $\mathbf{W}_{ij} = \mathbf{W}_{ij}^T \in \Re^{n\times n}$, $\mathbf{X} = \mathbf{X}^T \in \Re^{n\times n}$ *and* $\mathbf{\Lambda}_i = \mathbf{\Lambda}_i^T \in \Re^{n\times n}$ *such that the following BMIs and LMIs are satisfied.*

$$\mathbf{X} > 0;$$

$$\mathbf{J}_{ij} \geq 0 \; \forall \; i \; j;$$

$$\mathbf{K}_{ij} \geq 0 \; \forall \; i, \; j;$$

$$\mathbf{M}_{ij} \geq 0 \; \forall \; i, \; j;$$

$$\mathbf{W}_{ij} \geq 0 \; \forall \; i, \; j;$$

BMIs/LMIs in (4.18) to (4.22);

$$\begin{bmatrix} \mathbf{R} & \mathbf{S}^T \\ \mathbf{S} & \mathbf{T} \end{bmatrix} < 0;$$

and the feedback gains are designed as $\mathbf{G}_j = \mathbf{N}_j \mathbf{X}^{-1}$ *for all* j.

Remark 4.5. The BMI-based stability conditions in Theorem 4.1 apply to the FMB control system with imperfectly matched membership functions. It can be shown that the solution of the stability conditions in Theorem 2.1 is a particular case of the proposed BMI-based stability conditions. Considering there exists a solution $\mathbf{X}$ to the stability conditions in Theorem 2.1, we have $\mathbf{Q}_{ij} = \mathbf{A}_i\mathbf{X} + \mathbf{X}\mathbf{A}_i^T + \mathbf{B}_i\mathbf{N}_j + \mathbf{N}_j^T\mathbf{B}_i^T < 0$ for all i and j. Choosing the same $\mathbf{X}$ as the solution and $a_j = 1$ for all j for the stability conditions in Theorem 4.1, we have $\overline{\mathbf{Q}}_{ij} = \mathbf{Q}_{ij} < 0$ for all i and j. Then, we choose the slack matrices $\mathbf{J}_{ij} = \mathbf{K}_{ij} = \mathbf{M}_{ij} = \mathbf{W}_{ij} = \mathbf{0}$ for all i and j. As a result, (4.18) and (4.19) become $\mathbf{R}_{ii} > -\mathbf{V}_i$, $i = 1, 2, \cdots, p$, and $\mathbf{R}_{ij} + \mathbf{R}_{ij}^T \geq -\mathbf{V}_i - \mathbf{V}_j$, $j = 1, 2, \cdots, p$, $i < j$. By choosing $\mathbf{R}_{ii} = -\tau\mathbf{I}$ and $\mathbf{V}_i = \varsigma\mathbf{I}$ for all i where $\varsigma > \tau > 0$, the inequalities $\mathbf{R}_{ii} > -\mathbf{V}_i$ for all i become $-\tau\mathbf{I} > -\varsigma\mathbf{I}$ and are satisfied. Choosing $\mathbf{R}_{ij} = \mathbf{0}$ for all $i \neq j$, the inequalities of $\mathbf{R}_{ij} + \mathbf{R}_{ij}^T \geq -\mathbf{V}_i - \mathbf{V}_j$ are satisfied for $j = 1, 2, \cdots, p$, $i < j$. Hence, we have $\mathbf{R} = \begin{bmatrix} -\tau\mathbf{I} & \mathbf{0} & \cdots & \mathbf{0} \\ \mathbf{0} & -\tau\mathbf{I} & \cdots & \mathbf{0} \\ \vdots & \vdots & \vdots & \vdots \\ \mathbf{0} & \mathbf{0} & \cdots & -\tau\mathbf{I} \end{bmatrix} < 0$. Similarly, (4.21) and (4.22) become $\mathbf{T}_{ii} > \mathbf{\Lambda}_i$, $i = 1, 2, \cdots, p$, and $\mathbf{T}_{ij} + \mathbf{T}_{ij}^T \geq \mathbf{\Lambda}_i + \mathbf{\Lambda}_j$, $j = 1, 2, \cdots, p$, $i < j$, respectively. Choosing $\mathbf{T}_{ii} = -\tau\mathbf{I}$ and $\mathbf{\Lambda}_i = -\varsigma\mathbf{I}$ for all i, and $\mathbf{T}_{ij} = \mathbf{0}$ for all $i \neq j$, the inequalities $\mathbf{T}_{ii} > \mathbf{\Lambda}_i$ becoming $-\tau\mathbf{I} > -\varsigma\mathbf{I}$ and $\mathbf{T}_{ij}+\mathbf{T}_{ij}^T \geq \mathbf{\Lambda}_i+\mathbf{\Lambda}_j$ becoming $0 \geq -2\mathbf{I}$ are both satisfied. Hence, we have $\mathbf{T} = \begin{bmatrix} -\tau\mathbf{I} & \mathbf{0} & \cdots & \mathbf{0} \\ \mathbf{0} & -\tau\mathbf{I} & \cdots & \mathbf{0} \\ \vdots & \vdots & \vdots & \vdots \\ \mathbf{0} & \mathbf{0} & \cdots & -\tau\mathbf{I} \end{bmatrix} < 0$. As $\overline{\mathbf{Q}}_{ij} = \mathbf{Q}_{ij} < 0$ for all i and j, it can be seen from (4.20) that the inequalities become $\mathbf{S}_{ij} + \mathbf{S}_{ij}^T \geq \overline{\mathbf{Q}}_{ij} - \mathbf{\Lambda}_i + \mathbf{V}_j = \mathbf{Q}_{ij} + 2\varsigma\mathbf{I}$ for all i and j. Recalling that $\mathbf{Q}_{ij} < 0$ and choosing $\mathbf{S}_{ij} = \mathbf{0}$ for all i and j, there must exist a sufficiently small value of ς such that $0 \geq \mathbf{Q}_{ij} + 2\varsigma\mathbf{I}$ are satisfied. Consequently, we have $\begin{bmatrix} \mathbf{R} & \mathbf{S}^T \\ \mathbf{S} & \mathbf{T} \end{bmatrix} = \begin{bmatrix} \mathbf{R} & \mathbf{0} \\ \mathbf{0} & \mathbf{T} \end{bmatrix} < 0$, which implies the asymptotic stability of the FMB control system (2.10) and shows that the solution of the stability conditions in Theorem 2.1 is a particular of Theorem 4.1.

4.3.1 GA-Based Convex Programming Technique

The stability conditions in Theorem 4.1 are in terms of BMIs or LMIs. As a result, convex programming techniques cannot be employed to find the solution of the stability conditions. A BMI becomes an LMI if one of the variables of the BMI is fixed. This nice property enables the solution of the BMIs to be found numerically by combining the genetic algorithm (GA) and the convex programming techniques. Denote the BMI-based stability conditions as $\mathbf{L}(\mathbf{P_m}, \mathbf{P_s}) + z\mathbf{I} > 0$, where $\mathbf{P_s} = [a_1 \;\; a_2 \;\; \cdots \;\; a_p]$ and $\mathbf{P_m} = [\mathbf{X} \;\; \mathbf{M}_{11} \;\; \mathbf{M}_{12} \;\; \cdots \;\; \mathbf{M}_{pp} \;\; \mathbf{N}_1 \;\; \mathbf{N}_2 \;\; \cdots \;\; \mathbf{N}_p \;\; \mathbf{J}_{11} \;\; \mathbf{J}_{12} \;\; \cdots \;\; \mathbf{J}_{pp} \;\; \mathbf{K}_{11} \;\; \mathbf{K}_{12} \;\; \cdots \;\; \mathbf{K}_{pp} \;\; \mathbf{R}_{11} \;\; \mathbf{R}_{12} \;\; \cdots \;\; \mathbf{R}_{pp} \;\; \mathbf{S}_{11} \;\; \mathbf{S}_{12} \;\; \cdots \;\; \mathbf{S}_{pp} \;\; \mathbf{T}_{11} \;\; \mathbf{T}_{12} \;\; \cdots \;\; \mathbf{T}_{pp} \;\; \mathbf{W}_{11} \;\; \mathbf{W}_{12} \;\; \cdots \;\; \mathbf{W}_{pp} \;\; \mathbf{V}_1 \;\; \mathbf{V}_2 \;\; \cdots \;\; \mathbf{V}_p \;\; \mathbf{\Lambda}_1 \;\; \mathbf{\Lambda}_2 \;\; \cdots \;\; \mathbf{\Lambda}_p]$. If there exist $\mathbf{P_s}$ and $\mathbf{P_m}$ that satisfy $\mathbf{L}(\mathbf{P_m}, \mathbf{P_s}) + z\mathbf{I} > 0$ with a negative value of z, $\mathbf{P_s}$ and $\mathbf{P_m}$ are the solution of the BMI.

A GA-based convex programming technique as shown in Fig. 4.1 is proposed to solve the solution of the BMIs and is summarized as follows.

1. GA generates the potential solution of $\mathbf{P_s}$. It should be noted that when the value of $\mathbf{P_s}$ is kept constant, the BMI-based stability conditions become LMIs which can be solved using the convex programming technique. In general, the initial value of $\mathbf{P_s}$ is randomly generated.
2. The LMI solver searches for the solution $\mathbf{P_m}$ to the LMI condition of $\mathbf{L}(\mathbf{P_m}, \mathbf{P_s}) + z\mathbf{I} > 0$ (which becomes an LMI when $\mathbf{P_s}$ generated by GA in Step 1 is kept constant). The initial value of $\mathbf{P_m}$ is randomly generated or determined by the LMI solver.
3. If there exists a negative z such that $\mathbf{L}(\mathbf{P_m}, \mathbf{P_s}) + z\mathbf{I} > 0$, it implies that both $\mathbf{P_m}$ and $\mathbf{P_s}$ are the solutions of the BMI stability conditions.
4. If the stopping criterion is not met, return to Step 1.

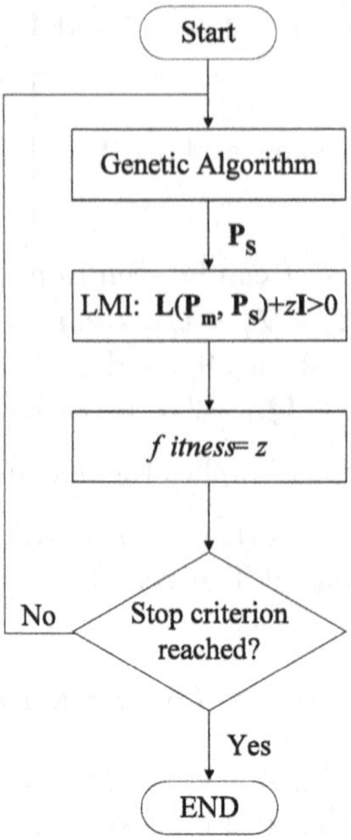

Fig. 4.1 Procedure of the GA-based convex programming technique.

Remark 4.6. For the GA-based convex programming process, z is taken as a fitness function to indicate the degree of satisfaction of both $\mathbf{P_m}$ and $\mathbf{P_s}$ to the BMI problem. A more negative value of z indicates a better solution $\mathbf{P_m}$ and $\mathbf{P_s}$. Consequently, the finding of solution is realized as a minimization problem (minimizing the fitness value of z). A stopping criterion should be set to stop the process, e.g. a pre-defined number of iterations or a feasible solution has been reached.

Example 4.1. Consider the same FMB control system in Example 2.1. The membership functions in Example 3.1 shown in Fig. 3.1 are used in this example. Based on the membership functions, it can be found that the values

Table 4.1 Vaues of σ_j, γ_j, δ_{ij} and ρ_{ij} for Example 4.1.

$\sigma_1 = 0.069182,\ \sigma_2 = 0.094927,\ \sigma_3 = 0.069182.$
$\gamma_1 = 0.069182,\ \gamma_2 = 0.074195,\ \gamma_3 = 0.069182.$
$\delta_{11} = 0.000000,\ \delta_{12} = \delta_{21} = 0.005459,\ \delta_{13} = \delta_{31} = 0.003054,$
$\delta_{22} = 0.000000,\ \delta_{23} = \delta_{32} = 0.005459,\ \delta_{33} = 0.000000.$
$\rho_{11} = 0.004786,\ \rho_{12} = \rho_{21} = 0.001137,\ \rho_{13} = \delta_{31} = 0.002253,$
$\rho_{22} = 0.009011,\ \rho_{23} = \rho_{32} = 0.001137,\ \rho_{33} = 0.004786.$

of σ_j, γ_j, δ_{ij} and ρ_{ij}, i, $j = 1, 2, 3$, listed in Table 4.1 satisfy the inequalities of (4.12) to (4.15).

The stability conditions in Theorem 4.1 are employed to check for the system stability for the TS fuzzy model parameters of $2 \leq a \leq 9$ and $0 \leq b \leq 8$. The real-coded GA with arithmetic crossover and non-uniform mutation [90] working with the MATLAB LMI toolbox is employed to search for the solution of the BMI stability conditions in Theorem 4.1. The lower and upper bounds of a_j, are chosen to be 10^{-3} and 5, respectively. With the chosen lower and upper bounds, the inequality of $\sum_{k=1}^{p} m_k(x_1(t))a_k > 0$ is satisfied. The parameters a_j, $j = 1, 2, 3$, form the chromosomes of the GA process and their initial values are randomly generated. The control parameters of the real-coded GA are tabulated in Table 4.2.

Table 4.2 Control parameters of the real-coded GA with arithmetic crossover and non-uniform mutation for Example 4.1.

Population size: 40
Number of iterations: 500
Probability of crossover: 0.8
Probability mutation: 0.5
Shape parameter: 1

The stability region is shown in Fig. 4.2 denoted by '∘'. To show the effectiveness of the parameter a_j, we choose $a_j = 1$ for $j = 1, 2, 3$ which make the stability conditions in Theorem 4.1 become LMIs. In this case, the proposed fuzzy controller is reduced to a traditional fuzzy controller (2.6) [122] with constant feedback gains. The stability region under $a_j = 1$ is shown in Fig. 4.2 denoted by '×'. It can be seen from Fig. 4.2 that the stability region given by the BMI-based stability conditions is larger. For comparison purposes, the stability conditions in Theorem 2.1 [122] applying to the FMB control systems with imperfectly matched membership functions are employed to check for the stability region for this example. However, no feasible solution is found. It is worth nothing that the stability conditions in Theorem 2.2 to Theorem 2.5 for FMB control systems with perfectly matched membership functions cannot be applied in this example.

Comparing with the traditional fuzzy controller (2.6), the proposed one (4.2) will increase the implementation cost owing to the time-varying nonlinear feedback gains that complicate the structure of the fuzzy controller. As a result, it is suggested to employ the LMI-based stability conditions in Theorem 4.1 with $a_j = 1$ for all j (that is reduced to Theorem 2.1) as the initial design. If a stable design cannot be achieved, the BMI-based stability conditions can then be employed.

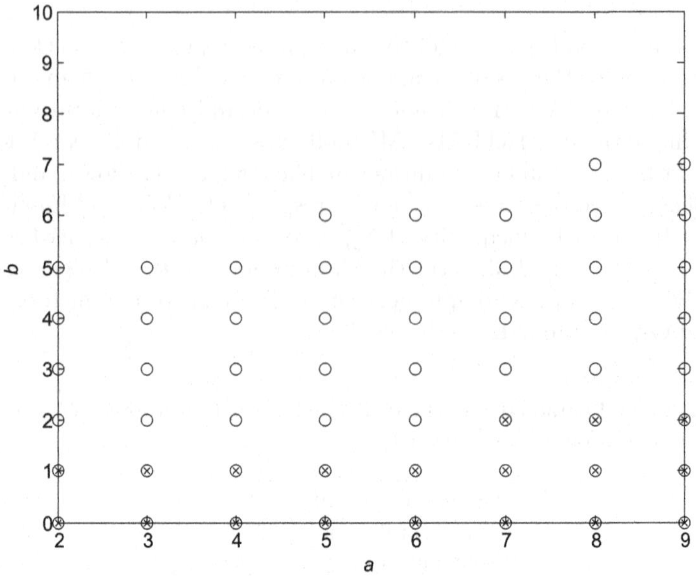

Fig. 4.2 Stability regions given by stability conditions in Theorem 4.1 with a_j determined by the GA-based convex programming technique ('$\circ$') and with $a_j = 1$ ('$\times$') for all j for Example 4.1.

4.4 BMI-Based Performance Conditions

Performance conditions in terms of BMIs are derived to guarantee the system performance under the consideration of system stability in this section. We consider the scalar performance index (3.14). From the performance index (3.14) and the fuzzy controller (4.3), and with the property of the membership functions (2.9), we have,

$$J = \int_0^\infty \begin{bmatrix} \mathbf{x}(t) \\ \mathbf{x}(t) \end{bmatrix}^T \begin{bmatrix} \mathbf{I} & \mathbf{0} \\ \mathbf{0} & \frac{\sum_{i=1}^p m_i \mathbf{G}_i}{\sum_{j=1}^p m_k a_k} \end{bmatrix}^T \begin{bmatrix} \mathbf{J}_1 & \mathbf{0} \\ \mathbf{0} & \mathbf{J}_2 \end{bmatrix} \begin{bmatrix} \mathbf{I} & \mathbf{0} \\ \mathbf{0} & \frac{\sum_{j=1}^p m_j \mathbf{G}_j}{\sum_{l=1}^p m_l a_l} \end{bmatrix} \begin{bmatrix} \mathbf{x}(t) \\ \mathbf{x}(t) \end{bmatrix} dt$$
$$= \frac{\sum_{i=1}^p \sum_{j=1}^p m_i m_j}{\sum_{k=1}^p \sum_{l=1}^p m_k m_l a_k a_l} \int_0^\infty \begin{bmatrix} \mathbf{x}(t) \\ \mathbf{x}(t) \end{bmatrix}^T \begin{bmatrix} a_i \mathbf{I} & \mathbf{0} \\ \mathbf{0} & \mathbf{G}_i \end{bmatrix}^T \begin{bmatrix} \mathbf{J}_1 & \mathbf{0} \\ \mathbf{0} & \mathbf{J}_2 \end{bmatrix}$$
$$\times \begin{bmatrix} a_j \mathbf{I} & \mathbf{0} \\ \mathbf{0} & \mathbf{G}_j \end{bmatrix} \begin{bmatrix} \mathbf{x}(t) \\ \mathbf{x}(t) \end{bmatrix} dt. \tag{4.24}$$

Let the performance index J satisfy the following condition.

$$J < \frac{\eta}{\sum_{k=1}^p \sum_{l=1}^p m_k m_l a_k a_l} \int_0^\infty \begin{bmatrix} \mathbf{x}(t) \\ \mathbf{x}(t) \end{bmatrix}^T \begin{bmatrix} \mathbf{X}^{-1} & \mathbf{0} \\ \mathbf{0} & \mathbf{X}^{-1} \end{bmatrix} \begin{bmatrix} \mathbf{x}(t) \\ \mathbf{x}(t) \end{bmatrix} dt \tag{4.25}$$

where η is a non-zero positive scalar to be chosen. It can be seen that if the inequality (4.25) is satisfied, the scalar performance index J is attenuated to a prescribed level governed by the value of η. From (4.24) and (4.25), recalling that $\mathbf{z}(t) = \mathbf{X}^{-1}\mathbf{x}(t)$ and the feedback gains of the fuzzy controller as $\mathbf{G}_j = \mathbf{N}_j\mathbf{X}^{-1}$, $j = 1, 2, \cdots, p$, we have,

$$\frac{1}{\sum_{k=1}^{p}\sum_{l=1}^{p} m_k m_l a_k a_l}\int_0^{\infty}\begin{bmatrix}\mathbf{z}(t)\\ \mathbf{z}(t)\end{bmatrix}^T \mathbf{W}\begin{bmatrix}\mathbf{z}(t)\\ \mathbf{z}(t)\end{bmatrix} dt < 0 \tag{4.26}$$

where

$$\mathbf{W} = \sum_{i=1}^{p}\sum_{j=1}^{p} m_i m_j \left(\begin{bmatrix} a_i\mathbf{X} & \mathbf{0}\\ \mathbf{0} & \mathbf{N}_i\end{bmatrix}^T \begin{bmatrix}\mathbf{J}_1 & \mathbf{0}\\ \mathbf{0} & \mathbf{J}_2\end{bmatrix}\begin{bmatrix} a_j\mathbf{X} & \mathbf{0}\\ \mathbf{0} & \mathbf{N}_j\end{bmatrix} - \eta\begin{bmatrix}\mathbf{X} & \mathbf{0}\\ \mathbf{0} & \mathbf{X}\end{bmatrix}\right). \tag{4.27}$$

To make sure that the inequality (4.26) is satisfied, the condition $\mathbf{W} < 0$ is required. It should be noted that the inequality $\mathbf{W} < 0$ cannot be formulated as an LMI problem with constant a_j. By Schur complement [120], $\mathbf{W} < 0$ is equivalent to the follow inequality.

$$\overline{\mathbf{W}} = \sum_{j=1}^{p} m_j \overline{\mathbf{W}}_j < 0 \tag{4.28}$$

where $\overline{\mathbf{W}}_j = \begin{bmatrix} -\eta\mathbf{X} & \mathbf{0} & a_j\mathbf{X} & \mathbf{0}\\ \mathbf{0} & -\eta\mathbf{X} & \mathbf{0} & \mathbf{N}_j^T\\ a_j\mathbf{X} & \mathbf{0} & -\mathbf{J}_1^{-1} & \mathbf{0}\\ \mathbf{0} & \mathbf{N}_j & \mathbf{0} & -\mathbf{J}_2^{-1}\end{bmatrix}$, $j = 1, 2, \cdots, p$. The inequalities $\overline{\mathbf{W}}_j < 0$ for all j (guaranteeing $\mathbf{W} < 0$) are regarded as the BMI-based performance conditions which have to be applied with the stability conditions in Theorem 4.1 to realize the system performance with pre-defined weighting matrices $\mathbf{J}_1$ and $\mathbf{J}_2$.

Theorem 4.2. *The system performance of a stable FMB control system (4.4), formed by the nonlinear plant represented by the fuzzy model (2.2) and the fuzzy controller (4.2) connected in a closed loop, satisfies the performance index J defined in (4.24) characterized by the pre-defined weighting matrices $\mathbf{J}_1 \geq 0$ and $\mathbf{J}_2 \geq 0$ that is attenuated to the level $\eta > 0$, if any stability conditions reported in this book and the LMIs $\overline{\mathbf{W}}_j < 0$ for all j are satisfied.*

Remark 4.7. The BMI-based stability and performance analysis can be further relaxed by adopting the SMF technique in Section 3.3 that incorporates the SMFs to the stability and performance conditions.

Remark 4.8. The BMI-based performance conditions in Theorem 4.2 are reduced to those in Theorem 3.2 with $a_j = 1$ for all j.

Example 4.2. The inverted pendulum on a cart shown in Example 3.2 is considered as the nonlinear plant. In this example, the stability condition in Theorem 4.1 and the performance conditions $\overline{\mathbf{W}}_j < 0$ for all j in Theorem 4.2 are employed to achieve a stable and well-performed FMB control system. The real-coded GA with arithmetic crossover and non-uniform mutation [90] working with the MATLAB LMI toolbox is employed to search for the solution of the BMI stability conditions in Theorem 4.1. The lower and upper bounds of a_j, are chosen to be 10^{-3} and 1, respectively. The parameters a_j, $j = 1, 2, 3$, form the chromosomes of the GA process and their initial values are randomly generated. The control parameters of the real-coded GA are the same as those in Table 4.2 except that the number of iterations is chosen to be 200. Based on these conditions, we obtain the feedback gains for the proposed fuzzy controllers (4.3) with different values of the weighting matrices $\mathbf{J}_1$ and $\mathbf{J}_2$, which are given in Table 4.3.

Table 4.3 Feedback gains for fuzzy controllers 1 to 3 in Example 4.2.

	Feedback Gains	η, a_j, $\mathbf{J}_1$ and $\mathbf{J}_2$
FC† 1	$\mathbf{G}_1 = \begin{bmatrix} 2064.8162 & 150.9920 & 28.5666 & 134.2425 \end{bmatrix}$ $\mathbf{G}_2 = \begin{bmatrix} 2655.0591 & 191.2585 & 36.2692 & 164.1681 \end{bmatrix}$	NA†† $a_1 = 1$ $a_2 = 1$
FC† 2	$\mathbf{G}_1 = \begin{bmatrix} 188.8736 & 20.5537 & 0.2143 & 27.3657 \end{bmatrix}$ $\mathbf{G}_2 = \begin{bmatrix} 453.2808 & 34.6134 & 0.4298 & 28.8520 \end{bmatrix}$	$\eta = 0.1$ $\mathbf{J}_2 = 1$ $\mathbf{J}_1 = \begin{bmatrix} 1 & 0 & 0 & 0 \\ 0 & 1 & 0 & 0 \\ 0 & 0 & 1 & 0 \\ 0 & 0 & 0 & 1 \end{bmatrix}$ $a_1 = 0.7900$ $a_2 = 0.4400$
FC† 3	$\mathbf{G}_1 = \begin{bmatrix} 201.3708 & 21.6676 & 0.5061 & 28.5817 \end{bmatrix}$ $\mathbf{G}_2 = \begin{bmatrix} 473.1369 & 36.2498 & 0.9967 & 30.4797 \end{bmatrix}$	$\eta = 0.1$ $\mathbf{J}_2 = 1$ $\mathbf{J}_1 = \begin{bmatrix} 1 & 0 & 0 & 0 \\ 0 & 1 & 0 & 0 \\ 0 & 0 & 5 & 0 \\ 0 & 0 & 0 & 1 \end{bmatrix}$ $a_1 = 0.8100$ $a_2 = 0.4500$

† FC stands for Fuzzy Controller.
†† LMI performance conditions are not considered.

The system responses and control signals of the inverted pendulum with fuzzy controllers 1 to 3 are shown in Fig. 4.3 and Fig. 4.4, respectively. It can be seen that fuzzy controller 1 designed without the performance conditions offers the best system performance in terms of rise and settling times at the cost of a large magnitude of control signal. By employing the weighting matrices of $\mathbf{J}_1$ and $\mathbf{J}_2$, the system responses and the control signals change accordingly. By employing $\mathbf{J}_2 = 1$, the control signal can be suppressed as shown in Fig. 4.4. The responses of the system states can be changed by specifying different values of weighting matrix $\mathbf{J}_1$. Referring to Fig. 4.3, the weighting matrix $\mathbf{J}_1$ with a higher weighting for $x_3(t)$ improves the system responses of $x_3(t)$ in terms of rise and settling times.

For comparison purposes, the traditional fuzzy controllers (with $a_j = 1$) are employed to control the inverted pendulum. By setting $a_j = 1$ for all j, the stability conditions in Theorem 4.1 and performance conditions in Theorem 4.2 become LMIs. With the MATLAB LMI toolbox, the feedback gains of the traditional fuzzy controllers subject to different values of the weighting matrices $\mathbf{J}_1$ and $\mathbf{J}_2$ are found and given in Table 4.2. It should be noted that no solution can be found for $\eta = 0.1$; hence, we choose η =0.2 to design the traditional fuzzy controllers in this example. It can be seen that the BMI-based stability and performance conditions are able to offer solution subject to some tightened design conditions. The system responses and control signals given by the traditional fuzzy controllers are shown in Fig. 4.5 and Fig. 4.6, respectively. It can be seen that $\mathbf{J}_1$ and $\mathbf{J}_2$ influence the system responses in a similar way. However, the proposed BMI-based fuzzy controllers are able to perform better in terms of rise and settling times because the feedback gains are achieved with a smaller value of η.

Table 4.4 Feedback gains for traditional fuzzy controllers 1 to 3 (with $a_j = 1$ for all j) in Example 4.2.

	Feedback Gains	η, $\mathbf{J}_1$ and $\mathbf{J}_2$
FC† 1	$\mathbf{G}_1 = \begin{bmatrix}2064.8162 & 150.9920 & 28.5666 & 134.2425\end{bmatrix}$ $\mathbf{G}_2 = \begin{bmatrix}2655.0591 & 191.2585 & 36.2692 & 164.1681\end{bmatrix}$	NA††
FC† 2	$\mathbf{G}_1 = \begin{bmatrix}363.6643 & 34.5143 & 0.2650 & 39.5054\end{bmatrix}$ $\mathbf{G}_2 = \begin{bmatrix}766.9431 & 57.6057 & 0.3615 & 45.5768\end{bmatrix}$	$\eta = 0.2$ $\mathbf{J}_2 = 1$ $\mathbf{J}_1 = \begin{bmatrix}1&0&0&0\\0&1&0&0\\0&0&1&0\\0&0&0&1\end{bmatrix}$
FC† 3	$\mathbf{G}_1 = \begin{bmatrix}365.2711 & 34.6056 & 0.5969 & 39.5254\end{bmatrix}$ $\mathbf{G}_2 = \begin{bmatrix}769.7314 & 57.7860 & 0.8160 & 45.6857\end{bmatrix}$	$\eta = 0.2$ $\mathbf{J}_2 = 1$ $\mathbf{J}_1 = \begin{bmatrix}1&0&0&0\\0&1&0&0\\0&0&5&0\\0&0&0&1\end{bmatrix}$

† FC stands for Fuzzy Controller.

†† LMI performance conditions are not considered.

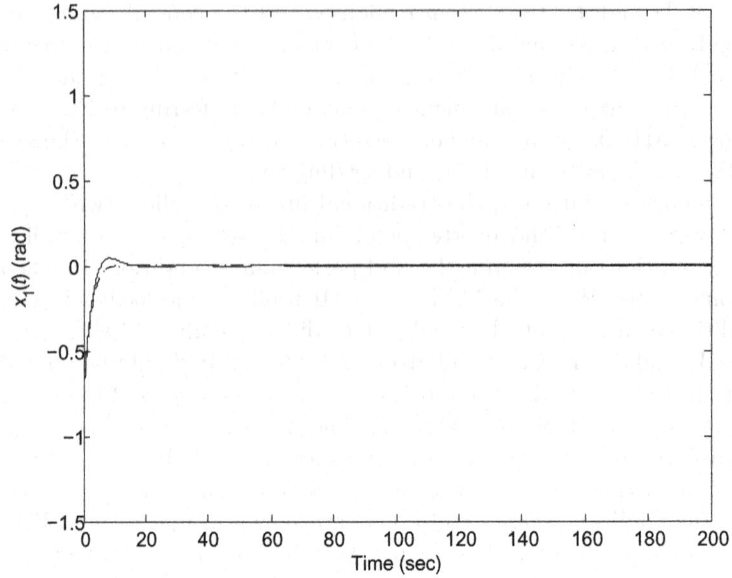

(a) $x_1(t)$.

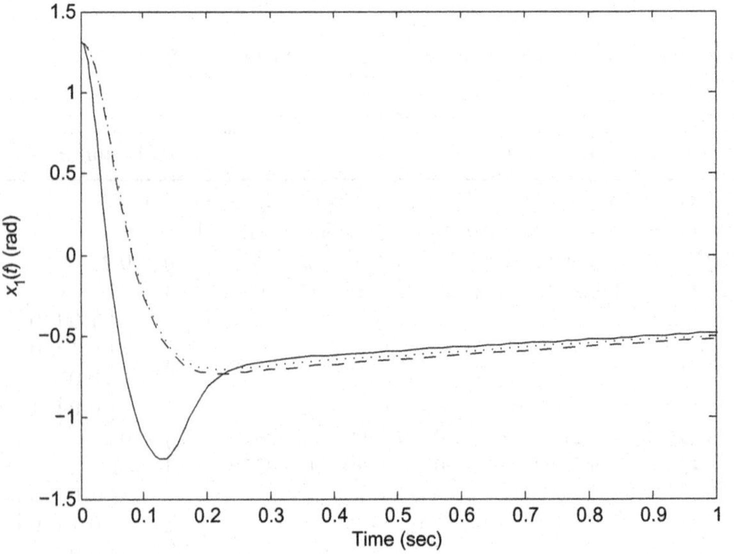

(b) $x_1(t)$ for $0s \leq t \leq 1s$.

Fig. 4.3 System state responses and control signals of the inverted pendulum with fuzzy controllers 1 (solid lines), 2 (dotted lines) and 3 (dash lines).

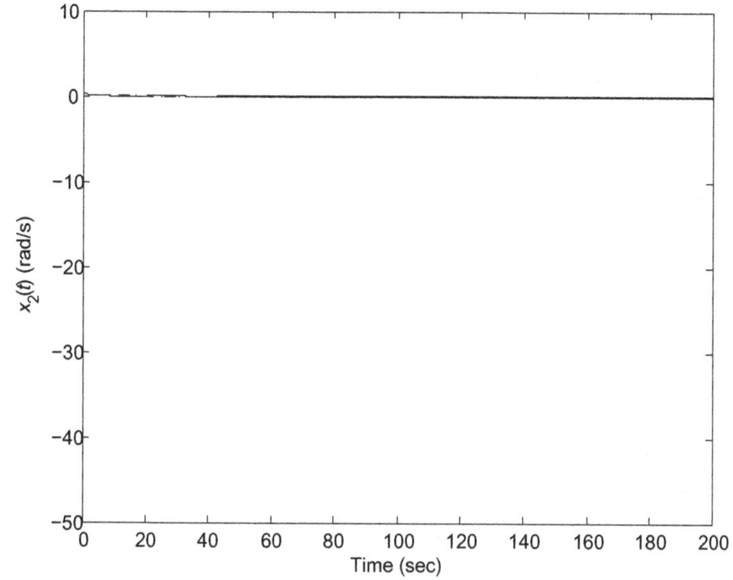

(c) $x_1(t)$.

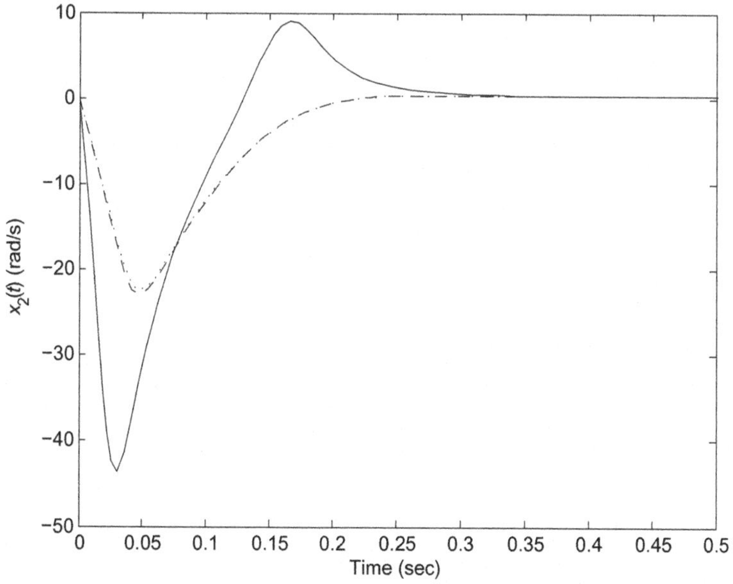

(d) $x_2(t)$ for $0s \leq t \leq 0.5s$.

Fig. 4.3 (*continued*)

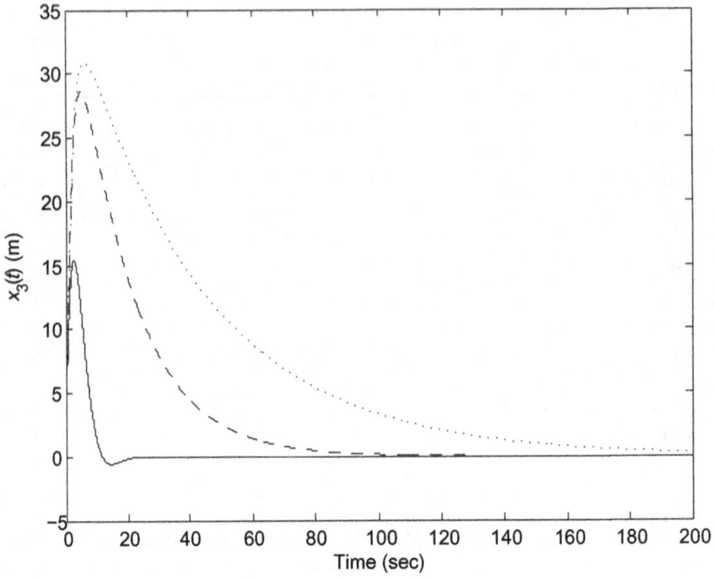

(e) $x_3(t)$.

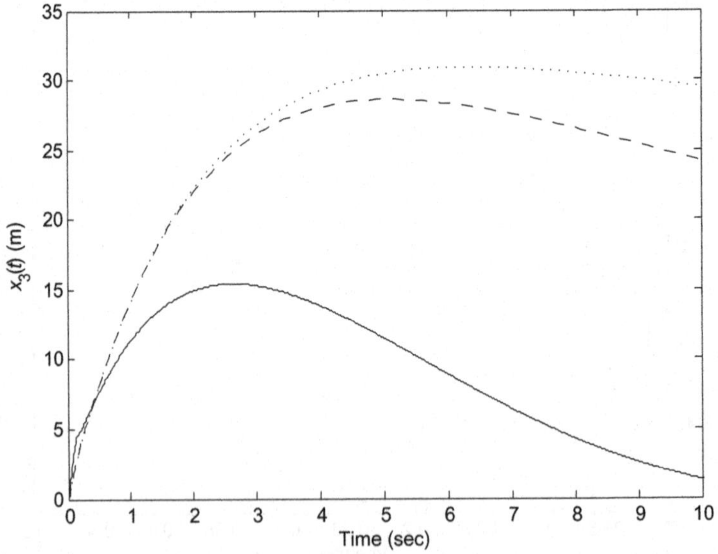

(f) $x_3(t)$ for $0s \leq t \leq 10s$.

Fig. 4.3 (*continued*)

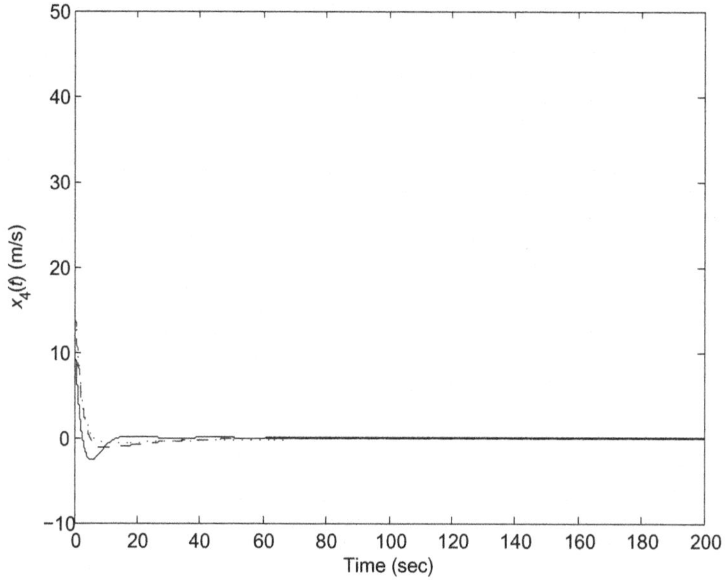

(g) $x_4(t)$.

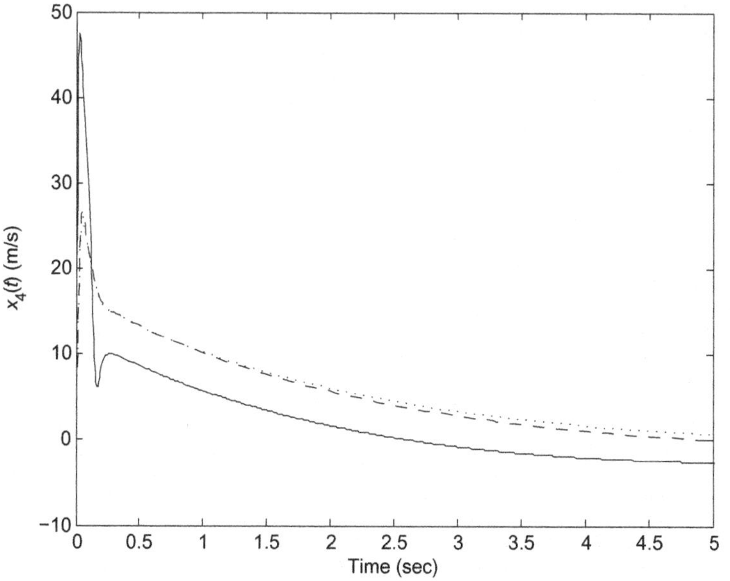

(h) $x_4(t)$ for $0s \leq t \leq 5s$.

Fig. 4.3 (*continued*)

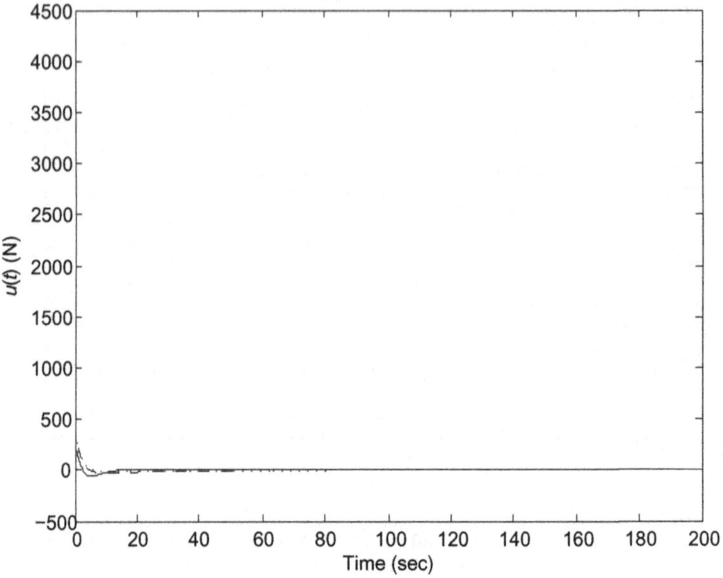

(a) $u(t)$.

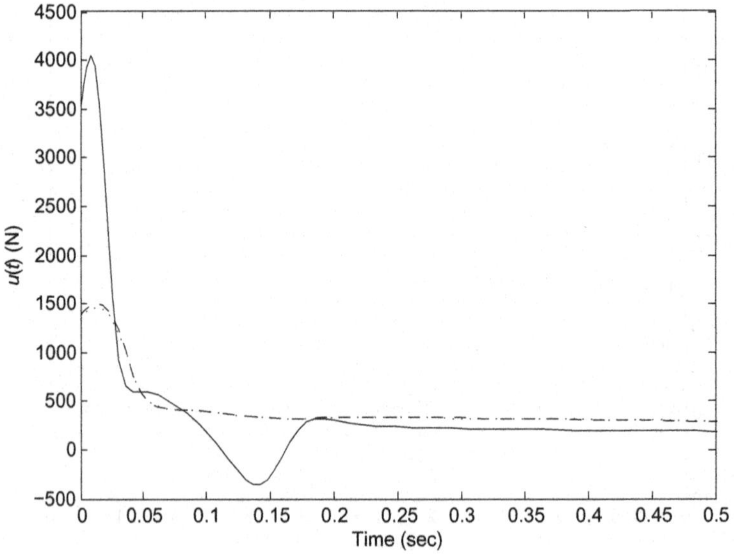

(b) $u(t)$ for $0s \leq t \leq 0.5s$.

Fig. 4.4 Control signals of the inverted pendulum with fuzzy controllers 1 (solid lines), 2 (dotted lines) and 3 (dash lines).

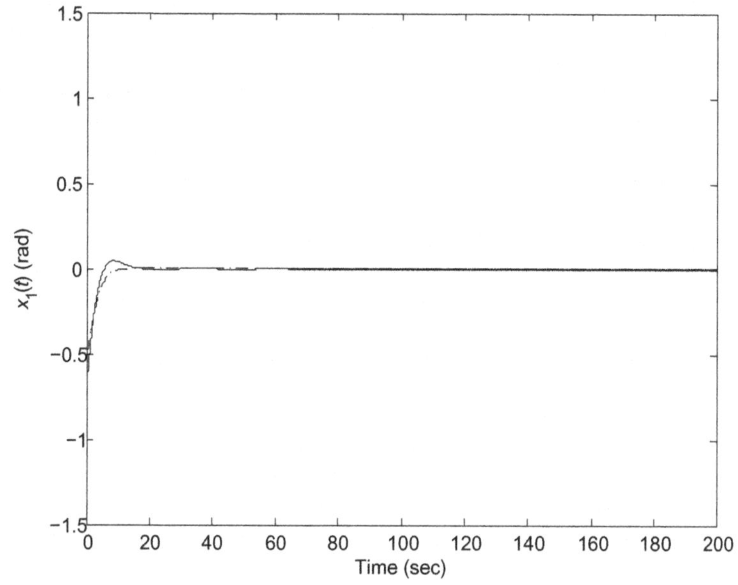

(a) $x_1(t)$.

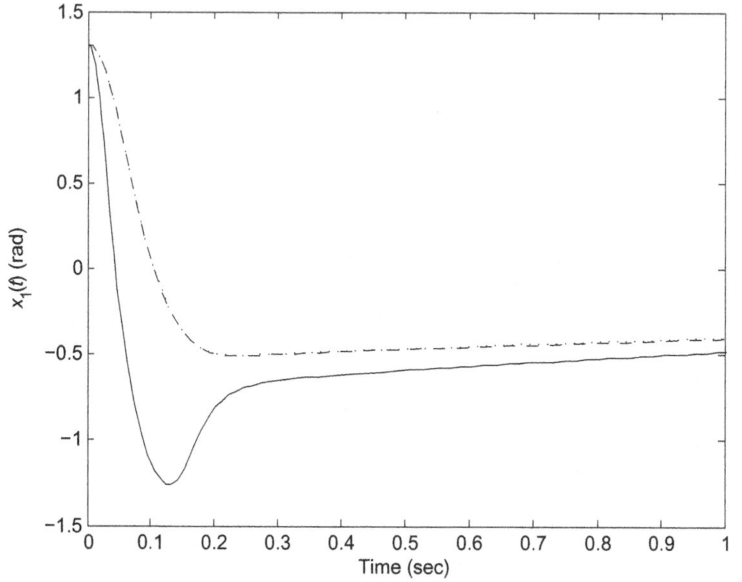

(b) $x_1(t)$ for $0s \leq t \leq 1s$.

Fig. 4.5 System state responses and control signals of the inverted pendulum with traditional fuzzy controllers 1 (solid lines), 2 (dotted lines) and 3 (dash lines).

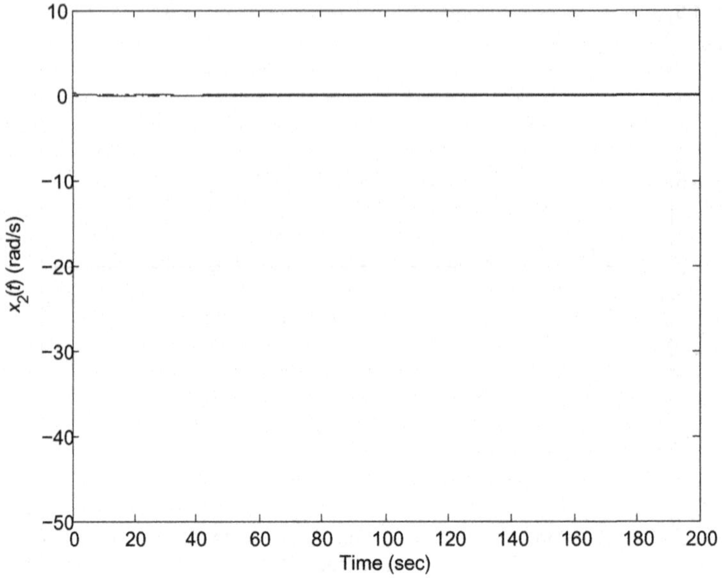

(c) $x_2(t)$.

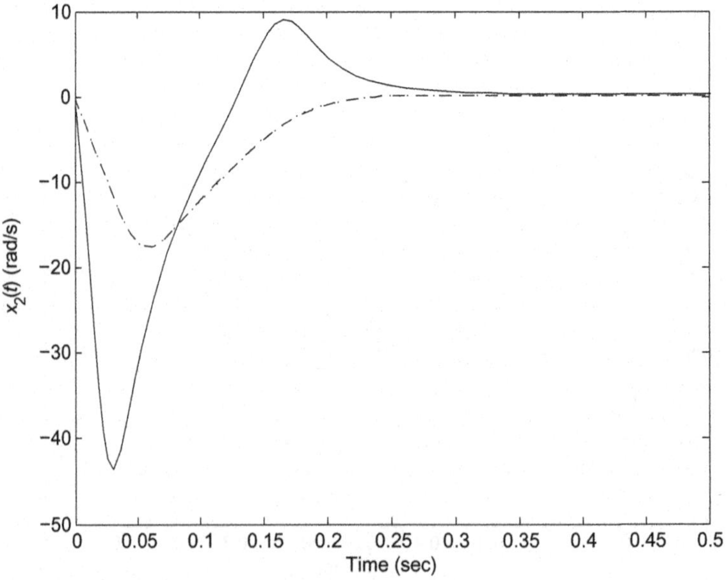

(d) $x_2(t)$ for $0s \leq t \leq 0.5s$.

Fig. 4.5 *(continued)*

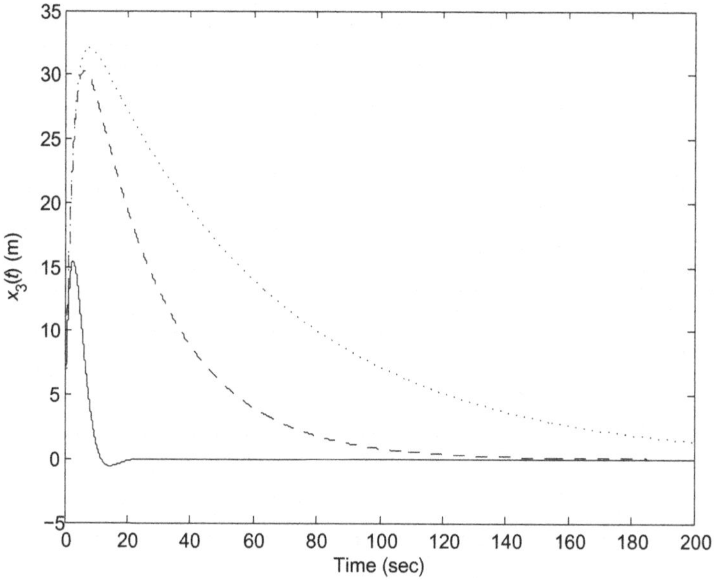

(e) $x_3(t)$.

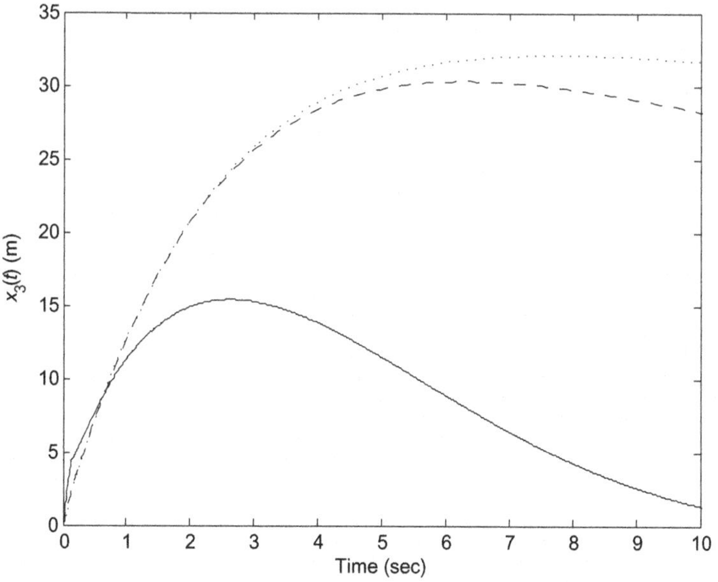

(f) $x_3(t)$ for $0s \leq t \leq 10s$.

Fig. 4.5 (*continued*)

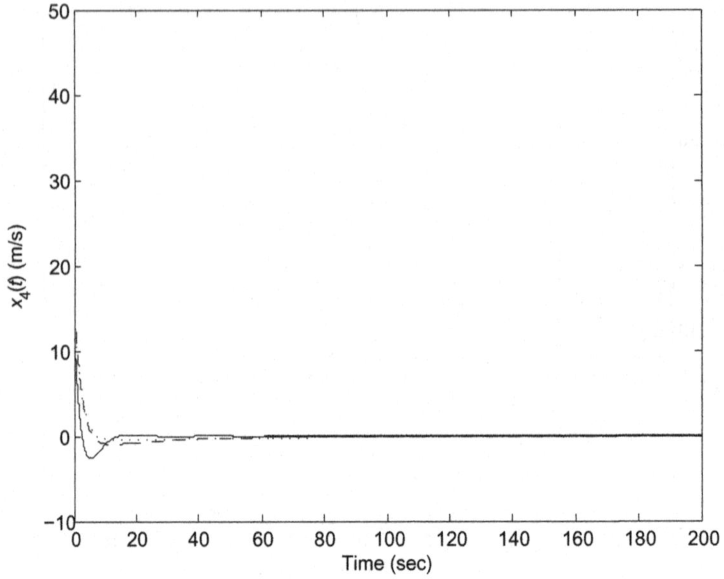

(g) $x_4(t)$.

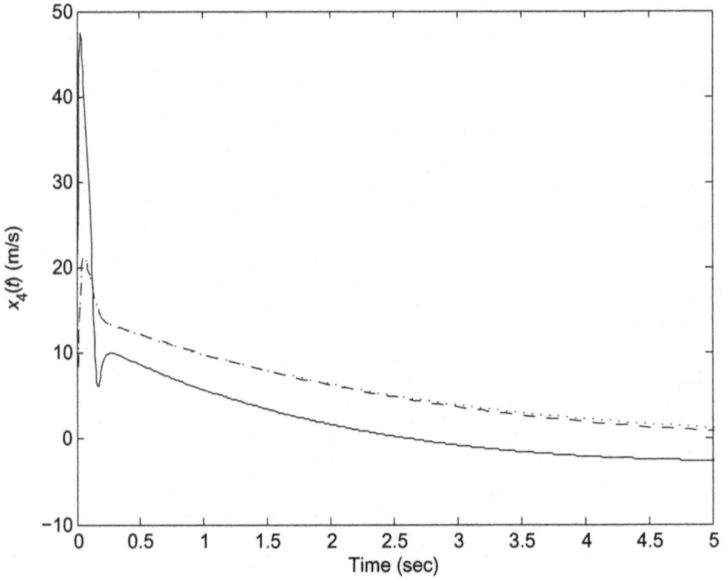

(h) $x_4(t)$ for $0s \leq t \leq 5s$.

Fig. 4.5 (*continued*)

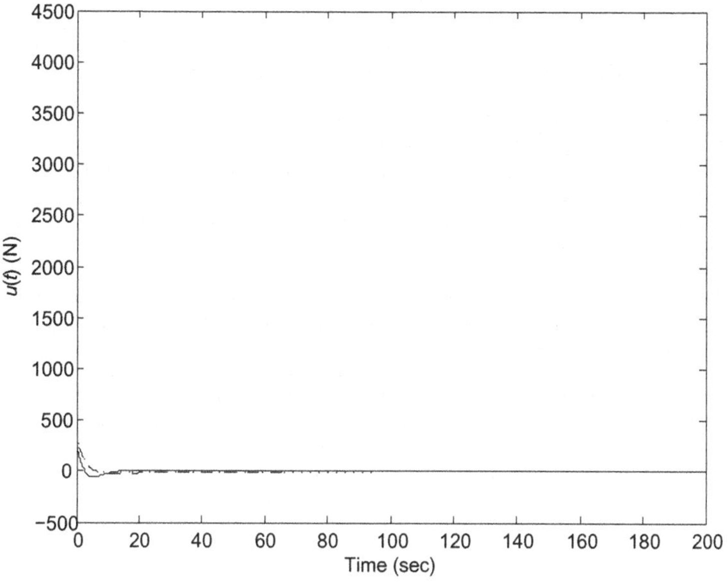

(a) $u(t)$.

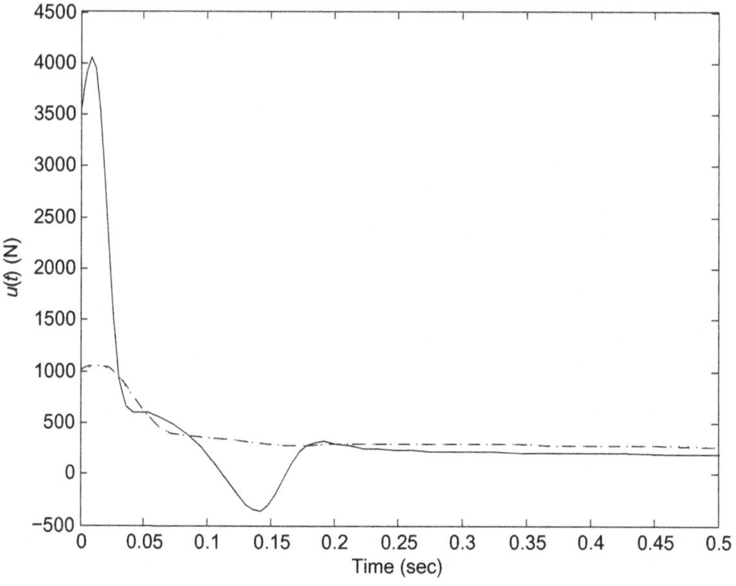

(b) $u(t)$ for $0s \leq t \leq 0.5s$.

Fig. 4.6 Control signals of the inverted pendulum with traditional fuzzy controllers 1 (solid lines), 2 (dotted lines) and 3 (dash lines).

4.5 Conclusion

The system stability of FMB control systems subject to imperfectly matched premise membership functions has been investigated. A fuzzy controller with nonlinear state feedback gains has been proposed to deal with the nonlinear plants. Comparing to the traditional fuzzy controller, the proposed one, with enhanced nonlinearity, exhibits stronger stabilizability for nonlinear control processes. Based on the Lyapunov stability theory, stability conditions in terms of BMIs have been derived to guarantee the system stability. Furthermore, based on a scalar performance index function, performance conditions in terms of BMIs have been derived. A GA-based convex programming technique has been proposed to search for the numerical solution of the BMI-based stability and performance conditions. Simulation examples have been given to show that the proposed BMI-based stability conditions are able to offer larger stability regions and the proposed BMI-based performance conditions can be applied to realize a well-performed FMB control system.

Chapter 5
Stability Analysis of FMB Control Systems Using PDLF

5.1 Introduction

The FMB control approach offers a systematic way to control nonlinear systems. Based on the TS fuzzy model, a nonlinear plant can be represented as a weighted sum of linear systems. This particular form offers a general and systematic framework to represent the nonlinear plant and provides an effective platform to facilitate stability analysis and controller synthesis. In the past two decades, FMB control systems have been extensively investigated, and flourish analysis results, particularly on the stability issue, have been obtained. Some relaxed LMI-based stability conditions were derived and introduced in Chapter 2.

The most popular tool for investigating the stability of FMB control systems is based on the Lyapunov stability theory. In general, FMB control systems are mainly investigated with parameter-independent and parameter-dependent Lyapunov functions. In Chapter 2, the analysis was conducted with a quadratic Lyapunov function candidate. As the matrix of the quadratic Lyapunov function does not depend on the system parameters/system states, it is referred to as parameter-independent Lyapunov function (PILF) in this book.

In [103], a fuzzy Lyapunov function (FLF), which is a kind of parameter-dependent Lyapunov function (PDLF), was proposed. As further information about the membership functions is considered and nonlinearity is introduced in the Lyapunov function, the FLF demonstrated a great potential to further relax the stability conditions as compared to PILF. However, for the continuous-time case, the FLF will produce the time derivatives of membership functions that complicate the stability analysis. (It was shown that the time derivatives of membership functions will disappear under a particular type of fuzzy models [96].) Furthermore, when the PDC fuzzy controller is considered [96, 103], the stability conditions cannot be formulated in terms of LMIs. As a result, the solution of the stability conditions cannot be found numerically by some convex programming techniques such as the MATLAB

H.-K. Lam and F.H.F. Leung: FMB Control Systems, STUDFUZZ 264, pp. 85–100.
springerlink.com

LMI toolbox. To hurdle the difficulty, a completing square technique was proposed in [103] to formulate the stability conditions in terms of LMIs. However, as redundant matrix terms are introduced in the stability analysis, conservative stability conditions are obtained.

In [59, 107], to further utilize the advantages of the FLF, a non-PDC fuzzy controller was proposed. Comparing to the PDC fuzzy controller, the non-PDC fuzzy controller [59, 107] offers greater design flexibility and shows greater potential to relax the stability conditions. Moreover, with the non-PDC fuzzy controller, the stability conditions can be formulated in terms of LMIs. The FLF was also employed to investigate the system stability of discrete-time FMB control systems [19, 36].

In general, the PDLF approach is good for general applications while the PILF approach is particularly good for some applications of which the bounds of the time derivatives of membership functions are known. As more information is included in the non-PDC fuzzy controller with PDLF, it has been reported in [59, 107] that the PDLF analysis approach is superior to the PILF approach in relaxing the stability conditions. As graphically illustrated in Fig. 5.1, the PDC fuzzy controller with PILF and non-PDC fuzzy controller with PDLF approaches are suitable for different classes of applications. It can be seen that the PDC fuzzy controller with PILF is a subset of the non-PDC fuzzy controller with PDLF. The latter approach is able to handle applications with time derivatives of membership functions available.

In this chapter, we further extend our fundamental PDLF work in [59] to investigate the FMB control systems subject to perfectly matched membership functions. By taking advantage of the PDLF, relaxed stability conditions are achieved by considering the characteristic of the fuzzy model. An improved non-PDC fuzzy controller is proposed to close the feedback loop. As the information about the time derivatives of membership functions is considered by the non-PDC fuzzy controller, it allows the introduction of slack matrices to facilitate the stability analysis. Based on the PDLF, stability conditions in terms of LMIs are derived to guarantee the system stability of the FMB control systems. It will be shown by simulation examples that when the time derivatives of membership functions are considered, the non-PDC stability conditions are more relaxed than the PDC ones. The proposed

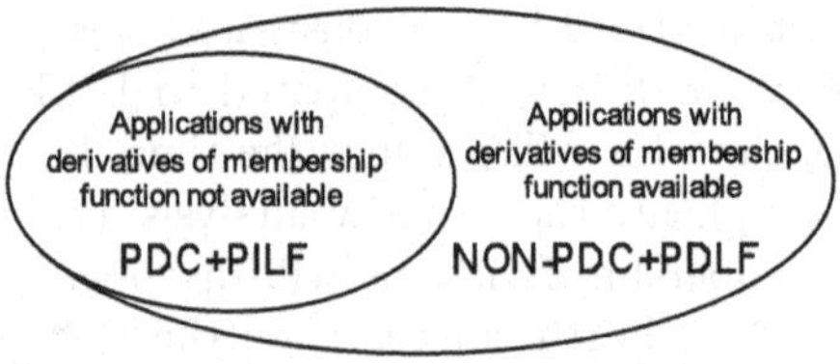

Fig. 5.1 A diagram illustrating the coexistence of PDC+PILF and Non-PDC+PDLF approaches.

non-PDC fuzzy controller provides an alternative solution to applications of which time derivatives of membership functions are available.

5.2 Non-PDC Fuzzy Controller

Consider the nonlinear plant represented by the TS fuzzy model (2.2). A non-PDC fuzzy controller [59, 107] with p rules is proposed to close the feedback loop and is described by the fuzzy rules of the following format:

$$\begin{aligned}&\text{Rule } j\text{: IF } f_1(\mathbf{x}(t)) \text{ is } M_1^j \text{ AND } \cdots \text{ AND } f_\Psi(\mathbf{x}(t)) \text{ is } M_\Psi^j \\ &\qquad \text{THEN } \mathbf{u}(t) = \big(\overline{\mathbf{F}}_j + \overline{\mathbf{G}}(\mathbf{x}(t))\big)\mathbf{P}(\mathbf{x}(t))^{-1}\mathbf{x}(t), j = 1, 2, \cdots, p \end{aligned} \tag{5.1}$$

where $\overline{\mathbf{G}}(\mathbf{x}(t)) = \sum_{j=1}^{p} \dot{w}_j(\mathbf{x}(t))\overline{\mathbf{G}}_j$, $\overline{\mathbf{F}}_j \in \Re^{m\times n}$, $\overline{\mathbf{G}}_j \in \Re^{m\times n}$, $j = 1, 2, \cdots, p$, are constant feedback gains to be determined; $\dot{w}_j(\mathbf{x}(t)) = \frac{d}{dt}w_j(\mathbf{x}(t))$; $0 < \mathbf{P}(\mathbf{x}(t)) = \mathbf{P}(\mathbf{x}(t))^T \in \Re^{n\times n}$ to be determined. The rest variables are defined in Sections 2.2 and 2.3.

The inferred non-PDC fuzzy controller is defined as follows.

$$\begin{aligned}\mathbf{u}(t) &= \sum_{j=1}^{p} w_j(\mathbf{x}(t))\big(\overline{\mathbf{F}}_j + \overline{\mathbf{G}}(\mathbf{x}(t))\big)\mathbf{P}(\mathbf{x}(t))^{-1}\mathbf{x}(t) \\ &= \sum_{j=1}^{p} w_j(\mathbf{x}(t))\overline{\mathbf{F}}_j\mathbf{P}(\mathbf{x}(t))^{-1}\mathbf{x}(t) + \sum_{j=1}^{p} \dot{w}_j(\mathbf{x}(t))\overline{\mathbf{G}}_j\mathbf{P}(\mathbf{x}(t))^{-1}\mathbf{x}(t)\end{aligned} \tag{5.2}$$

It should be noted that the non-PDC fuzzy controller (5.2) is only applicable for a class of applications of which the time derivatives of membership functions are available [59, 107]. Comparing to the traditional PDC-fuzzy controller (2.6), the non-PDC fuzzy controller (5.2) contains additional information (time derivatives of the membership functions) to facilitate the stability analysis.

Remark 5.1. It should be noted that the non-PDC fuzzy controller proposed in this chapter shares the same premises with the fuzzy model.

Remark 5.2. The non-PDC fuzzy controller (5.2) is reduced to the traditional one (2.6) when we choose $\mathbf{P}(\mathbf{x}(t))$ as a constant matrix and $\overline{\mathbf{G}}_j = \mathbf{0}$ for all j.

5.3 Stability Analysis Based on PDLF

The system stability of the FMB control system with perfectly matched premise membership functions is investigated using a PDLF. Two non-PDC fuzzy controllers are proposed to handle the nonlinear plants in the form of (2.2). For brevity, $w_i(\mathbf{x}(t))$ and $\dot{w}_i(\mathbf{x}(t))$ are denoted as w_i and $\dot{w}_i$, respectively. The property of the membership functions (2.9) is utilized in the system analysis.

In the following, the MFSD analysis technique in Chapter 3 is employed to investigate the system stability. The relationship between the membership functions of the fuzzy model and fuzzy controllers is used to introduce some slack matrix variables in stability analysis through the S-procedure [6]. The boundary information of the membership functions is brought to the stability conditions and thus offers some relaxed stability conditions.

5.3.1 Non-PDC Fuzzy Controller

To facilitate the stability analysis, we consider the scalars ρ and σ_j that satisfy the inequalities of $\overline{w}_j = w_j + \rho\dot{w}_j + \sigma_j \geq 0$ for all j and the operating domain of $\mathbf{x}(t)$. The scalars ρ and σ_j can be found numerically when the form of w_j and $\dot{w}_j$ are known; otherwise, the proposed non-PDC fuzzy control approach cannot be applied. Based on the proposed inequalities, the non-PDC fuzzy controller (5.2) is rewritten as the following form.

$$\mathbf{u}(t) = \sum_{j=1}^{p} w_j \mathbf{F}_j \mathbf{P}(\mathbf{x}(t))^{-1}\mathbf{x}(t) + \sum_{j=1}^{p} \overline{w}_j \mathbf{G}_j \mathbf{P}(\mathbf{x}(t))^{-1}\mathbf{x}(t) \qquad (5.3)$$

where $\mathbf{F}_j \in \Re^{m \times n}$ and $\mathbf{G}_j \in \Re^{m \times n}$ are constant feedback gains to be determined. Reshuffling the terms of the non-PDC fuzzy controller (5.3), it becomes

$$\begin{aligned}\mathbf{u}(t) &= \sum_{j=1}^{p} w_j \mathbf{F}_j \mathbf{P}(\mathbf{x}(t))^{-1}\mathbf{x}(t) + \sum_{j=1}^{p} (w_j + \rho\dot{w}_j + \sigma_j) \mathbf{G}_j \mathbf{P}(\mathbf{x}(t))^{-1}\mathbf{x}(t) \\ &= \sum_{j=1}^{p} w_j \Big(\mathbf{F}_j + \mathbf{G}_j + \sum_{k=1}^{p} \sigma_k \mathbf{G}_k\Big) \mathbf{P}(\mathbf{x}(t))^{-1}\mathbf{x}(t) \\ &+ \sum_{j=1}^{p} \dot{w}_j \rho \mathbf{G}_j \mathbf{P}(\mathbf{x}(t))^{-1}\mathbf{x}(t). \end{aligned} \qquad (5.4)$$

Comparing (5.2) and (5.4) term by term, the non-PDC fuzzy controllers in the form of (5.2), (5.3) and (5.4) are equivalent by choosing $\overline{\mathbf{F}}_j = \mathbf{F}_j + \mathbf{G}_j + \sum_{k=1}^{p} \sigma_k \mathbf{G}_k$ and $\overline{\mathbf{G}}_j = \rho \mathbf{G}_j$ for all j.

Remark 5.3. Comparing to the traditional PDC-fuzzy controller (2.6), the non-PDC fuzzy controller (5.3) has an enhanced nonlinear feedback compensation ability because of $\mathbf{P}(\mathbf{x}(t))^{-1}$, which makes the non-PDC fuzzy controller be more capable of compensating the nonlinearity of the plant. Furthermore, the inequalities of $\overline{w}_j >= 0$ for all j relate the membership functions and their time derivatives by the scalars ρ and σ_j that offer further information of the nonlinear plant to facilitate the stability analysis.

5.3.1.1 Stability Analysis of FMB Control Systems with Non-PDC Fuzzy Controller

Considering the fuzzy model (2.2) and the non-PDC fuzzy controller (5.3) connected in a closed loop, the FMB control system is obtained as follows.

$$\begin{aligned}\dot{\mathbf{x}}(t) &= \sum_{i=1}^{p} w_i \Big(\mathbf{A}_i \mathbf{x}(t) + \mathbf{B}_i \Big(\sum_{j=1}^{p} w_j \mathbf{F}_j \mathbf{P}(\mathbf{x}(t))^{-1}\mathbf{x}(t) \\ &\quad + \sum_{j=1}^{p} \overline{w}_j \mathbf{G}_j \mathbf{P}(\mathbf{x}(t))^{-1}\mathbf{x}(t)\Big)\Big) \\ &= \sum_{i=1}^{p}\sum_{j=1}^{p} w_i w_j (\mathbf{A}_i + \mathbf{B}_i \mathbf{F}_j \mathbf{P}(\mathbf{x}(t))^{-1})\mathbf{x}(t) \\ &\quad + \sum_{i=1}^{p}\sum_{j=1}^{p} w_i \overline{w}_j \mathbf{B}_i \mathbf{G}_j \mathbf{P}(\mathbf{x}(t))^{-1}\mathbf{x}(t) \end{aligned} \tag{5.5}$$

To investigate the system stability of the FMB control system (5.5), we consider the following PDLF candidate.

$$V(t) = \mathbf{x}(t)^T \mathbf{P}(\mathbf{x}(t))^{-1}\mathbf{x}(t) \tag{5.6}$$

where $\mathbf{P}(\mathbf{x}(t)) = \mathbf{P}(\mathbf{x}(t))^T = \sum_{k=1}^{p}\sum_{l=1}^{p} w_k w_l \mathbf{P}_{k,l} > 0$ and $\mathbf{P}_{k,l} = \mathbf{P}_{l,k}^T \in \Re^{n\times n}$, $k, l = 1, 2, \cdots, p$.

Remark 5.4. It is required that $\mathbf{P}(\mathbf{x}(t)) > 0$ which implies non-singularity. Based on the matrix property, $\mathbf{P}(\mathbf{x}(t)) > 0$ implies that the inverse of $\mathbf{P}(\mathbf{x}(t))$ exists and $\mathbf{P}(\mathbf{x}(t))^{-1} > 0$.

Remark 5.5. The proposed PDLF is reduced to $\mathbf{P}(\mathbf{x}(t)) = \mathbf{P}(\mathbf{x}(t))^T = \sum_{k=1}^{p} w_k \mathbf{P}_k > 0$ used in [107] when $\mathbf{P}_k = \mathbf{P}_{k,l}$ are considered for all l.

Before proceeding further, the following lemma is introduced to facilitate the stability analysis.

Lemma 5.1. *For any invertible matrix* $\mathbf{P}(\mathbf{x}(t))$*, the time derivative of* $\mathbf{P}(\mathbf{x}(t))$ *is given by,*

$$\dot{\mathbf{P}}(\mathbf{x}(t)) = -\mathbf{P}(\mathbf{x}(t))\dot{\mathbf{P}}(\mathbf{x}(t))^{-1}\mathbf{P}(\mathbf{x}(t))$$

Proof. Considering $\mathbf{P}(\mathbf{x}(t))\mathbf{P}(\mathbf{x}(t))^{-1} = \mathbf{I}$, we have $\frac{d}{dt}\mathbf{P}(\mathbf{x}(t))\mathbf{P}(\mathbf{x}(t))^{-1} = \dot{\mathbf{P}}(\mathbf{x}(t))\mathbf{P}(\mathbf{x}(t))^{-1} + \mathbf{P}(\mathbf{x}(t))\dot{\mathbf{P}}(\mathbf{x}(t))^{-1} = \mathbf{0}$ Reshuffling the terms in the last equation, Lemma 5.1 is achieved.

Denote $\mathbf{z}(t) = \mathbf{P}(\mathbf{x}(t))^{-1}\mathbf{x}(t)$. From (5.5), (5.6) and Lemma 5.1, we have

$$\begin{aligned}
\dot{V}(t) &= \dot{\mathbf{x}}(t)^T\mathbf{P}(\mathbf{x}(t))^{-1}\mathbf{x}(t) + \mathbf{x}(t)^T\mathbf{P}(\mathbf{x}(t))^{-1}\dot{\mathbf{x}}(t) + \mathbf{x}(t)^T\frac{d}{dt}\mathbf{P}(\mathbf{x}(t))^{-1}\mathbf{x}(t) \\
&= \mathbf{x}(t)^T\Big(\sum_{i=1}^{p}\sum_{j=1}^{p} w_i w_j(\mathbf{A}_i + \mathbf{B}_i\mathbf{F}_j\mathbf{P}(\mathbf{x}(t))^{-1}) \\
&\quad + \sum_{i=1}^{p}\sum_{j=1}^{p} w_i \overline{w}_j \mathbf{B}_i\mathbf{G}_j\mathbf{P}(\mathbf{x}(t))^{-1}\Big)^T\mathbf{P}(\mathbf{x}(t))^{-1}\mathbf{x}(t) \\
&\quad + \mathbf{x}(t)^T\mathbf{P}(\mathbf{x}(t))^{-1}\Big(\sum_{i=1}^{p}\sum_{j=1}^{p} w_i w_j(\mathbf{A}_i + \mathbf{B}_i\mathbf{F}_j\mathbf{P}(\mathbf{x}(t))^{-1}) \\
&\quad + \sum_{i=1}^{p}\sum_{j=1}^{p} w_i \overline{w}_j \mathbf{B}_i\mathbf{G}_j\mathbf{P}(\mathbf{x}(t))^{-1}\Big)\mathbf{x}(t) \\
&\quad - \mathbf{x}(t)^T\mathbf{P}(\mathbf{x}(t))^{-1}\dot{\mathbf{P}}(\mathbf{x}(t))\mathbf{P}(\mathbf{x}(t))^{-1}\mathbf{x}(t) \\
&= \sum_{i=1}^{p}\sum_{j=1}^{p} w_i w_j \mathbf{z}(t)^T\big(\mathbf{P}(\mathbf{x}(t))\mathbf{A}_i^T + \mathbf{A}_i\mathbf{P}(\mathbf{x}(t)) + \mathbf{F}_j^T\mathbf{B}_i^T + \mathbf{B}_i\mathbf{F}_j\big)\mathbf{z}(t) \\
&\quad + \sum_{i=1}^{p}\sum_{j=1}^{p} w_i \overline{w}_j \mathbf{z}(t)^T\big(\mathbf{G}_j^T\mathbf{B}_i^T + \mathbf{B}_i\mathbf{G}_j\big)\mathbf{z}(t) \\
&\quad - \mathbf{z}(t)^T\sum_{i=1}^{p}\sum_{j=1}^{p}\big(\dot{w}_i w_j\mathbf{P}_{i,j} + w_i\dot{w}_j\mathbf{P}_{i,j}\big)\mathbf{z}(t).
\end{aligned} \tag{5.7}$$

Using the fact that $w_i w_j = w_j w_i$ for all i and j, we have $\sum_{i=1}^{p}\sum_{j=1}^{p} w_i w_j \mathbf{P}_{j,i} = \sum_{i=1}^{p}\sum_{j=1}^{p} w_j w_i \mathbf{P}_{j,i} = \sum_{i=1}^{p}\sum_{j=1}^{p} w_i w_j \mathbf{P}_{i,j}$, which will be used in the following analysis. Expanding $\mathbf{P}(\mathbf{x}(t))$, (5.7) becomes

$$\begin{aligned}
\dot{V}(t) &= \sum_{i=1}^{p}\sum_{j=1}^{p}\sum_{k=1}^{p} w_i w_j w_k \mathbf{z}(t)^T \big(\mathbf{P}_{j,k}\mathbf{A}_i^T + \mathbf{A}_i\mathbf{P}_{j,k} + \mathbf{F}_j^T\mathbf{B}_i^T + \mathbf{B}_i\mathbf{F}_j\big)\mathbf{z}(t) \\
&\quad + \sum_{i=1}^{p}\sum_{j=1}^{p} w_i \overline{w}_j \mathbf{z}(t)^T \big(\mathbf{G}_j^T\mathbf{B}_i^T + \mathbf{B}_i\mathbf{G}_j\big)\mathbf{z}(t) \\
&\quad - \sum_{i=1}^{p}\sum_{j=1}^{p} w_i(\rho\dot{w}_j + w_j - w_j + \sigma_j - \sigma_j)\mathbf{z}(t)^T \frac{1}{\rho}(\mathbf{P}_{i,j} + \mathbf{P}_{j,i})\mathbf{z}(t) \\
&= \sum_{i=1}^{p}\sum_{j=1}^{p}\sum_{k=1}^{p} w_i w_j w_k \mathbf{z}(t)^T \big(\mathbf{P}_{j,k}\mathbf{A}_i^T + \mathbf{A}_i\mathbf{P}_{j,k} + \mathbf{F}_j^T\mathbf{B}_i^T + \mathbf{B}_i\mathbf{F}_j\big)\mathbf{z}(t) \\
&\quad + \sum_{i=1}^{p}\sum_{j=1}^{p} w_i \overline{w}_j \mathbf{z}(t)^T \big(\mathbf{G}_j^T\mathbf{B}_i^T + \mathbf{B}_i\mathbf{G}_j\big)\mathbf{z}(t) \\
&\quad - \sum_{i=1}^{p}\sum_{j=1}^{p} w_i \overline{w}_j \mathbf{z}(t)^T \frac{1}{\rho}(\mathbf{P}_{i,j} + \mathbf{P}_{j,i})\mathbf{z}(t) \\
&\quad + \sum_{i=1}^{p}\sum_{j=1}^{p} w_i w_j \mathbf{z}(t)^T \frac{1}{\rho}(\mathbf{P}_{i,j} + \mathbf{P}_{j,i})\mathbf{z}(t) \\
&\quad + \sum_{i=1}^{p} w_i \mathbf{z}(t)^T \sum_{r=1}^{p} \frac{\sigma_r}{\rho}(\mathbf{P}_{i,r} + \mathbf{P}_{r,i})\mathbf{z}(t) \\
&= \sum_{i=1}^{p}\sum_{j=1}^{p}\sum_{k=1}^{p} w_i w_j w_k \mathbf{z}(t)^T \Big(\mathbf{P}_{j,k}\big(\mathbf{A}_i + \frac{1}{\rho}\mathbf{I}\big)^T + \big(\mathbf{A}_i + \frac{1}{\rho}\mathbf{I}\big)\mathbf{P}_{j,k} \\
&\quad + \mathbf{F}_j^T\mathbf{B}_i^T + \mathbf{B}_i\mathbf{F}_j + \sum_{r=1}^{p} \frac{\sigma_r}{\rho}(\mathbf{P}_{i,r} + \mathbf{P}_{r,i})\Big)\mathbf{z}(t) \\
&\quad + \sum_{i=1}^{p}\sum_{j=1}^{p} w_i \overline{w}_j \mathbf{z}(t)^T \Big(\mathbf{G}_j^T\mathbf{B}_i^T + \mathbf{B}_i\mathbf{G}_j - \frac{1}{\rho}(\mathbf{P}_{i,j} + \mathbf{P}_{j,i})\Big)\mathbf{z}(t) \qquad (5.8)
\end{aligned}$$

Considering arbitrary matrices of $\mathbf{\Lambda}_i = \mathbf{\Lambda}_i^T \in \Re^{n\times n}$, $i = 1, 2, \cdots, p$, and based on the property of the membership functions of (2.3), we have $\sum_{i=1}^{p} w_i = 1$ and $\sum_{i=1}^{p} \dot{w}_i = 0$ which lead to $\sum_{i=1}^{p}\sum_{j=1}^{p} w_i\dot{w}_j\mathbf{\Lambda}_i = \mathbf{0}$. It can be written as $\sum_{i=1}^{p}\sum_{j=1}^{p} w_i(\rho\dot{w}_j + w_j - w_j + \sigma_j - \sigma_j)\frac{1}{\rho}\mathbf{\Lambda}_i = \sum_{i=1}^{p}\sum_{j=1}^{p} w_i(\overline{w}_j - w_j - \sigma_j)\frac{1}{\rho}\mathbf{\Lambda}_i = \sum_{i=1}^{p}\sum_{j=1}^{p} w_i\overline{w}_j\frac{1}{\rho}\mathbf{\Lambda}_i - \sum_{i=1}^{p}\sum_{j=1}^{p} w_i(w_j + \sigma_j)\frac{1}{\rho}\mathbf{\Lambda}_i = \sum_{i=1}^{p}\sum_{j=1}^{p} w_i\overline{w}_j\frac{1}{\rho}\mathbf{\Lambda}_i - \sum_{i=1}^{p}\sum_{j=1}^{p} w_i w_j(\frac{1}{\rho} + \sum_{k=1}^{p}\frac{\sigma_k}{\rho})\mathbf{\Lambda}_i = \mathbf{0}$. Adding the last equation to (5.8), we have,

$$\dot{V}(t) = \sum_{i=1}^{p}\sum_{j=1}^{p}\sum_{k=1}^{p} w_i w_j w_k \mathbf{z}(t)^T \Big(\mathbf{P}_{j,k}(\mathbf{A}_i + \frac{1}{\rho}\mathbf{I})^T + (\mathbf{A}_i + \frac{1}{\rho}\mathbf{I})\mathbf{P}_{j,k}$$
$$+ \mathbf{F}_j^T\mathbf{B}_i^T + \mathbf{B}_i\mathbf{F}_j + \sum_{r=1}^{p}\frac{\sigma_r}{\rho}(\mathbf{P}_{i,r} + \mathbf{P}_{r,i}) - (\frac{1}{\rho} + \sum_{r=1}^{p}\frac{\sigma_r}{\rho})\mathbf{\Lambda}_i\Big)\mathbf{z}(t)$$
$$+ \sum_{i=1}^{p}\sum_{j=1}^{p} w_i \overline{w}_j \mathbf{z}(t)^T \Big(\mathbf{G}_j^T\mathbf{B}_i^T + \mathbf{B}_i\mathbf{G}_j - \frac{1}{\rho}(\mathbf{P}_{i,j} + \mathbf{P}_{j,i}) + \frac{1}{\rho}\mathbf{\Lambda}_i\Big)\mathbf{z}(t) \tag{5.9}$$

Denote $\mathbf{\Xi} = \sum_{i=1}^{p}\sum_{j=1}^{p}\sum_{k=1}^{p} w_i w_j w_k \mathbf{\Xi}_{ijk}$ where

$$\mathbf{\Xi}_{ijk} = \mathbf{P}_{j,k}(\mathbf{A}_i + \frac{1}{\rho}\mathbf{I})^T + (\mathbf{A}_i + \frac{1}{\rho}\mathbf{I})\mathbf{P}_{j,k} + \mathbf{F}_j^T\mathbf{B}_i^T + \mathbf{B}_i\mathbf{F}_j$$
$$+ \sum_{r=1}^{p}\frac{\sigma_r}{\rho}(\mathbf{P}_{i,r} + \mathbf{P}_{r,i}) - (\frac{1}{\rho} + \sum_{r=1}^{p}\frac{\sigma_r}{\rho})\mathbf{\Lambda}_i \ \forall\, i,\ j,\ k. \tag{5.10}$$

By following the analysis procedure in [24], we have,

$$\mathbf{\Xi} = \sum_{i=1}^{p} w_i^3 \mathbf{\Xi}_{iii} + \sum_{i=1}^{p}\sum_{j=1, j\neq i}^{p} w_i^2 w_j (\mathbf{\Xi}_{iij} + \mathbf{\Xi}_{iji} + \mathbf{\Xi}_{jii})$$
$$+ \sum_{i=1}^{p-2}\sum_{j=i+1}^{p-1}\sum_{k=j+1}^{p} w_i w_j w_k (\mathbf{\Xi}_{ijk} + \mathbf{\Xi}_{ikj} + \mathbf{\Xi}_{jik} + \mathbf{\Xi}_{kij} + \mathbf{\Xi}_{jki} + \mathbf{\Xi}_{kji}). \tag{5.11}$$

Define $\mathbf{Y}_{iii} = \mathbf{Y}_{iii}^T \in \Re^{n\times n}$, $i = 1, 2, \cdots, p$, $\mathbf{Y}_{iij} = \mathbf{Y}_{jii}^T \in \Re^{n\times n}$, $\mathbf{Y}_{iji} = \mathbf{Y}_{iji}^T \in \Re^{n\times n}$, $i, j = 1, 2, \cdots, p$; $i \neq j$, $\mathbf{Y}_{ijk} = \mathbf{Y}_{kji}^T \in \Re^{n\times n}$, $\mathbf{Y}_{ikj} = \mathbf{Y}_{jki}^T \in \Re^{n\times n}$ and $\mathbf{Y}_{jik} = \mathbf{Y}_{kij}^T \in \Re^{n\times n}$, $i = 1, 2, \cdots, p$-2; $j = i$+1, i+2, $\cdots$, p-1; $k = j$+1, j+2, $\cdots$, p

Let

$$\mathbf{Y}_{iii} \geq \mathbf{\Xi}_{iii} \ \forall\, i \tag{5.12}$$

$$\mathbf{Y}_{iij} + \mathbf{Y}_{iji} + \mathbf{Y}_{jii} \geq \mathbf{\Xi}_{iij} + \mathbf{\Xi}_{iji} + \mathbf{\Xi}_{jii} \ \forall\, i,\ j\ ; i \neq j \tag{5.13}$$

$$\mathbf{Y}_{ijk} + \mathbf{Y}_{ikj} + \mathbf{Y}_{jik} + \mathbf{Y}_{kij} + \mathbf{Y}_{jki} + \mathbf{Y}_{kji} \geq \mathbf{\Xi}_{ijk} + \mathbf{\Xi}_{ikj} + \mathbf{\Xi}_{jik} + \mathbf{\Xi}_{kij}$$
$$+ \mathbf{\Xi}_{jki} + \mathbf{\Xi}_{kji}, i = 1, 2, \cdots, p-2; j = i+1, i+2, \cdots, p-1; k = j+1, j+2, \cdots, p \tag{5.14}$$

From (5.12) to (5.14), we have

$$\begin{aligned}\boldsymbol{\Xi} \leq &\sum_{i=1}^{p} w_i^3 \mathbf{Y}_{iii} + \sum_{i=1}^{p} \sum_{j=1, j\neq i}^{p} w_i^2 w_j (\mathbf{Y}_{iij} + \mathbf{Y}_{iji} + \mathbf{Y}_{jii}) \\ &+ \sum_{i=1}^{p-2} \sum_{j=i+1}^{p-1} \sum_{k=j+1}^{p} w_i w_j w_k (\mathbf{Y}_{ijk} + \mathbf{Y}_{ikj} + \mathbf{Y}_{jik} + \mathbf{Y}_{kij} + \mathbf{Y}_{jki} + \mathbf{Y}_{kji}).\end{aligned} \tag{5.15}$$

From (5.9) to (5.15), considering $\mathbf{G}_j^T \mathbf{B}_i^T + \mathbf{B}_i \mathbf{G}_j - \frac{1}{\rho}(\mathbf{P}_{i,j} + \mathbf{P}_{j,i}) + \frac{1}{\rho}\boldsymbol{\Lambda}_i < 0$ for all i and j, we have

$$\begin{aligned}\dot{V}(t) \leq &\sum_{i=1}^{p} w_i^3 \mathbf{z}(t)^T \mathbf{Y}_{iii} \mathbf{z}(t) + \sum_{i=1}^{p} \sum_{j=1, j\neq i}^{p} w_i^2 w_j \mathbf{z}(t)^T (\mathbf{Y}_{iij} + \mathbf{Y}_{iji} + \mathbf{Y}_{jii}) \mathbf{z}(t) \\ &+ \sum_{i=1}^{p-2} \sum_{j=i+1}^{p-1} \sum_{k=j+1}^{p} w_i w_j w_k \mathbf{z}(t)^T (\mathbf{Y}_{ijk} + \mathbf{Y}_{ikj} + \mathbf{Y}_{jik} + \mathbf{Y}_{kij} \\ &+ \mathbf{Y}_{jki} + \mathbf{Y}_{kji}) \mathbf{z}(t) \\ = &\sum_{k=1}^{p} w_k \mathbf{r}(t)^T \mathbf{Y}_k \mathbf{r}(t)\end{aligned} \tag{5.16}$$

where $\mathbf{r}(t) = \begin{bmatrix} w_1 \mathbf{z}(t) \\ w_2 \mathbf{z}(t) \\ \vdots \\ w_p \mathbf{z}(t) \end{bmatrix}$, $\mathbf{Y}_k = \begin{bmatrix} \mathbf{Y}_{1k1} & \mathbf{Y}_{1k2} & \cdots & \mathbf{Y}_{1kp} \\ \mathbf{Y}_{2k1} & \mathbf{Y}_{2k2} & \cdots & \mathbf{Y}_{2kp} \\ \vdots & \vdots & \vdots & \vdots \\ \mathbf{Y}_{pk1} & \mathbf{Y}_{pk2} & \cdots & \mathbf{Y}_{pkp} \end{bmatrix}$.

Remark 5.6. Referring to the PDLF (5.6), it is required that $\mathbf{P}(\mathbf{x}(t)) = \sum_{k=1}^{p} \sum_{l=1}^{p} \mathbf{P}_{k,l} > 0$, which can be written as

$$\begin{bmatrix} w_1 \mathbf{z}(t) \\ w_2 \mathbf{z}(t) \\ \vdots \\ w_p \mathbf{z}(t) \end{bmatrix}^T \begin{bmatrix} \mathbf{P}_{1,1} & \mathbf{P}_{1,2} & \cdots & \mathbf{P}_{1,p} \\ \mathbf{P}_{2,1} & \mathbf{P}_{2,2} & \cdots & \mathbf{P}_{2,p} \\ \vdots & \vdots & \vdots & \vdots \\ \mathbf{P}_{p,1} & \mathbf{P}_{p,2} & \cdots & \mathbf{P}_{p,p} \end{bmatrix} \begin{bmatrix} w_1 \mathbf{z}(t) \\ w_2 \mathbf{z}(t) \\ \vdots \\ w_p \mathbf{z}(t) \end{bmatrix} > 0.$$

Thus, $\mathbf{P}(\mathbf{x}(t)) > 0$ is achieved by $\begin{bmatrix} \mathbf{P}_{1,1} & \mathbf{P}_{1,2} & \cdots & \mathbf{P}_{1,p} \\ \mathbf{P}_{2,1} & \mathbf{P}_{2,2} & \cdots & \mathbf{P}_{2,p} \\ \vdots & \vdots & \vdots & \vdots \\ \mathbf{P}_{p,1} & \mathbf{P}_{p,2} & \cdots & \mathbf{P}_{p,p} \end{bmatrix} > 0.$

From (5.6) and (5.16), based on the the Lyapunov stability theory, $V(t) > 0$ and $\dot{V}(t) < 0$ for $\mathbf{z}(t) \neq \mathbf{0}$ ($\mathbf{x}(t) \neq \mathbf{0}$) implying the asymptotic stability of the FMB control system (4.4), i.e., $\mathbf{x}(t) \to 0$ when time $t \to \infty$, can be achieved if the stability conditions summarized in the following theorem are satisfied.

Theorem 5.1. *The FMB control system (5.5), formed by the nonlinear plant represented by the fuzzy model (2.2) and the fuzzy controller (5.3) connected in a closed loop, is asymptotically stable if there exist pre-defined scalars* ρ *and and* σ_i, $i = 1, 2, \cdots, p$, *satisfying* $w_i(\mathbf{x}(t)) + \rho\dot{w}_i(\mathbf{x}(t)) + \sigma_i \geq 0$ *and there exist matrices* $\mathbf{F}_j \in \Re^{m\times n}$, $\mathbf{G}_j \in \Re^{m\times n}$, $j = 1, 2, \cdots, p$, $\mathbf{P}_{k,l} = \mathbf{P}_{l,k}^T \in \Re^{n\times n}$, $k, l = 1, 2, \cdots, p$, $\mathbf{Y}_{iii} = \mathbf{Y}_{iii}^T \in \Re^{n\times n}$, $i = 1, 2, \cdots, p$, $\mathbf{Y}_{iij} = \mathbf{Y}_{jii}^T \in \Re^{n\times n}$, $\mathbf{Y}_{iji} = \mathbf{Y}_{iji}^T \in \Re^{n\times n}$, $i, j = 1, 2, \cdots, p$; $i \neq j$, $\mathbf{Y}_{ijk} = \mathbf{Y}_{kji}^T \in \Re^{n\times n}$, $\mathbf{Y}_{ikj} = \mathbf{Y}_{jki}^T \in \Re^{n\times n}$, $\mathbf{Y}_{jik} = \mathbf{Y}_{kij}^T \in \Re^{n\times n}$, $i = 1, 2, \cdots, p\text{-}2$; $j = i+1, i+2, \cdots, p\text{-}1$; $k = j+1, j+2, \cdots, p$ *and* $\mathbf{\Lambda}_i = \mathbf{\Lambda}_i \in \Re^{n\times n}$, $i = 1, 2, \cdots, p$, *such that the following LMIs are satisfied.*

LMIs of (5.12) to (5.14);

$$\begin{bmatrix} \mathbf{P}_{1,1} & \mathbf{P}_{1,2} & \cdots & \mathbf{P}_{1,p} \\ \mathbf{P}_{2,1} & \mathbf{P}_{2,2} & \cdots & \mathbf{P}_{2,p} \\ \vdots & \vdots & \vdots & \vdots \\ \mathbf{P}_{p,1} & \mathbf{P}_{p,2} & \cdots & \mathbf{P}_{p,p} \end{bmatrix} > 0;$$

$$\mathbf{G}_j^T\mathbf{B}_i^T + \mathbf{B}_i\mathbf{G}_j - \frac{1}{\rho}(\mathbf{P}_{i,j} + \mathbf{P}_{j,i}) + \frac{1}{\rho}\mathbf{\Lambda}_i < 0 \; \forall \; i, \; j.$$

5.3.2 An Improved Non-PDC Fuzzy Controller

It is found that the stability analysis can be improved by including the membership function w_k in the second term of the non-PDC fuzzy controller of (5.3). Hence, we propose an improved non-PDC fuzzy controller in the following form:

$$\mathbf{u}(t) = \sum_{j=1}^{p} w_j\mathbf{F}_j\mathbf{P}(\mathbf{x}(t))^{-1}\mathbf{x}(t) + \sum_{j=1}^{p}\sum_{k=1}^{p} \overline{w}_j w_k \mathbf{H}_{jk}\mathbf{P}(\mathbf{x}(t))^{-1}\mathbf{x}(t) \quad (5.17)$$

where $\mathbf{H}_{jk} \in \Re^{m\times n}$, $j, k = 1, 2, \cdots, p$, are feedback gains to be determined.

Remark 5.7. Comparing to the non-PDC fuzzy controller (5.3), the extra membership functions w_k in the second term of (5.17) is able to further enhance the nonlinearity for feedback compensation. However, it complicates the controller structure that may increases the implementation cost. It is thus suggested that the non-PDC fuzzy controller (5.3) is employed for initial design. If no feasible design is achieved, the one in (5.17) can be employed.

In the following, we consider the FMB control system formed by connecting the fuzzy model (2.2) and the improved non-PDC fuzzy controller (5.17) in a closed loop that is shown as follows.

$$\begin{aligned}\dot{\mathbf{x}}(t) &= \sum_{i=1}^{p} w_i \Big(\mathbf{A}_i\mathbf{x}(t) + \mathbf{B}_i\Big(\sum_{j=1}^{p} w_j\mathbf{F}_j\mathbf{P}(\mathbf{x}(t))^{-1}\mathbf{x}(t) \\ &\quad + \sum_{j=1}^{p}\sum_{k=1}^{p} \overline{w}_j w_k \mathbf{H}_{jk}\mathbf{P}(\mathbf{x}(t))^{-1}\mathbf{x}(t)\Big)\Big) \\ &= \sum_{i=1}^{p}\sum_{j=1}^{p} w_i w_j (\mathbf{A}_i + \mathbf{B}_i\mathbf{F}_j\mathbf{P}(\mathbf{x}(t))^{-1})\mathbf{x}(t) \\ &\quad + \sum_{i=1}^{p}\sum_{j=1}^{p}\sum_{k=1}^{p} w_i\overline{w}_j w_k \mathbf{B}_i\mathbf{H}_{jk}\mathbf{P}(\mathbf{x}(t))^{-1}\mathbf{x}(t) \end{aligned} \tag{5.18}$$

We consider the PDLF candidate (5.6) to investigate the system stability of (5.18). By introducing the slack matrices $\mathbf{\Lambda}_{ik} = \mathbf{\Lambda}_{ki} \in \Re^{n\times n}$, we have $\sum_{i=1}^{p}\sum_{k=1}^{p} w_i w_k \mathbf{\Lambda}_{ik} \sum_{j=1}^{p} \dot{w}_j = \sum_{i=1}^{p}\sum_{j=1}^{p}\sum_{k=1}^{p} w_i \dot{w}_j w_k \mathbf{\Lambda}_{ik} = \mathbf{0}$. From the (5.6) and (5.18), recalling that $\mathbf{z}(t) = \mathbf{P}(\mathbf{x}(t))^{-1}\mathbf{x}(t)$ and following the same line of derivation procedure as in the previous section, we have

$$\begin{aligned}\dot{V}(t) &= \sum_{i=1}^{p}\sum_{j=1}^{p}\sum_{k=1}^{p} w_i w_j w_k \mathbf{z}(t)^T \Big(\mathbf{P}_{j,k}(\mathbf{A}_i + \frac{1}{\rho}\mathbf{I})^T + (\mathbf{A}_i + \frac{1}{\rho}\mathbf{I})\mathbf{P}_{j,k} \\ &\quad + \mathbf{F}_j^T\mathbf{B}_i^T + \mathbf{B}_i\mathbf{F}_j + \sum_{r=1}^{p}\frac{\sigma_r}{\rho}(\mathbf{P}_{i,r} + \mathbf{P}_{r,i}) - (\frac{1}{\rho} + \sum_{r=1}^{p}\frac{\sigma_r}{\rho})\mathbf{\Lambda}_{ik}\Big)\mathbf{z}(t) \\ &\quad + \sum_{i=1}^{p}\sum_{j=1}^{p}\sum_{k=1}^{p} w_i\overline{w}_j w_k \mathbf{z}(t)^T \Big(\mathbf{H}_{jk}^T\mathbf{B}_i^T + \mathbf{B}_i\mathbf{H}_{jk} - \frac{1}{\rho}(\mathbf{P}_{i,j} + \mathbf{P}_{j,i}) \\ &\quad + \frac{1}{\rho}\mathbf{\Lambda}_{ik}\Big)\mathbf{z}(t). \end{aligned} \tag{5.19}$$

Denote

$$\begin{aligned}\mathbf{\Xi}_{ijk} &= \mathbf{P}_{j,k}(\mathbf{A}_i + \frac{1}{\rho}\mathbf{I})^T + (\mathbf{A}_i + \frac{1}{\rho}\mathbf{I})\mathbf{P}_{j,k} + \mathbf{F}_j^T\mathbf{B}_i^T + \mathbf{B}_i\mathbf{F}_j \\ &\quad + \sum_{r=1}^{p}\frac{\sigma_r}{\rho}(\mathbf{P}_{i,r} + \mathbf{P}_{r,i}) - (\frac{1}{\rho} + \sum_{r=1}^{p}\frac{\sigma_r}{\rho})\mathbf{\Lambda}_{ik} \;\forall\, i,\, j,\, k \end{aligned} \tag{5.20}$$

which satisfy (5.12) to (5.14). Introducing matrices $\mathbf{S}_{ijk} = \mathbf{S}_{kji} \in \Re^{n\times n}$, we consider the following inequalities.

$$\mathbf{S}_{iji} > \mathbf{H}_{ji}^T\mathbf{B}_i^T + \mathbf{B}_i\mathbf{H}_{ji} - \frac{1}{\rho}(\mathbf{P}_{i,j} + \mathbf{P}_{j,i}) + \frac{1}{\rho}\mathbf{\Lambda}_{ii} \;\forall\, i,\, j \tag{5.21}$$

$$\mathbf{S}_{ijk} + \mathbf{S}_{ijk}^T \geq \mathbf{H}_{jk}^T\mathbf{B}_i^T + \mathbf{B}_i\mathbf{H}_{jk} - \frac{1}{\rho}(\mathbf{P}_{i,j} + \mathbf{P}_{j,i}) + \frac{1}{\rho}\mathbf{\Lambda}_{ik} + \mathbf{H}_{ji}^T\mathbf{B}_k^T + \mathbf{B}_k\mathbf{H}_{ji} - \frac{1}{\rho}(\mathbf{P}_{k,j} + \mathbf{P}_{j,k}) + \frac{1}{\rho}\mathbf{\Lambda}_{ki}\ \forall\, j,\ k;\ i < k \tag{5.22}$$

From (5.12) to (5.14), (5.21) to (5.22), (5.19) becomes

$$\begin{aligned}\dot{V}(t) &\leq \sum_{k=1}^{p} w_k\mathbf{r}(t)^T\mathbf{Y}_k\mathbf{r}(t) + \sum_{i=1}^{p}\sum_{j=1}^{p}\overline{w}_j w_i^2\mathbf{z}(t)^T\mathbf{S}_{iji}\mathbf{z}(t) \\ &\quad + \sum_{j=1}^{p}\sum_{i<k}\overline{w}_j w_i w_k\mathbf{z}(t)^T(\mathbf{S}_{ijk} + \mathbf{S}_{ijk}^T)\mathbf{z}(t) \\ &= \sum_{i=1}^{p} w_k\mathbf{r}(t)^T\mathbf{Y}_k\mathbf{r}(t) + \sum_{k=1}^{p}\overline{w}_k\mathbf{r}(t)^T\mathbf{S}_k\mathbf{r}(t)\end{aligned} \tag{5.23}$$

where $\mathbf{S}_k = \begin{bmatrix} \mathbf{S}_{1k1} & \mathbf{S}_{1k2} & \cdots & \mathbf{S}_{1kp} \\ \mathbf{S}_{2k1} & \mathbf{S}_{2k2} & \cdots & \mathbf{S}_{2kp} \\ \vdots & \vdots & \vdots & \vdots \\ \mathbf{S}_{pk1} & \mathbf{S}_{pk2} & \cdots & \mathbf{S}_{pkp} \end{bmatrix}$.

From (5.6) and (5.23), based on the the Lyapunov stability theory, $V(t) > 0$ and $\dot{V}(t) < 0$ for $\mathbf{z}(t) \neq \mathbf{0}$ ($\mathbf{x}(t) \neq \mathbf{0}$) implying the asymptotic stability of the FMB control system (5.5), i.e., $\mathbf{x}(t) \to 0$ when time $t \to \infty$, can be achieved if the stability conditions summarized in the following theorem are satisfied.

Theorem 5.2. *The FMB control system (5.5), formed by the nonlinear plant represented by the fuzzy model (2.2) and the fuzzy controller (5.17) connected in a closed loop, is asymptotically stable if there exist pre-defined scalars ρ and and σ_i, $i = 1, 2, \cdots, p$, satisfying $w_i(\mathbf{x}(t)) + \rho\dot{w}_i(\mathbf{x}(t)) + \sigma_i \geq 0$ and there exist matrices $\mathbf{F}_j \in \Re^{m\times n}$, $\mathbf{H}_{jk} \in \Re^{m\times n}$, $j, k = 1, 2, \cdots, p$, $\mathbf{P}_{k,l} = \mathbf{P}_{l,k}^T \in \Re^{n\times n}$, $k, l = 1, 2, \cdots, p$, $\mathbf{Y}_{iii} = \mathbf{Y}_{iii}^T \in \Re^{n\times n}$, $i = 1, 2, \cdots, p$, $\mathbf{Y}_{iij} = \mathbf{Y}_{jii}^T \in \Re^{n\times n}$, $\mathbf{Y}_{iji} = \mathbf{Y}_{iji}^T \in \Re^{n\times n}$, $i, j = 1, 2, \cdots, p$; $i \neq j$, $\mathbf{Y}_{ijk} = \mathbf{Y}_{kji}^T \in \Re^{n\times n}$, $\mathbf{Y}_{ikj} = \mathbf{Y}_{jki}^T \in \Re^{n\times n}$, $\mathbf{Y}_{jik} = \mathbf{Y}_{kij}^T \in \Re^{n\times n}$, $i = 1, 2, \cdots, p\text{-}2$; $j = i+1, i+2, \cdots, p\text{-}1$; $k = j+1, j+2, \cdots, p$ and $\mathbf{\Lambda}_{ik} = \mathbf{\Lambda}_{ki} \in \Re^{n\times n}$, $i, k = 1, 2, \cdots, p$, such that the following LMIs are satisfied.*

LMIs of (5.12) to (5.14), (5.21), (5.22);

$$\mathbf{Y}_i < 0;$$

$$\mathbf{S}_i < 0\ \forall\, i;$$

$$\begin{bmatrix} \mathbf{P}_{1,1} & \mathbf{P}_{1,2} & \cdots & \mathbf{P}_{1,p} \\ \mathbf{P}_{2,1} & \mathbf{P}_{2,2} & \cdots & \mathbf{P}_{2,p} \\ \vdots & \vdots & \vdots & \vdots \\ \mathbf{P}_{p,1} & \mathbf{P}_{p,2} & \cdots & \mathbf{P}_{p,p} \end{bmatrix} > 0.$$

Remark 5.8. The proposed PDLF-based non-PDC fuzzy control approach can be applied subject to the following conditions.

1. The FMB control systems are subject to perfectly matched premised membership functions. It means the fuzzy controller has to share the same membership functions as those of the TS fuzzy model.
2. The forms of the membership functions and their time derivatives have to be known and do not depend on the control input, $\mathbf{u}(t)$, in order to find the scalars ρ and σ_j that provide information of the membership functions to facilitate the stability analysis.

Remark 5.9. In this chapter, the stability analysis requires the inequalities of $w_i(\mathbf{x}(t)) + \rho\dot{w}_i(\mathbf{x}(t)) + \sigma_i \geq 0$, $i = 1, 2, \cdots, p$, to be satisfied. The scalars of ρ and σ_i bring the MFB information to the stability analysis. Hence, the stability analysis is considered as an MFSD analysis approach.

The solution of the stability conditions in Theorem 2.4 is a particular case of the solution of the proposed stability conditions in Theorem 5.1 and Theorem 5.2. It can be shown that if we choose $\mathbf{P}_{j,j} = \mathbf{P}$ for all j (where $\mathbf{P}$ is a constant symmetric positive definite matrix), $\mathbf{P}_{j,k} = \mathbf{0}$ for all $j \neq k$, $\mathbf{G}_j = \mathbf{0}$ for all j, and $\mathbf{\Lambda}_i = \mathbf{0}$ for all i (for Theorem 5.1) or $\mathbf{\Lambda}_{ij} = \mathbf{0}$ for all i and j (for Theorem 5.2); the stability conditions in Theorem 5.1 and Theorem 5.2 are reduced to the stability conditions in Theorem 2.4. It was reported that Theorem 2.1 to Theorem 2.3 are particular cases of Theorem 2.4. Hence, it can be concluded that the solutions of the particular cases of Theorem 2.1 to Theorem 2.3 are also the solutions of the proposed stability conditions in Theorem 5.1 and Theorem 5.2. Further relaxed stability conditions can be achieved by using the analysis procedure with the Polya's Theorem [97].

Example 5.1. Consider the TS fuzzy plant model in the form of (2.2) with $\mathbf{x}(t) = \begin{bmatrix} x_1(t) & x_2(t) & x_3(t) \end{bmatrix}^T$, $\mathbf{A}_1 = \begin{bmatrix} 1.59 & -7.29 & 0 \\ 0.01 & 0 & 0 \\ 0 & -0.17 & 0 \end{bmatrix}$, $\mathbf{A}_2 = \begin{bmatrix} 0.02 & -4.64 & 0 \\ 0.35 & 0.21 & 0 \\ 0 & -0.78 & 0 \end{bmatrix}$, $\mathbf{A}_3 = \begin{bmatrix} -a & -4.33 & 0 \\ 0 & -0.05 & 0 \\ 0 & 0 & -0.21 \end{bmatrix}$, $\mathbf{B}_1 = \begin{bmatrix} 1 \\ 0 \\ 0 \end{bmatrix}$, $\mathbf{B}_2 = \begin{bmatrix} 8 \\ 0 \\ 0 \end{bmatrix}$, $\mathbf{B}_3 = \begin{bmatrix} -b+6 \\ -1 \\ 0 \end{bmatrix}$, where $2 \leq a \leq 12$ and $2 \leq b \leq 24$ are constant parameters.

The membership functions are defined as $w_1(x_3(t)) = \mu_{M_1^1}(x_3(t)) = 1 - \frac{1}{1+e^{-\frac{x_3(t)+1}{0.8}}}$, $w_2(x_3(t)) = \mu_{M_1^2}(x_3(t)) = 1 - w_1(x_3(t)) - w_3(x_3(t))$, $w_3(x_3(t)) = \mu_{M_1^3}(x_3(t)) = \frac{1}{1+e^{-\frac{x_3(t)-1}{0.8}}}$. Based on the membership functions and the operating region of the TS fuzzy model, the time derivatives of the membership functions are obtained as $\dot{w}_1(x_3(t)) = -\frac{1.25e^{-\frac{x_3(t)+1}{0.8}}\dot{x}_3(t)}{\left(1+e^{-\frac{x_3(t)+1}{0.8}}\right)^2}$, $\dot{w}_2(x_3(t)) = -\dot{w}_1(x_3(t)) - \dot{w}_3(x_3(t))$ and $\dot{w}_3(x_3(t)) = \frac{1.25e^{-\frac{x_3(t)-1}{0.8}}\dot{x}_3(t)}{\left(1+e^{-\frac{x_3(t)-1}{0.8}}\right)^2}$ where $\dot{x}_3(t) = \big(-0.17w_1(x_3(t)) - 0.78w_2(x_3(t))\big)x_2(t) - 0.21w_3(x_3(t))x_3(t)$ as given by the TS fuzzy model. It can be seen that the time derivatives of the membership functions are independent of the control signal $\mathbf{u}(t)$ as required by the conditions stated in Remark 5.8. Based on the membership functions and their derivatives, it can be found numerically that $\rho = 1$, $\sigma_1 = -0.012993$, $\sigma_2 = -0.093097$ and $\sigma_3 = -0.001244$ satisfy the inequalities of $w_i(x_3(t)) + \rho\dot{w}_i(x_3(t)) + \sigma_i \geq 0$, $i = 1, 2, 3$.

In this example, the proposed non-PDC fuzzy controllers (5.3) and (5.17) are employed to close the feedback loop of the TS fuzzy model. The stability conditions in Theorem 5.1 and Theorem 5.2 are employed respectively to check for the stability regions. The stability regions corresponding to the non-PDC fuzzy controllers (5.3) and (5.17) are shown in Fig. 5.2 denoted by '×' and '∘', respectively. It can be seen that the stability conditions in Theorem 5.2 offer a larger stability region. However, the non-PDC fuzzy controller (5.17) has a more complex structure than the one of (5.3). Hence, in the overlapping stability region in Fig. 5.2, it is suggested to employ the non-PDC fuzzy controller (5.3) for the control process to lower the structural complexity of the controller.

For comparison purposes, the stability conditions in [59, 107] are employed to check for the stability regions, which are shown in Fig. 5.3 denoted by '×' and '∘' respectively. Furthermore, the published stability conditions in Theorem 2.4 and Theorem 2.5 (the degree of fuzzy summation [97] is chosen as $d = 4$) are also employed to check for the stability regions which are shown in Fig. 5.4. As the stability conditions in Theorem 2.1 to Theorem 2.3 are particular cases of those in Theorem 2.4, hence, the stability regions offered by them are not shown. It can be seen that the proposed stability conditions in this chapter offer larger stability regions than the published ones.

Remark 5.10. It has been shown in Fig. 5.2 to Fig. 5.4 that the proposed stability conditions in Theorem 5.1 and Theorem 5.2 are more relaxed than the published ones. However, the proposed non-PDC controllers have comparatively more complicated structures. Hence, it is recommended that the published stability conditions under the PDC design and PILF analysis approach should be employed first to design the fuzzy controller for simpler structure and lower implementation cost. If a feasible design cannot be realized, the proposed stability conditions can then be employed.

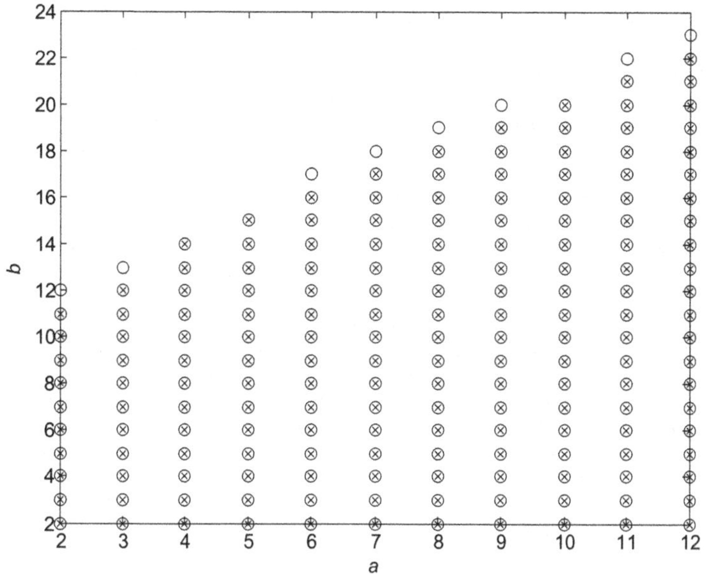

Fig. 5.2 Stability regions given by Theorem 5.1 ('×') and Theorem 5.2 ('○').

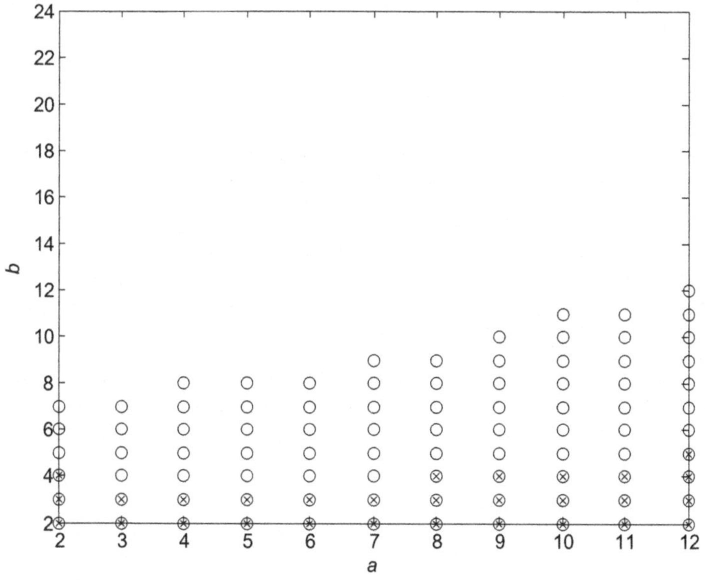

Fig. 5.3 Stability regions given by stability conditions in [107] ('×') and [59] ('○').

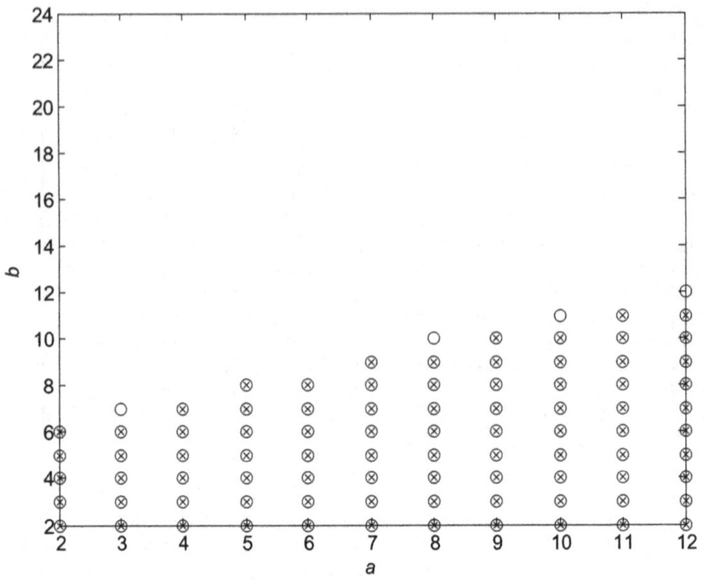

Fig. 5.4 Stability regions given by Theorem 2.4 ('×') and Theorem 2.5 ('○').

5.4 Conclusion

The system stability of FMB control systems subject to perfectly matched premised membership functions has been investigated under the PDLF approach. Two types of non-PDC fuzzy controllers have been proposed to control the nonlinear plants. The proposed controllers employ both the membership functions and their derivatives for the feedback compensation of the plant nonlinearity. Furthermore, the information of the membership function boundary has been utilized to facilitate the stability analysis. LMI-based stability conditions have been obtained to achieve stable FMB control systems. It has been shown that the solution of the stability conditions under the PDC design and MFSI analysis approach discussed in Chapter 2 is a particular case of the proposed ones in this chapter. Through simulation examples, it has been demonstrated that the proposed stability conditions under the non-PDC design and PDLF analysis approach are more relaxed than some published ones.

Chapter 6
Regional Switching FMB Control Systems

6.1 Introduction

In the previous chapters, the stability analysis of FMB control systems subject to perfectly and imperfectly matched premise membership functions was investigated. Various techniques, such as the MFSI and MFSD approaches with PILF/PDLF, were introduced for relaxing the stability conditions. Under these approaches, a single fuzzy controller is employed for the control process.

However, if the nonlinearity of the plant is strong enough, it is possible that a single fuzzy controller is not able to compensate the nonlinearity and stabilize the plant. In this chapter, a regional switching fuzzy control technique was proposed [64, 106] to handle nonlinear plants with strong nonlinearity. The basic idea is illustrated in Fig. 6.1. The operating region of the nonlinear plant is divided into a number of operating sub-regions. Referring to Fig. 6.1, it is assumed that the system has 2 system states, namely $x_1(t)$ and $x_2(t)$. The operating region is denoted by the dash square, which contains 4 sub-regions denoted as R_1 to R_4. Corresponding to each sub-region, a local TS fuzzy model is employed to describe the system dynamics of the plant. Based on each local TS fuzzy model, a local fuzzy controller is designed accordingly. As the nonlinearity for each operating sub-region is weaker than that for the full operating region, it is more likely to come up with a stable design in each sub-region. When the FMB control system is working in R_i, $i = 1, 2, 3, 4$, the i-th local fuzzy controller is employed to control the system. The resulting switching FMB control approach offers a potential to further relax the stability conditions. The switching FMB control approach was investigated in [21, 22, 128]. By using the fuzzy combination technique [62], the abrupt change of the control signal due to the switching activity is smoothed out. Some other switching fuzzy control techniques can be found in [57, 63, 65].

In [21, 22, 62, 64, 106, 128], the stability analysis is subject to the following limitations: 1) it is required that a switching TS fuzzy model is needed for the stability analysis and controller synthesis. In some situations, it is

H.-K. Lam and F.H.F. Leung: FMB Control Systems, STUDFUZZ 264, pp. 101–122.
springerlink.com

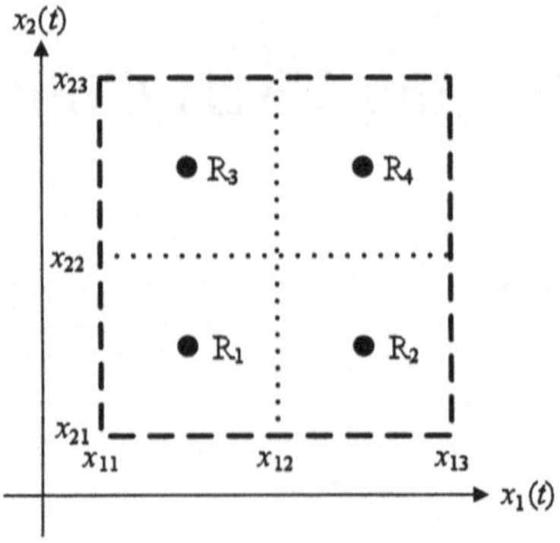

Fig. 6.1 A diagram illustrating the concept of regional switching FMB control approach.

inconvenient or difficult to obtain the switching TS fuzzy model. For example, when input-output data are needed for constructing the TS fuzzy model, it may be difficult to obtain the data for the nonlinear plant that works in a particular region, especially when a large number of small operating sub-regions are considered. 2) The stability analysis is conducted under perfectly matched premise membership functions. Hence, the regional switching FMB control approach [21, 22, 62, 64, 106, 128] cannot be applied for nonlinear plants subject to parameter uncertainties appearing in the premise membership functions of the fuzzy model. In this chapter, another regional switching FMB control approach is proposed to deal with nonlinear plants with strong nonlinearity. The proposed approach is able to circumvent the aforementioned limitations by using the regional information together with the consideration of membership functions.

6.2 Regional Switching Fuzzy Controller

Considering that the operating domain is divided to c operating sub-regions, a fuzzy controller for the operating sub-region j is proposed to control the nonlinear plant represented by the fuzzy model (2.2). The k-th rule for the j-th regional fuzzy controller is of the following form:

$$\begin{aligned}&\text{Rule } k\text{: IF } g_{j1}(\mathbf{x}(t)) \text{ is } N_1^{jk} \text{ AND } \cdots \text{ AND } g_{j\Omega}(\mathbf{x}(t)) \text{ is } N_\Omega^{jk}\\&\qquad \text{THEN } \mathbf{u}_j(t) = \mathbf{G}_{jk}\mathbf{x}(t), j = 1, 2, \ldots, c; k = 1, 2, \ldots, p\end{aligned} \tag{6.1}$$

where N_β^{jk} is the fuzzy set of rule k corresponding to the function $g_{j\beta}(\mathbf{x}(t))$, $\beta = 1, 2, \cdots, \Omega$; $\mathbf{G}_{jk} \in \Re^{m\times n}$, is the constant feedback gain of rule k for the j-th regional fuzzy controller; $\mathbf{u}_j(t) \in \Re^{m\times n}$ is the input vector for the j-th sub-operating domain. The rest variables are defined in Section 2.3. The inferred j-th regional switching (RS) fuzzy controller is defined as follows.

$$\mathbf{u}_j(t) = \sum_{k=1}^{p} m_{jk}(\mathbf{x}(t))\mathbf{G}_{jk}\mathbf{x}(t) \tag{6.2}$$

where

$$m_{jk}(\mathbf{x}(t)) \geq 0 \;\forall\; j, k, \sum_{k=1}^{p} m_{jk}(\mathbf{x}(t)) = 1 \;\forall\; j, \tag{6.3}$$

$$m_{jk}(\mathbf{x}(t)) = \frac{\prod_{l=1}^{\Omega} \mu_{N_l^{jk}}(g_{jl}(\mathbf{x}(t)))}{\sum_{k=1}^{p}\prod_{l=1}^{\Omega} \mu_{N_l^{jk}}(g_{jl}(\mathbf{x}(t)))} \;\forall\; j, \tag{6.4}$$

$m_{jk}(\mathbf{x}(t))$, $j = 1, 2, \cdots, c$, $k = 1, 2, \cdots, p$, are the normalized grades of membership function, $\mu_{N_\beta^{jk}}(g_\beta(\mathbf{x}(t)))$, $\beta = 1, 2, \cdots, \Omega$, $k = 1, 2, \cdots, p$, are the membership functions corresponding to the fuzzy set N_β^{jk}.

The j-th regional fuzzy controller defined in (6.2) is applied when the nonlinear plant is working in the j-th operating sub-region, which is characterized by the constraints $\underline{h}_\varsigma \leq h_\varsigma(\mathbf{x}(t)) \leq \overline{h}_\varsigma$, where $\varsigma = 1, 2, \cdots, \Phi$; Φ is an integer denoting the number of constraints in each operating sub-region; $\underline{h}_\varsigma$ and $\overline{h}_\varsigma$ being scalar constants are the boundaries of the j-th operating sub-region. For example, referring to Fig. 6.1, we choose $h_1(\mathbf{x}(t))$ as $x_1(t)$ and $h_2(\mathbf{x}(t))$ as $x_2(t)$. The 2 constraints of $x_{11} \leq x_1(t) \leq x_{12}$ and $x_{21} \leq x_2(t) \leq x_{22}$ characterize the sub-region R_1, where x_{11}, x_{12}, x_{21} and x_{22} are defined in Fig. 6.1.

Define a scalar function $v_j(\mathbf{x}(t))$ which takes the value of 1 if $h_\varsigma(\mathbf{x}(t))$ indicates that the system states are inside the j-th operating sub-region; otherwise, $v_j(\mathbf{x}(t)) = 0$. As there is no overlapping between operating sub-regions, it exhibits the property of $\sum_{j=1}^{c} v_j(\mathbf{x}(t)) = 1$. With this concept, from (6.2), the overall regional switching fuzzy controller is defined as follows.

$$\begin{aligned}\mathbf{u}(t) &= \sum_{j=1}^{c} v_j(\mathbf{x}(t))\mathbf{u}_j(t) \\ &= \sum_{j=1}^{c}\sum_{k=1}^{p} v_j(\mathbf{x}(t))m_{jk}(\mathbf{x}(t))\mathbf{G}_{jk}\mathbf{x}(t)\end{aligned} \tag{6.5}$$

Remark 6.1. With the function $v_j(\mathbf{x}(t))$, when the system is working in the j-th operating sub-region, the corresponding j-th regional fuzzy controller will be activated to perform the control process.

Remark 6.2. The regional switching fuzzy controller is reduced to the traditional fuzzy controller (2.6) when we choose $c = 1$ (i.e. the full operating region is considered). Consequently, comparing (2.6) with (6.5), we have $m_k(\mathbf{x}(t)) = m_{jk}(\mathbf{x}(t))$ and $\mathbf{G}_k = \mathbf{G}_{jk}$ for all j.

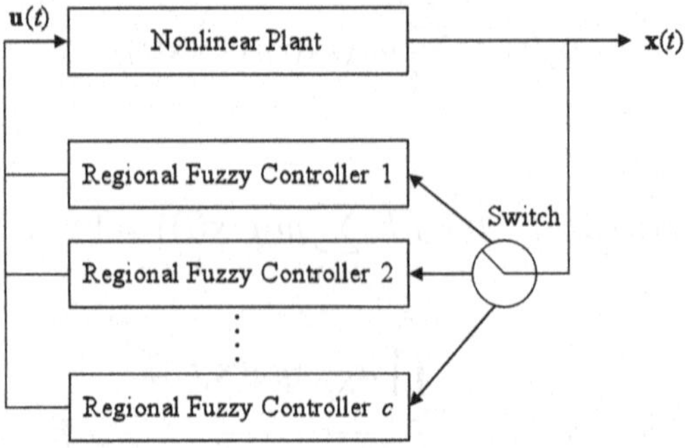

Fig. 6.2 A diagram illustrating the concept of regional switching FMB control approach.

A block diagram illustrating the regional switching fuzzy control scheme is shown in Fig. 6.2.

6.3 Stability Analysis of Regional Switching FMB Control Systems

Considering the fuzzy model (2.2) and the RS fuzzy controller (6.5) connected in a closed loop, the RS FMB control system is obtained as follows.

$$\begin{aligned}\dot{\mathbf{x}}(t) &= \sum_{i=1}^{p} w_i(\mathbf{x}(t))\big(\mathbf{A}_i\mathbf{x}(t) + \mathbf{B}_i \sum_{j=1}^{c}\sum_{k=1}^{p} v_j(\mathbf{x}(t)) m_{jk}(\mathbf{x}(t))\mathbf{G}_{jk}\mathbf{x}(t)\big) \\ &= \sum_{i=1}^{p}\sum_{j=1}^{c}\sum_{k=1}^{p} w_i(\mathbf{x}(t)) v_j(\mathbf{x}(t)) m_{jk}(\mathbf{x}(t))(\mathbf{A}_i + \mathbf{B}_i\mathbf{G}_{jk})\mathbf{x}(t) \end{aligned} \quad (6.6)$$

For brevity, $w_i(\mathbf{x}(t))$, $v_j(\mathbf{x}(t))$ and $m_{jk}(\mathbf{x}(t))$ are denoted as w_i, v_j and m_{jk}, respectively. In the following analysis, as given by the property of these variables, the equality $\sum_{i=1}^{p} w_i(\mathbf{x}(t)) = \sum_{j=1}^{c} v_j(\mathbf{x}(t)) = \sum_{k=1}^{p} m_{jk}(\mathbf{x}(t)) = \sum_{i=1}^{p}\sum_{j=1}^{c}\sum_{k=1}^{p} w_i(\mathbf{x}(t))v_j(\mathbf{x}(t))m_{jk}(\mathbf{x}(t)) = 1$ will be utilized. To investigate the system stability of the regional switching FMB control system (6.6), we consider the quadratic Lyapunov function candidate (3.1). Define $\mathbf{X} = \mathbf{P}^{-1}$, $\mathbf{z}(t) = \mathbf{X}^{-1}\mathbf{x}(t)$ and the feedback gains of the fuzzy controller as $\mathbf{G}_{jk} = \mathbf{N}_{jk}\mathbf{X}^{-1}$, $j = 1, 2, \cdots, c$, $k = 1, 2, \cdots, p$, where $\mathbf{N}_{jk} \in \Re^{m\times n}$.

From (3.1) and (6.6), we have

$$\dot{V}(t) = \dot{\mathbf{x}}(t)^T\mathbf{P}\mathbf{x}(t) + \mathbf{x}(t)^T\mathbf{P}\dot{\mathbf{x}}(t)$$
$$= \sum_{i=1}^{p}\sum_{j=1}^{c}\sum_{k=1}^{p} w_i v_j m_{jk}\mathbf{x}(t)^T\left((\mathbf{A}_i + \mathbf{B}_i\mathbf{G}_{jk})^T\mathbf{P} + \mathbf{P}(\mathbf{A}_i + \mathbf{B}_i\mathbf{G}_{jk})\right)\mathbf{x}(t)$$
$$= \sum_{i=1}^{p}\sum_{j=1}^{c}\sum_{k=1}^{p} w_i v_j m_{jk}\mathbf{z}(t)^T\mathbf{Q}_{ijk}\mathbf{z}(t) \tag{6.7}$$

where $\mathbf{Q}_{ijk} = \mathbf{A}_i\mathbf{X} + \mathbf{X}\mathbf{A}_i^T + \mathbf{B}_i\mathbf{N}_{jk} + \mathbf{N}_{jk}^T\mathbf{B}_i^T$ for all i, j and k.

Remark 6.3. When $c = 1$, the RS FMB control system is reduced to (2.10). Then, two cases have been investigated. Case 1 (Imperfectly matched premise membership functions): when the number of rules of the fuzzy model and the fuzzy controller are different, the system stability of the RS FMB control system (6.6) is guaranteed by the stability conditions in Theorem 2.1. Case 2 (Perfectly matched premise membership functions): when the fuzzy model and the fuzzy controller share the same set of premise rules and membership functions, the stability conditions in Theorem 2.2 to Theorem 2.5 were derived to examine the system stability and facilitate the controller synthesis.

In the following, we will investigate the stability of the RS FMB control systems (6.6) based on the Lyapunov stability theory for the cases of imperfectly and perfectly matched premise membership functions.

6.3.1 Imperfectly Matched Premise Membership Functions Based Inequalities

It should be noted that we cannot produce relaxed stability conditions by directly following the stability analysis approach for Theorem 2.2 to Theorem 2.5 as the regional information is not considered. To bring the regional information into the stability analysis, we propose some MFSI and MFSD inequalities to introduce some slack matrices.

Based on the property of the membership functions that $\sum_{k=1}^{p} w_k = \sum_{k=1}^{p} m_{jk} = 1$, we have $\sum_{k=1}^{p}(w_k - m_{jk}) = 0$ for all j. Introducing slack matrices $\mathbf{\Lambda}_{ij} = \mathbf{\Lambda}_{ij}^T \in \Re^{n\times n}$ and $\mathbf{V}_{jk} = \mathbf{V}_{jk}^T \in \Re^{n\times n}$, we have the following MFSI inequalities.

$$\sum_{i=1}^{p}\sum_{k=1}^{p} w_i(w_k - m_{jk})\mathbf{\Lambda}_{ij} = \mathbf{0} \tag{6.8}$$

$$\sum_{i=1}^{p}\sum_{k=1}^{p} (w_i - m_{ji})m_{jk}\mathbf{V}_{jk} = \mathbf{0} \tag{6.9}$$

As shown in Chapter 3, the boundary information of the membership functions plays an important role for the stability analysis. Introducing slack matrices $0 \leq \mathbf{M}_{ijk} = \mathbf{M}_{kji}^T \in \Re^{n \times n}$ and $0 \leq \mathbf{W}_{ijk} = \mathbf{W}_{kji}^T \in \Re^{n \times n}$, i, $k = 1, 2, \cdots, p$; $j = 1, 2, \cdots, c$, we consider the following MFSD inequalities for the j-th operating sub-region.

$$\sum_{i=1}^{p}\sum_{k=1}^{p}(w_i w_k - w_i m_{jk} - m_{ji} w_k + m_{ji} m_{jk} - \rho_{ijk})\mathbf{M}_{ijk} \geq 0 \;\forall\, j \tag{6.10}$$

$$\sum_{i=1}^{p}\sum_{k=1}^{p}(\sigma_{ijk} - (w_i w_k - w_i m_{jk} - m_{ji} w_k + m_{ji} m_{jk}))\mathbf{W}_{ijk} \geq 0 \;\forall\, j \tag{6.11}$$

where ρ_{ijk} and σ_{ijk} are constant scalars to be determined and satisfy the inequalities $w_i w_k - w_i m_{jk} - m_{ji} w_k + m_{ji} m_{jk} - \rho_{ijk} \geq 0$ and $\sigma_{ijk} - (w_i w_k - w_i m_{jk} - m_{ji} w_k + m_{ji} m_{jk})) \geq 0$, respectively, for all i, j and k.

Furthermore, for the j-th operating sub-region, it is assumed that the membership functions satisfy the following inequalities.

$$\delta_{ji} \leq w_i(\mathbf{x}(t)) \leq \varepsilon_{ji} \;\forall\, i,\, j \tag{6.12}$$

$$\eta_{jk} \leq m_{jk}(\mathbf{x}(t)) \leq \gamma_{jk} \;\forall\, j,\, k \tag{6.13}$$

where the scalar constants of δ_{ji}, ε_{ji}, η_{jk} and γ_{jk} are the lower or upper membership-function boundaries. Consequently, introducing slack matrices $0 < \mathbf{H}_{ijk} = \mathbf{H}_{ijk}^T \in \Re^{n \times n}$, $0 < \mathbf{J}_{ijk} = \mathbf{J}_{ijk}^T \in \Re^{n \times n}$, $0 < \mathbf{K}_{ijk} = \mathbf{K}_{ijk}^T \in \Re^{n \times n}$ and $0 < \mathbf{L}_{ijk} = \mathbf{L}_{ijk}^T \in \Re^{n \times n}$ for all i, j and k, we can construct the following regional MFSD inequalities for the j-th operating sub-region.

$$\sum_{i=1}^{p}\sum_{k=1}^{p}(w_i - \delta_{ji})(w_k - \delta_{jk})\mathbf{H}_{ijk} \geq \mathbf{0} \;\forall\, j \tag{6.14}$$

$$\sum_{i=1}^{p}\sum_{k=1}^{p}(\varepsilon_{ji} - w_i)(\varepsilon_{jk} - w_k)\mathbf{J}_{ijk} \geq \mathbf{0} \;\forall\, j \tag{6.15}$$

$$\sum_{i=1}^{p}\sum_{k=1}^{p}(m_{ji} - \eta_{ji})(m_{jk} - \eta_{jk})\mathbf{K}_{ijk} \geq \mathbf{0} \;\forall\, j \tag{6.16}$$

$$\sum_{i=1}^{p}\sum_{k=1}^{p}(\gamma_{ji} - m_{ji})(\gamma_{jk} - m_{jk})\mathbf{L}_{ijk} \geq \mathbf{0} \;\forall\, j \tag{6.17}$$

Remark 6.4. The matrix and scalar variables in (6.10) to (6.17) contain the regional and boundary information of the FMB control system to facilitate the stability analysis. It should be noted that there must exist the lower and upper bounds (ρ_{ijk}, σ_{ijk}, δ_{ji}, ε_{ji}, η_{ji} and γ_{ji}) such that the inequalities of (6.10) to (6.17) are satisfied. As the form of the membership functions of $w_i(\mathbf{x}(t))$ and $m_{jk}(\mathbf{x}(t))$ is known, the lower and upper bounds can be found analytically or numerically.

6.3.2 Stability Analysis with Imperfectly Matched Premise Membership Functions

Adding the inequalities of (6.10) and (6.11) to (6.11), we have the follows.

$$\begin{aligned}
\dot{V}(t) \leq & \sum_{i=1}^{p}\sum_{j=1}^{c}\sum_{k=1}^{p} w_i v_j m_{jk}\mathbf{z}(t)^T\mathbf{Q}_{ijk}\mathbf{z}(t) \\
& + \sum_{i=1}^{p}\sum_{j=1}^{c}\sum_{k=1}^{p} v_j(w_i w_k - w_i m_{jk} - m_{ji} w_k + m_{ji} m_{jk} - \rho_{ijk})\mathbf{z}(t)^T\mathbf{M}_{ijk}\mathbf{z}(t) \\
& + \sum_{i=1}^{p}\sum_{j=1}^{c}\sum_{k=1}^{p} v_j(\sigma_{ijk} - (w_i w_k - w_i m_{jk} - m_{ji} w_k + m_{ji} m_{jk}))\mathbf{z}(t)^T\mathbf{W}_{ijk}\mathbf{z}(t) \\
& + \sum_{i=1}^{p}\sum_{j=1}^{c}\sum_{k=1}^{p} w_i v_j(w_k - m_{jk})\mathbf{z}(t)^T\boldsymbol{\Lambda}_{ij}\mathbf{z}(t) \\
& + \sum_{i=1}^{p}\sum_{j=1}^{c}\sum_{k=1}^{p} (w_i - m_{ji}) v_j m_{jk}\mathbf{z}(t)^T\mathbf{V}_{jk}\mathbf{z}(t) \\
= & \sum_{i=1}^{p}\sum_{j=1}^{c}\sum_{k=1}^{p} w_i v_j m_{jk}\mathbf{z}(t)^T(\mathbf{Q}_{ijk} - \mathbf{M}_{ijk} - \mathbf{M}_{kji} + \mathbf{W}_{ijk} + \mathbf{W}_{kji} \\
& - \boldsymbol{\Lambda}_{ij} + \mathbf{V}_{jk})\mathbf{z}(t) + \sum_{i=1}^{p}\sum_{j=1}^{c}\sum_{k=1}^{p} w_i v_j w_k \mathbf{z}(t)^T\big(\mathbf{M}_{ijk} - \mathbf{W}_{ijk} \\
& + \boldsymbol{\Lambda}_{ij} + \sum_{r=1}^{p}\sum_{s=1}^{p}(\sigma_{rjs}\mathbf{W}_{rjs} - \rho_{rjs}\mathbf{M}_{rjs})\big)\mathbf{z}(t) \\
& + \sum_{i=1}^{p}\sum_{j=1}^{c}\sum_{k=1}^{p} m_{ji} v_j m_{jk}\mathbf{z}(t)^T(\mathbf{M}_{ijk} - \mathbf{W}_{ijk} - \mathbf{V}_{jk})\mathbf{z}(t) \qquad (6.18)
\end{aligned}$$

Furthermore, reshuffling the terms in (6.18) and considering the inequalities of (6.14) to (6.17), (6.18) becomes the following.

$$\begin{aligned}
\dot{V}(t) \leq & \sum_{i=1}^{p}\sum_{j=1}^{c}\sum_{k=1}^{p} w_i v_j m_{jk}\mathbf{z}(t)^T(\mathbf{Q}_{ijk} - \mathbf{M}_{ijk} - \mathbf{M}_{kji} + \mathbf{W}_{ijk} + \mathbf{W}_{kji} \\
& - \mathbf{\Lambda}_{ij} + \mathbf{V}_{jk})\mathbf{z}(t) + \sum_{i=1}^{p}\sum_{j=1}^{c}\sum_{k=1}^{p} w_i v_j w_k \mathbf{z}(t)^T \Big(\mathbf{M}_{ijk} - \mathbf{W}_{ijk} \\
& + \mathbf{\Lambda}_{ij} + \mathbf{H}_{ijk} + \mathbf{J}_{ijk} + \sum_{r=1}^{p}\sum_{s=1}^{p}(\sigma_{rjs}\mathbf{W}_{rjs} - \rho_{rjs}\mathbf{M}_{rjs} + \delta_{jr}\delta_{js}\mathbf{H}_{rjs} \\
& + \varepsilon_{jr}\varepsilon_{js}\mathbf{J}_{rjs} + \eta_{jr}\eta_{js}\mathbf{K}_{rjs} + \gamma_{jr}\gamma_{js}\mathbf{L}_{rjs}) - \sum_{r=1}^{p}\big(\delta_{jr}(\mathbf{H}_{ijr} + \mathbf{H}_{rji}) \\
& + \varepsilon_{jr}(\mathbf{J}_{ijr} + \mathbf{J}_{rji})\big)\Big)\mathbf{z}(t) + \sum_{i=1}^{p}\sum_{j=1}^{c}\sum_{k=1}^{p} m_{ji} v_j m_{jk}\mathbf{z}(t)^T\Big(\mathbf{M}_{ijk} - \mathbf{W}_{ijk} \\
& - \mathbf{V}_{jk} + \mathbf{K}_{ijk} + \mathbf{L}_{ijk} - \sum_{r=1}^{p}\big(\eta_{jr}(\mathbf{K}_{ijr} + \mathbf{K}_{rji}) + \gamma_{jr}(\mathbf{L}_{ijr} + \mathbf{L}_{rji})\big)\Big)\mathbf{z}(t)
\end{aligned} \tag{6.19}$$

Introducing matrices $\mathbf{R}_{ijk} = \mathbf{R}_{kji}^T \in \Re^{n \times n}$, $\mathbf{S}_{ijk} \in \Re^{n \times n}$ and $\mathbf{T}_{ijk} = \mathbf{T}_{kji}^T \in \Re^{n \times n}$ for all i, j and k, we consider the following inequalities.

$$\begin{aligned}
\mathbf{R}_{iji} > & \mathbf{M}_{iji} - \mathbf{W}_{iji} - \mathbf{V}_{ji} + \mathbf{K}_{iji} + \mathbf{L}_{iji} \\
& - \sum_{r=1}^{p}\big(\eta_{jr}(\mathbf{K}_{ijr} + \mathbf{K}_{rji}) + \gamma_{jr}(\mathbf{L}_{ijr} + \mathbf{L}_{rji})\big) \ \forall\, i,\ j
\end{aligned} \tag{6.20}$$

$$\begin{aligned}
\mathbf{R}_{ijk} + \mathbf{R}_{ijk}^T \geq & \mathbf{M}_{ijk} - \mathbf{W}_{ijk} - \mathbf{V}_{jk} + \mathbf{K}_{ijk} + \mathbf{L}_{ijk} \\
& - \sum_{r=1}^{p}\big(\eta_{jr}(\mathbf{K}_{ijr} + \mathbf{K}_{rji}) + \gamma_{jr}(\mathbf{L}_{ijr} + \mathbf{L}_{rji})\big) \\
& + \mathbf{M}_{kji} - \mathbf{W}_{kji} - \mathbf{V}_{ji} + \mathbf{K}_{kji} + \mathbf{L}_{kji} \\
& - \sum_{r=1}^{p}\big(\eta_{jr}(\mathbf{K}_{kjr} + \mathbf{K}_{rjk}) + \gamma_{jr}(\mathbf{L}_{kjr} + \mathbf{L}_{rjk})\big) \ \forall\, j, k;\ i < k
\end{aligned} \tag{6.21}$$

$$\mathbf{S}_{ijk} + \mathbf{S}_{ijk}^T \geq \mathbf{Q}_{ijk} - \mathbf{M}_{ijk} - \mathbf{M}_{kji} + \mathbf{W}_{ijk} + \mathbf{W}_{kji} - \mathbf{\Lambda}_{ij} + \mathbf{V}_{jk} \ \forall\, i,\ j,\ k \tag{6.22}$$

$$\mathbf{T}_{iji} > \mathbf{M}_{iji} - \mathbf{W}_{iji} + \boldsymbol{\Lambda}_{ij} + \mathbf{H}_{iji} + \mathbf{J}_{iji} + \sum_{r=1}^{p}\sum_{s=1}^{p}(\sigma_{rjs}\mathbf{W}_{rjs} - \rho_{rjs}\mathbf{M}_{rjs} + \delta_{jr}\delta_{js}\mathbf{H}_{rjs} + \varepsilon_{jr}\varepsilon_{js}\mathbf{J}_{rjs} + \eta_{jr}\eta_{js}\mathbf{K}_{rjs} + \gamma_{jr}\gamma_{js}\mathbf{L}_{rjs}) - \sum_{r=1}^{p}\big(\delta_{jr}(\mathbf{H}_{ijr} + \mathbf{H}_{rji}) + \varepsilon_{jr}(\mathbf{J}_{ijr} + \mathbf{J}_{rji})\big)\ \forall\, i,\ j \tag{6.23}$$

$$\begin{aligned}\mathbf{T}_{ijk} + \mathbf{T}_{ijk}^{T} \geq\ & \mathbf{M}_{ijk} - \mathbf{W}_{ijk} + \boldsymbol{\Lambda}_{ij} + \mathbf{H}_{ijk} + \mathbf{J}_{ijk} + 2\sum_{r=1}^{p}\sum_{s=1}^{p}(\sigma_{rjs}\mathbf{W}_{rjs} \\ & - \rho_{rjs}\mathbf{M}_{rjs} + \delta_{jr}\delta_{js}\mathbf{H}_{rjs} + \varepsilon_{jr}\varepsilon_{js}\mathbf{J}_{rjs} + \eta_{jr}\eta_{js}\mathbf{K}_{rjs} \\ & + \gamma_{jr}\gamma_{js}\mathbf{L}_{rjs}) - \sum_{r=1}^{p}\big(\delta_{jr}(\mathbf{H}_{ijr} + \mathbf{H}_{rji}) + \varepsilon_{jr}(\mathbf{J}_{ijr} + \mathbf{J}_{rji})\big) \\ & + \mathbf{M}_{kji} - \mathbf{W}_{kji} + \boldsymbol{\Lambda}_{kj} + \mathbf{H}_{kji} + \mathbf{J}_{kji} \\ & - \sum_{r=1}^{p}\big(\delta_{jr}(\mathbf{H}_{kjr} + \mathbf{H}_{rjk}) + \varepsilon_{jr}(\mathbf{J}_{kjr} + \mathbf{J}_{rjk})\big)\ \forall\, j, k;\ i < k\end{aligned} \tag{6.24}$$

From (6.19) to (6.24), it follows that,

$$\begin{aligned}\dot{V}(t) \leq & \sum_{i=1}^{p}\sum_{j=1}^{c} m_{ji}^2 v_j \mathbf{z}(t)^T \mathbf{R}_{iji}\mathbf{z}(t) + \sum_{k=1}^{p}\sum_{j=1}^{c}\sum_{i<k} m_{ji}v_j m_{jk}\mathbf{z}(t)^T(\mathbf{R}_{ijk} + \mathbf{R}_{ijk}^T)\mathbf{z}(t) \\ & + \sum_{k=1}^{p}\sum_{j=1}^{c}\sum_{k=1}^{p} w_i v_j m_{jk}\mathbf{z}(t)^T(\mathbf{S}_{ijk} + \mathbf{S}_{ijk}^T)\mathbf{z}(t) \\ & + \sum_{i=1}^{p}\sum_{j=1}^{c} w_i^2 v_j \mathbf{z}(t)^T \mathbf{T}_{iji}\mathbf{z}(t) + \sum_{k=1}^{p}\sum_{j=1}^{c}\sum_{i<k} w_i v_j w_k \mathbf{z}(t)^T(\mathbf{T}_{ijk} + \mathbf{T}_{ijk}^T)\mathbf{z}(t) \\ = & \sum_{j=1}^{c} v_j \begin{bmatrix}\mathbf{r}_j(t) \\ \mathbf{s}(t)\end{bmatrix}^T \begin{bmatrix}\mathbf{R}_j & \mathbf{S}_j^T \\ \mathbf{S}_j & \mathbf{T}_j\end{bmatrix} \begin{bmatrix}\mathbf{r}_j(t) \\ \mathbf{s}(t)\end{bmatrix}.\end{aligned} \tag{6.25}$$

where $\mathbf{r}_j(t) = \begin{bmatrix} m_{j1}\mathbf{z}(t) \\ m_{j2}\mathbf{z}(t) \\ \vdots \\ m_{jp}\mathbf{z}(t)\end{bmatrix}$, $\mathbf{s}(t) = \begin{bmatrix} w_1\mathbf{z}(t) \\ w_2\mathbf{z}(t) \\ \vdots \\ w_p\mathbf{z}(t)\end{bmatrix}$, $\mathbf{R}_j = \begin{bmatrix}\mathbf{R}_{1j1} & \mathbf{R}_{1j2} & \cdots & \mathbf{R}_{1jp} \\ \mathbf{R}_{2j1} & \mathbf{R}_{2j2} & \cdots & \mathbf{R}_{2jp} \\ \vdots & \vdots & \vdots & \vdots \\ \mathbf{R}_{pj1} & \mathbf{R}_{pj2} & \cdots & \mathbf{R}_{pjp}\end{bmatrix}$,

$\mathbf{S}_j = \begin{bmatrix}\mathbf{S}_{1j1} & \mathbf{S}_{1j2} & \cdots & \mathbf{S}_{1jp} \\ \mathbf{S}_{2j1} & \mathbf{S}_{2j2} & \cdots & \mathbf{S}_{2p} \\ \vdots & \vdots & \vdots & \vdots \\ \mathbf{S}_{p1} & \mathbf{S}_{pj2} & \cdots & \mathbf{S}_{pjp}\end{bmatrix}$ and $\mathbf{T}_j = \begin{bmatrix}\mathbf{T}_{1j1} & \mathbf{T}_{1j2} & \cdots & \mathbf{T}_{1jp} \\ \mathbf{T}_{2j1} & \mathbf{T}_{2j2} & \cdots & \mathbf{T}_{2jp} \\ \vdots & \vdots & \vdots & \vdots \\ \mathbf{T}_{pj1} & \mathbf{T}_{pj2} & \cdots & \mathbf{T}_{pjp}\end{bmatrix}$.

From (3.1) and (6.25), based on the the Lyapunov stability theory, $V(t) > 0$ and $\dot{V}(t) < 0$ for $\mathbf{z}(t) \neq \mathbf{0}$ ($\mathbf{x}(t) \neq \mathbf{0}$) implying the asymptotic stability of the FMB control system (6.6) with imperfectly matched membership functions, i.e., $\mathbf{x}(t) \to 0$ when time $t \to \infty$, can be achieved if the stability conditions summarized in the following theorem are satisfied.

Theorem 6.1. *The RS FMB control system (6.6) with imperfectly matched premise membership functions, formed by the nonlinear plant represented by the fuzzy model (2.2) and the RS fuzzy controller (6.5) connected in a closed loop, is asymptotically stable if there exist predefined constant scalars* ρ_{ijk}, σ_{ijk}, δ_{ji}, ε_{ji}, η_{ji} *and* γ_{ji}, $i, k = 1, 2, \cdots, p$, $j = 1, 2, \cdots, c$, *satisfying the inequalities* $w_i(\mathbf{x}(t))w_k(\mathbf{x}(t)) - w_i(\mathbf{x}(t))m_{jk}(\mathbf{x}(t)) - m_{ji}(\mathbf{x}(t))w_k(\mathbf{x}(t)) + m_{ji}(\mathbf{x}(t))m_{jk}(\mathbf{x}(t)) - \rho_{ijk} \geq 0$, $\sigma_{ijk} - (w_i(\mathbf{x}(t))w_k(\mathbf{x}(t)) - w_i(\mathbf{x}(t))m_{jk}(\mathbf{x}(t)) - m_{ji}(\mathbf{x}(t))w_k(\mathbf{x}(t)) + m_{ji}(\mathbf{x}(t))m_{jk}(\mathbf{x}(t)))) \geq 0$, $\delta_{ji} \leq w_i(\mathbf{x}(t)) \leq \varepsilon_{ji}$ *and* $\eta_{jk} \leq m_{jk}(\mathbf{x}(t)) \leq \gamma_{jk}$ *for all* i, j, k *and* $\mathbf{x}(t)$, *and there exist matrices* $\mathbf{H}_{ijk} = \mathbf{H}_{ijk}^T \in \Re^{n\times n}$, $\mathbf{J}_{ijk} = \mathbf{J}_{ijk}^T \in \Re^{n\times n}$, $\mathbf{K}_{ijk} = \mathbf{K}_{ijk}^T \in \Re^{n\times n}$, $\mathbf{L}_{ijk} = \mathbf{L}_{ijk}^T \in \Re^{n\times n}$, $\mathbf{M}_{ijk} = \mathbf{M}_{kji}^T \in \Re^{n\times n}$, $\mathbf{N}_{jk} = \mathbf{N}_{jk}^T \in \Re^{m\times n}$, $\mathbf{R}_{ijk} = \mathbf{R}_{kji}^T \in \Re^{n\times n}$, $\mathbf{S}_{ijk} \in \Re^{n\times n}$, $\mathbf{T}_{ijk} = \mathbf{T}_{kji}^T \in \Re^{n\times n}$, $\mathbf{V}_{jk} = \mathbf{V}_{jk}^T \in \Re^{n\times n}$, $\mathbf{W}_{ijk} = \mathbf{W}_{kji}^T \in \Re^{n\times n}$, $\mathbf{X} = \mathbf{X}^T \in \Re^{n\times n}$ *and* $\mathbf{\Lambda}_{ij} = \mathbf{\Lambda}_{ij}^T \in \Re^{n\times n}$ *such that the following LMIs are satisfied.*

$$\mathbf{X} > 0;$$

$$\mathbf{H}_{ijk} \geq 0 \,\forall\, i,\, j,\, k;$$

$$\mathbf{J}_{ijk} \geq 0 \,\forall\, i,\, j,\, k;$$

$$\mathbf{K}_{ijk} \geq 0 \,\forall\, i,\, j,\, k;$$

$$\mathbf{L}_{ijk} \geq 0 \,\forall\, i,\, j,\, k;$$

$$\mathbf{M}_{ijk} \geq 0 \,\forall\, i,\, j,\, k;$$

$$\mathbf{W}_{ijk} \geq 0 \,\forall\, i,\, j,\, k;$$

LMIs of (6.20) to (6.24);

$$\begin{bmatrix} \mathbf{R}_j & \mathbf{S}_j^T \\ \mathbf{S}_j & \mathbf{T}_j \end{bmatrix} < 0 \,\forall\, j;$$

and the feedback gains are designed as $\mathbf{G}_{jk} = \mathbf{N}_{jk}\mathbf{X}^{-1}$ *for all* j *and* k.

Remark 6.5. When only one operating region is considered, the RS FMB control system (6.6) is reduced to $\dot{\mathbf{x}}(t) = \sum_{i=1}^{p}\sum_{j=1}^{p} w_i m_j(\mathbf{A}_i + \mathbf{B}_i\mathbf{G}_j)\mathbf{x}(t)$. The time derivative of the quadratic Lyapunov function (6.7) becomes $\dot{V}(t) = \sum_{i=1}^{p}\sum_{k=1}^{p} w_i m_k \mathbf{z}(t)^T\mathbf{Q}_{ik}\mathbf{z}(t)$ where $\mathbf{Q}_{ik} = \mathbf{A}_i\mathbf{X} + \mathbf{X}\mathbf{A}_i^T + \mathbf{B}_i\mathbf{N}_k + \mathbf{N}_k^T\mathbf{B}_i^T$. From the stability conditions in Theorem 2.1, the FMB control system with imperfectly matched premise membership functions is asymptotically stable if $\mathbf{Q}_{ik} < 0$ for all i and k. It can be shown that the solution of the stability conditions in Theorem 2.1 (i.e., $\mathbf{Q}_{ik} < 0$ for all i and k) is also the solution of the proposed stability conditions in Theorem 6.1. Considering only one operating region, the subscript j is omitted in the following. Choosing $\mathbf{H}_{ik} = \mathbf{J}_{ik} = \mathbf{K}_{ik} = \mathbf{L}_{ik} = \mathbf{M}_{ik} = \mathbf{W}_{ik} = \mathbf{0}$ and $\mathbf{V}_i = -\mathbf{\Lambda}_i = \tau\mathbf{I}$ where τ is a non-zero positive scalar, (6.20) to (6.24) are reduced to $\mathbf{R}_{ii} > -\tau\mathbf{I}$; $\mathbf{R}_{ik} + \mathbf{R}_{ik}^T \geq -2\tau\mathbf{I}$, i, $k = 1, 2, \cdots, p$; $i < k$; $\mathbf{S}_{ik} + \mathbf{S}_{ik}^T > \mathbf{Q}_{ik} + 2\tau\mathbf{I}$, i, $k = 1, 2, \cdots, p$; $\mathbf{T}_{ii} > -\tau\mathbf{I}$, $i = 1, 2$, $\mathbf{T}_{ik} + \mathbf{T}_{ik}^T > -2\tau\mathbf{I}$, $k = 1, 2, \cdots, p$; $i < k$, respectively. As $\mathbf{Q}_{ik} < 0$ for all i and k, by choosing $\mathbf{S}_{ik} + \mathbf{S}_{ik}^T = \mathbf{0}$ and a sufficiently small value of τ, we have $0 > \mathbf{Q}_{ik} + 2\tau\mathbf{I}$. Furthermore, we choose $\mathbf{S}_{ik} + \mathbf{S}_{ik}^T = \mathbf{0} \geq -2\tau\mathbf{I}$ and $\mathbf{T}_{ik} + \mathbf{T}_{ik}^T = \mathbf{0} \geq -2\tau\mathbf{I}$. As a result, (6.25) becomes $\dot{V}(t) \leq \begin{bmatrix} \mathbf{r}(t) \\ \mathbf{s}(t) \end{bmatrix}^T \begin{bmatrix} \mathbf{R} & \mathbf{S}^T \\ \mathbf{S} & \mathbf{T} \end{bmatrix} \begin{bmatrix} \mathbf{r}(t) \\ \mathbf{s}(t) \end{bmatrix} = \begin{bmatrix} \mathbf{r}(t) \\ \mathbf{s}(t) \end{bmatrix}^T \begin{bmatrix} \mathbf{R} & \mathbf{0} \\ \mathbf{0} & \mathbf{T} \end{bmatrix} \begin{bmatrix} \mathbf{r}(t) \\ \mathbf{s}(t) \end{bmatrix} = \mathbf{r}(t)^T\mathbf{R}\mathbf{r}(t) + \mathbf{s}(t)^T\mathbf{T}\mathbf{s}(t)$ where $\mathbf{R} = \begin{bmatrix} \mathbf{R}_{11} & 0 & \cdots & 0 \\ 0 & \mathbf{R}_{22} & \cdots & 0 \\ \vdots & \vdots & \vdots & \vdots \\ 0 & 0 & \cdots & \mathbf{R}_{pp} \end{bmatrix}$ and $\mathbf{T} = \begin{bmatrix} \mathbf{T}_{11} & 0 & \cdots & 0 \\ 0 & \mathbf{T}_{22} & \cdots & 0 \\ \vdots & \vdots & \vdots & \vdots \\ 0 & 0 & \cdots & \mathbf{T}_{pp} \end{bmatrix}$. It can be seen that $\mathbf{R} = \mathbf{T} < 0$ can be achieved by properly choosing $\mathbf{R}_{ii} > -\tau\mathbf{I}$ and $\mathbf{T}_{ii} > -\tau\mathbf{I}$ for all i, which leads to $\dot{V}(t) < 0$ for $\mathbf{z}(t) \neq \mathbf{0}$ ($\mathbf{x}(t) \neq \mathbf{0}$) that further implies the asymptotic stability of the FMB control system. Hence, the stability conditions in Theorem 2.1 are covered by the proposed stability conditions in Theorem 6.1.

Example 6.1. Consider the same TS fuzzy model in Example 2.1. The membership functions of the TS fuzzy model are defined in Example 3.1 which are shown graphically in Fig. 6.3.

To design the RS fuzzy controller, the operating region is divided into 3 sub-regions as shown in Fig. 6.3. Corresponding to the j-th sub-region, a sub-region j fuzzy controller is proposed with the following three rules.

$$\begin{aligned} &\text{Rule } k\text{: IF } x_1(t) \text{ is } N_1^{jk} \\ &\qquad\text{THEN } \mathbf{u}(t) = \mathbf{G}_{jk}\mathbf{x}(t), j, k = 1, 2, 3 \end{aligned} \tag{6.26}$$

where N_1^{jk} is a fuzzy set of rule k corresponding to the system state $x_1(t)$ in the j-th sub-region; $\mathbf{G}_{jk} \in \Re^{m \times n}$, j, $k = 1, 2, 3$, are constant feedback gains to be determined.

The membership functions of the fuzzy controller in the sub-region j are defined in Example 3.1 which are shown graphically in Fig. 6.3. In this example, for simplicity, the membership functions for all sub-regional fuzzy controllers are the same. In general, different membership functions can be applied in practice. Based on the chosen membership functions, it can be found that the values of ρ_{ijk}, σ_{ijk}, δ_{ji}, ε_{ji}, η_{ji} and γ_{ji} listed in Table 6.1 satisfy the inequalities of (6.10) to (6.13). Based on the parameters in Table 6.1, the stability conditions in Theorem 6.1 are employed to check for the stability region of the RS FMB fuzzy control system by using the MATLAB LMI toolbox. The stability region is shown in Fig. 6.4 indicated by 'o' for the system parameters of the TS fuzzy model in the range of $2 \leq a \leq 9$,

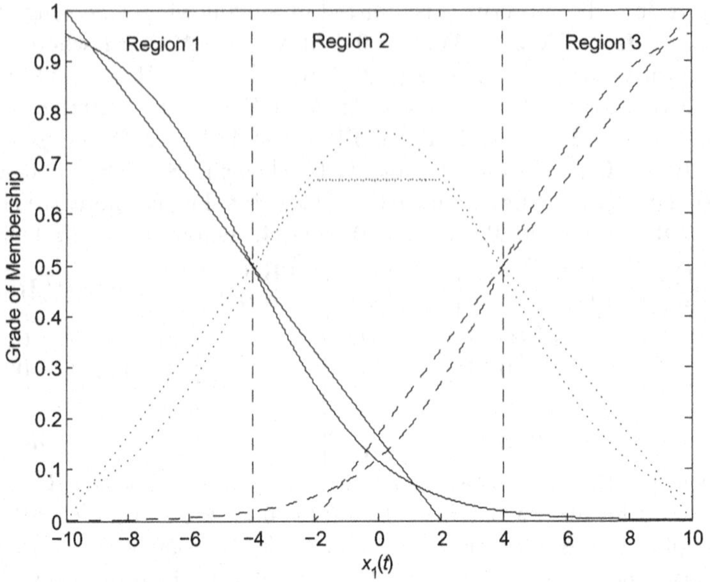

Fig. 6.3 Membership functions of the fuzzy model: $w_1(x_1(t))$ (left triangle in solid line), $w_2(x_1(t))$ (trapezoid in dotted line) and $w_3(x_1(t))$ (right triangle in dash line). Membership functions of the fuzzy controller: $m_{j1}(x_1(t))$ (left z shape in solid line), $m_{j2}(x_1(t))$ (bell shape in dotted line) and $m_{j3}(x_1(t))$ (right s shape in dash line), $j = 1, 2, 3$.

$2 \le b \le 62$. For comparison purposes, the stability conditions in Theorem 2.1 and Theorem 3.3, [2, 55] are employed to check for the stability region. However, no stability region can be found. It should be noted that the stability conditions in Theorem 2.2 to Theorem 2.5 for perfectly matched premise membership functions cannot be applied in this example.

Furthermore, we consider the case that all regional fuzzy controllers sharing the same feedback gains and membership functions in Example 3.1 as a result of the design in this example. From Remark 6.2, the regional switching fuzzy controller is reduced to the traditional fuzzy controller. Under this case, the stability region is shown in Fig. 6.4 indicated by '×'. It can be seen that the proposed stability conditions are able to produce a stable design for the FMB system in a large domain. In the overlapping stability region (indicated by both 'o' and '×' simultaneously at each point) in Fig. 6.4, it is recommended to apply the traditional fuzzy controller in order to lower the implementation cost. For the stability region indicated by 'o' only, the regional switching fuzzy controller can be employed to realize a stable FMB control system.

Table 6.1 Information of operating sub-regions for Example 6.1.

Operating sub-region	ρ_{ijk}, σ_{ijk}, δ_{ji}, ε_{ji}, η_{ji} and γ_{ji}
Region 1: $x_1(t) < -4$	$\rho_{111} = 0.0000$; $\rho_{112} = \rho_{211} = -0.0051$; $\rho_{212} = 0.0000$; $\rho_{113} = \rho_{311} = 0.0000$; $\rho_{312} = \rho_{213} = -0.0006$; $\rho_{313} = 0.0000$; $\sigma_{111} = 0.0048$; $\sigma_{112} = \sigma_{211} = 0.0000$; $\sigma_{212} = 0.0055$; $\sigma_{113} = \sigma_{311} = 0.0005$; $\sigma_{312} = \sigma_{213} = 0.0000$; $\sigma_{313} = 0.0003$; $\delta_{11} = 0.5000$; $\varepsilon_{11} = 1.0000$; $\delta_{12} = 0.0000$; $\varepsilon_{12} = 0.5000$; $\delta_{13} = 0.0000$; $\varepsilon_{13} = 0.0000$; $\eta_{11} = 0.5000$; $\gamma_{11} = 0.9536$; $\eta_{12} = 0.0465$; $\gamma_{12} = 0.04820$; $\eta_{13} = 0.0009$; $\gamma_{13} = 0.0180$;
Region 2: $-4 \leq x_1(t) < 4$	$\rho_{121} = 0.0000$; $\rho_{122} = \rho_{221} = -0.0055$; $\rho_{222} = 0.0000$; $\rho_{123} = \rho_{321} = -0.0031$; $\rho_{322} = \rho_{223} = -0.0055$; $\rho_{323} = 0.0000$; $\sigma_{121} = 0.0048$; $\sigma_{122} = \sigma_{221} = 0.0011$; $\sigma_{222} = 0.0090$; $\sigma_{123} = \sigma_{321} = 0.0023$; $\sigma_{322} = \sigma_{223} = 0.0011$; $\sigma_{323} = 0.0048$; $\delta_{21} = 0.0000$; $\varepsilon_{21} = 0.5000$; $\delta_{22} = 0.5000$; $\varepsilon_{22} = 0.6667$; $\delta_{23} = 0.0000$; $\varepsilon_{23} = 0.5000$; $\eta_{21} = 0.0180$; $\gamma_{21} = 0.5000$; $\eta_{22} = 0.4820$; $\gamma_{22} = 0.7616$; $\eta_{23} = 0.0180$; $\gamma_{23} = 0.5000$;
Region 3: $x_1(t) > 4$	$\rho_{131} = 0.0000$; $\rho_{132} = \rho_{231} = -0.0006$; $\rho_{232} = 0.0000$; $\rho_{133} = \rho_{331} = 0.0000$; $\rho_{332} = \rho_{233} = -0.0051$; $\rho_{333} = 0.0000$; $\sigma_{131} = 0.0003$; $\sigma_{132} = \sigma_{231} = 0.0000$; $\sigma_{232} = 0.0055$; $\sigma_{133} = \sigma_{331} = 0.0005$; $\sigma_{332} = \sigma_{233} = 0.0000$; $\sigma_{333} = 0.0048$; $\delta_{31} = 0.0000$; $\varepsilon_{31} = 0.0000$; $\delta_{32} = 0.0000$; $\varepsilon_{32} = 0.5000$; $\delta_{33} = 0.5000$; $\varepsilon_{33} = 1.0000$; $\eta_{31} = 0.0009$; $\gamma_{31} = 0.0180$; $\eta_{32} = 0.0465$; $\gamma_{32} = 0.4820$; $\eta_{33} = 0.5000$; $\gamma_{33} = 0.9526$;

6.3.3 Perfectly Matched Premise Membership Functions Based Inequalities

In this section, we consider the RS FMB control system with perfectly matched premise membership functions of which $w_i = m_{ji}$ for all i and j. Similar to the analysis approach for the RS FMB control system with imperfectly matched premise membership functions, some MFSD inequalities are introduced. Removing all terms with m_{ji} and/or m_{jk} in (6.10) and (6.11), we have the following MFSD inequalities.

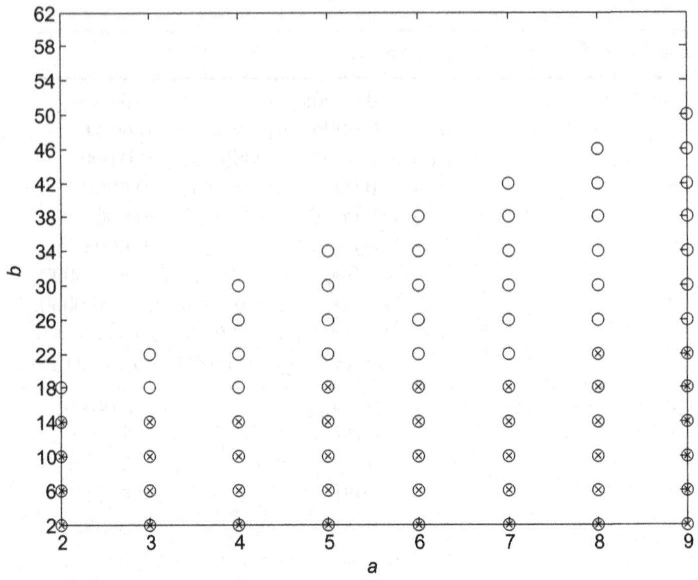

Fig. 6.4 Stability region indicated by 'o' given by the stability conditions in Theorem 6.1 for Example 6.1.

$$\sum_{i=1}^{p}\sum_{k=1}^{p}(w_i w_k - \rho_{ijk})\mathbf{M}_{ijk} \geq 0 \; \forall \, j \tag{6.27}$$

$$\sum_{i=1}^{p}\sum_{k=1}^{p}(\sigma_{ijk} - w_i w_k)\mathbf{W}_{ijk} \geq 0 \; \forall \, j \tag{6.28}$$

where ρ_{ijk} and σ_{ijk} are constant scalars to be determined and satisfy $w_i w_k - \rho_{ijk} \geq 0$ and $\sigma_{ijk} - w_i w_k \geq 0$, respectively, for all i, j and k.

6.3.4 Stability Analysis with Perfectly Matched Premise Membership Functions

From (6.14), (6.15), (6.27) and (6.28), with $m_{jk} = w_i$ for all j and k, it follows from (6.7) that

$$
\begin{aligned}
\dot{V}(t) \leq & \sum_{i=1}^{p}\sum_{j=1}^{c}\sum_{k=1}^{p} w_i v_j w_k \mathbf{z}(t)^T \mathbf{Q}_{ijk}\mathbf{z}(t) \\
& + \sum_{i=1}^{p}\sum_{j=1}^{c}\sum_{k=1}^{p} v_j(w_i w_k - \rho_{ijk})\mathbf{z}(t)^T \mathbf{M}_{ijk}\mathbf{z}(t) \\
& + \sum_{i=1}^{p}\sum_{j=1}^{c}\sum_{k=1}^{p} v_j(\sigma_{ijk} - w_i w_k)\mathbf{z}(t)^T \mathbf{W}_{ijk}\mathbf{z}(t) \\
& + \sum_{i=1}^{p}\sum_{j=1}^{c}\sum_{k=1}^{p} (w_i w_k - w_i\delta_{jk} - \delta_{ji}w_k + \delta_{ji}\delta_{jk})\mathbf{z}(t)^T \mathbf{H}_{ijk}\mathbf{z}(t) \\
& + \sum_{i=1}^{p}\sum_{j=1}^{c}\sum_{k=1}^{p} (w_i w_k - w_i\varepsilon_{jk} - \varepsilon_{ji}w_k + \varepsilon_{ji}\varepsilon_{jk})\mathbf{z}(t)^T \mathbf{J}_{ijk}\mathbf{z}(t) \\
= & \sum_{i=1}^{p}\sum_{j=1}^{c}\sum_{k=1}^{p} w_i v_j w_k \mathbf{z}(t)^T \Big(\mathbf{Q}_{ijk} + \mathbf{M}_{ijk} - \mathbf{W}_{ijk} + \mathbf{H}_{ijk} + \mathbf{J}_{ijk} \\
& + \sum_{r=1}^{p}\sum_{s=1}^{p}(-\rho_{rjs}\mathbf{M}_{rjs} + \sigma_{rjs}\mathbf{W}_{rjs} + \delta_{jr}\delta_{js}\mathbf{H}_{rjs} + \varepsilon_{jr}\varepsilon_{js}\mathbf{J}_{rjs}) \\
& - \sum_{r=1}^{p}\big(\delta_{jr}(\mathbf{H}_{ijr} + \mathbf{H}_{rji}) + \varepsilon_{jr}(\mathbf{J}_{ijr} + \mathbf{J}_{rji})\big)\Big)\mathbf{z}(t)
\end{aligned}
\tag{6.29}
$$

Introducing matrices $\mathbf{R}_{ijk} = \mathbf{R}_{kji}^T \in \Re^{n\times n}$ for all i, j and k, we consider the following inequalities.

$$
\begin{aligned}
\mathbf{R}_{iji} > & \mathbf{Q}_{iji} + \mathbf{M}_{iji} - \mathbf{W}_{iji} + \mathbf{H}_{iji} + \mathbf{J}_{iji} \\
& + \sum_{r=1}^{p}\sum_{s=1}^{p}(-\rho_{rjs}\mathbf{M}_{rjs} + \sigma_{rjs}\mathbf{W}_{rjs} + \delta_{jr}\delta_{js}\mathbf{H}_{rjs} + \varepsilon_{jr}\varepsilon_{js}\mathbf{J}_{rjs}) \\
& - \sum_{r=1}^{p}\big(\delta_{jr}(\mathbf{H}_{ijr} + \mathbf{H}_{rji}) + \varepsilon_{jr}(\mathbf{J}_{ijr} + \mathbf{J}_{rji})\big)\ \forall\, i,\ j
\end{aligned}
\tag{6.30}
$$

$$
\begin{aligned}
\mathbf{R}_{ijk} + \mathbf{R}_{ijk}^T \geq & \mathbf{Q}_{ijk} + \mathbf{M}_{ijk} - \mathbf{W}_{ijk} + \mathbf{H}_{ijk} + \mathbf{J}_{ijk} \\
& + \mathbf{Q}_{kji} + \mathbf{M}_{kji} - \mathbf{W}_{kji} + \mathbf{H}_{kji} + \mathbf{J}_{kji} \\
& + 2\sum_{r=1}^{p}\sum_{s=1}^{p}(-\rho_{rjs}\mathbf{M}_{rjs} + \sigma_{rjs}\mathbf{W}_{rjs} + \delta_{jr}\delta_{js}\mathbf{H}_{rjs} + \varepsilon_{jr}\varepsilon_{js}\mathbf{J}_{rjs}) \\
& - \sum_{r=1}^{p}\big(\delta_{jr}(\mathbf{H}_{ijr} + \mathbf{H}_{rjk} + \mathbf{H}_{kjr} + \mathbf{H}_{rji}) \\
& + \varepsilon_{jr}(\mathbf{J}_{ijr} + \mathbf{J}_{rjk} + \mathbf{J}_{kjr} + \mathbf{J}_{rji})\big)\ \forall\, j, k;\ i < k
\end{aligned}
\tag{6.31}
$$

From (6.30) and (6.31), it follows that,

$$\begin{aligned}\dot{V}(t) &\leq \sum_{i=1}^{p}\sum_{j=1}^{c} w_{ji}^2 v_j \mathbf{z}(t)^T \mathbf{R}_{iji}\mathbf{z}(t) \\ &+ \sum_{k=1}^{p}\sum_{j=1}^{c}\sum_{i<k} w_{ji}v_j w_{jk}\mathbf{z}(t)^T(\mathbf{R}_{ijk}+\mathbf{R}_{ijk}^T)\mathbf{z}(t) \\ &= \sum_{j=1}^{c} v_j \mathbf{s}(t)^T \mathbf{R}_j \mathbf{s}(t). \end{aligned} \tag{6.32}$$

From (3.1) and (6.32), based on the the Lyapunov stability theory, $V(t) > 0$ and $\dot{V}(t) < 0$ for $\mathbf{z}(t) \neq \mathbf{0}$ ($\mathbf{x}(t) \neq \mathbf{0}$) implies the asymptotic stability of the FMB control system (6.6) with perfectly matched membership functions, i.e., $\mathbf{x}(t) \to 0$ when time $t \to \infty$, if the stability conditions summarized in the following theorem are satisfied.

Theorem 6.2. *The RS FMB control system (6.6) with perfectly matched premise membership functions (i.e., $m_{ji} = w_i$ for all i), formed by the nonlinear plant represented by the fuzzy model (2.2) and the RS fuzzy controller (6.5) connected in a closed loop, is asymptotically stable if there exist predefined constant scalars ρ_{ijk}, σ_{ijk}, δ_{ji} and ε_{ji}, $i, k = 1, 2, \cdots, p$, satisfying the inequalities $w_i(\mathbf{x}(t))w_k(\mathbf{x}(t)) - \rho_{ijk} \geq 0$, $\sigma_{ijk} - w_i(\mathbf{x}(t))w_k(\mathbf{x}(t)) \geq 0$ and $\delta_{ji} \leq w_i(\mathbf{x}(t)) \leq \varepsilon_{ji}$ for all i, j, k and $\mathbf{x}(t)$, and there exist matrices $\mathbf{H}_{ijk} = \mathbf{H}_{ijk}^T \in \Re^{n\times n}$, $\mathbf{J}_{ijk} = \mathbf{J}_{ijk}^T \in \Re^{n\times n}$, $\mathbf{M}_{ijk} = \mathbf{M}_{kji}^T \in \Re^{n\times n}$, $\mathbf{N}_{jk} = \mathbf{N}_{jk}^T \in \Re^{m\times n}$, $\mathbf{R}_{ijk} = \mathbf{R}_{kji}^T \in \Re^{n\times n}$, $\mathbf{W}_{ijk} = \mathbf{W}_{kji}^T \in \Re^{n\times n}$ and $\mathbf{X} = \mathbf{X}^T \in \Re^{n\times n}$ such that the following LMIs are satisfied.*

$$\mathbf{X} > 0;$$

$$\mathbf{H}_{ijk} \geq 0 \,\forall\, i,\ j,\ k;$$

$$\mathbf{J}_{ijk} \geq 0 \,\forall\, i,\ j,\ k;$$

$$\mathbf{M}_{ijk} \geq 0 \,\forall\, i,\ j,\ k;$$

$$\mathbf{W}_{ijk} \geq 0 \,\forall\, i,\ j,\ k;$$

LMIs of (6.30) and (6.31);

$$\mathbf{R}_j < 0 \,\forall\, j;$$

and the feedback gains are designed as $\mathbf{G}_{jk} = \mathbf{N}_{jk}\mathbf{X}^{-1}$ for all j and k.

Remark 6.6. The stability conditions in Theorem 2.3 is a particular case of Theorem 6.2. It can be shown easily that by choosing $\mathbf{G}_j = \mathbf{G}_{jk}$ for all j and k, $\mathbf{H}_{ijk} = \mathbf{J}_{ijk} = \mathbf{M}_{ijk} = \mathbf{W}_{ijk} = \mathbf{0}$, the stability conditions in Theorem 6.2 is reduced to those in Theorem 2.3. The stability conditions in Theorem 6.2

can be further relaxed by considering the analysis approach using the Polya's Theorem briefly discussed in Chapter 2.

Remark 6.7. When the traditional fuzzy controller in Remark 6.2 is considered, it can be deduced from Remark 6.5 that the stability conditions in Theorem 6.2 potentially are more relaxed than the published stability conditions in Theorem 2.3, as the regional information of the membership functions is considered.

Remark 6.8. Different membership functions constraints will produce different stability analysis results by bringing different level of membership function information to the stability analysis. Some polynomial constraints were investigated in [99] which can also be employed in the stability analysis for relaxing the stability conditions in this chapter.

Example 6.2. Consider the same RS FMB control system in Example 6.1. In this example, the RS fuzzy controller shares the same membership functions of the TS fuzzy model. The membership functions of the TS fuzzy model are defined in 3.1 and divided into 3 regions which are shown graphically in Fig. 6.3. Based on the membership functions, it can be found that the values

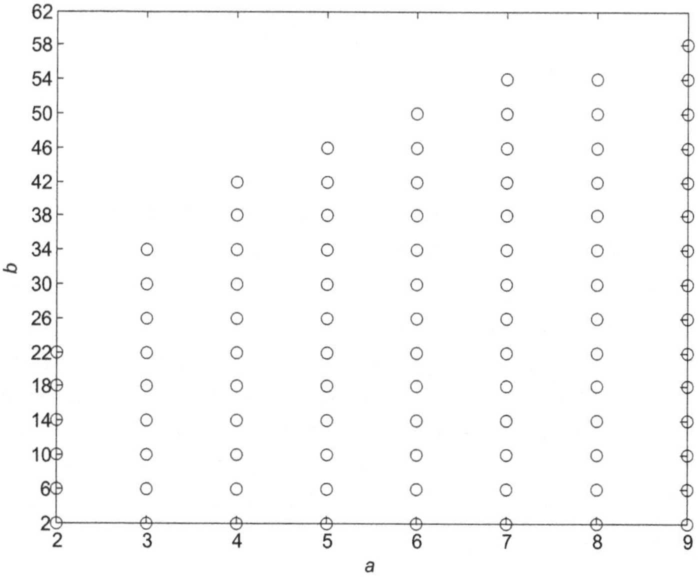

Fig. 6.5 Stability region indicated by 'o' given by the stability conditions in Theorem 6.2 for Example 6.2.

Table 6.2 Information of operating sub-regions for Example 6.2.

Operating sub-region	ρ_{ijk}, σ_{ijk}, δ_{ji} and ε_{ji}
Region 1: $x_1(t) < -4$	$\rho_{111} = 0.2500$; $\rho_{112} = \rho_{211} = 0.0000$; $\rho_{212} = 0.0000$; $\rho_{113} = \rho_{311} = 0.0000$; $\rho_{312} = \rho_{213} = 0.0000$; $\rho_{313} = 0.0000$; $\sigma_{111} = 1.0000$; $\sigma_{112} = \sigma_{211} = 0.2500$; $\sigma_{212} = 0.2500$; $\sigma_{113} = \sigma_{311} = 0.0000$; $\sigma_{312} = \sigma_{213} = 0.0000$; $\sigma_{313} = 0.0000$; $\delta_{11} = 0.5000$; $\varepsilon_{11} = 1.0000$; $\delta_{12} = 0.0000$; $\varepsilon_{12} = 0.5000$; $\delta_{13} = 0.0000$; $\varepsilon_{13} = 0.0000$;
Region 2: $-4 \le x_1(t) < 4$	$\rho_{121} = 0.0000$; $\rho_{122} = \rho_{221} = 0.0000$; $\rho_{222} = 0.2500$; $\rho_{123} = \rho_{321} = 0.0000$; $\rho_{322} = \rho_{223} = 0.0000$; $\rho_{323} = 0.0000$; $\sigma_{121} = 0.2500$; $\sigma_{122} = \sigma_{221} = 0.2500$; $\sigma_{222} = 0.4444$; $\sigma_{123} = \sigma_{321} = 0.0278$; $\sigma_{322} = \sigma_{223} = 0.2500$; $\sigma_{323} = 0.2500$; $\delta_{21} = 0.0000$; $\varepsilon_{21} = 0.5000$; $\delta_{22} = 0.5000$; $\varepsilon_{22} = 0.6667$; $\delta_{23} = 0.0000$; $\varepsilon_{23} = 0.5000$;
Region 3: $x_1(t) > 4$	$\rho_{131} = 0.0000$; $\rho_{132} = \rho_{231} = 0.0000$; $\rho_{232} = 0.0000$; $\rho_{133} = \rho_{331} = 0.0025$; $\rho_{332} = \rho_{233} = 0.000$; $\rho_{333} = 0.2500$; $\sigma_{131} = 0.0000$; $\sigma_{132} = \sigma_{231} = 0.0000$; $\sigma_{232} = 0.2500$; $\sigma_{133} = \sigma_{331} = 0.0000$; $\sigma_{332} = \sigma_{233} = 0.2500$; $\sigma_{333} = 1.0000$; $\delta_{31} = 0.0000$; $\varepsilon_{31} = 0.0000$; $\delta_{32} = 0.0000$; $\varepsilon_{32} = 0.5000$; $\delta_{33} = 0.5000$; $\varepsilon_{33} = 1.0000$;

of ρ_{ijk}, σ_{ijk}, δ_{ji} and ε_{ji} listed in Table 6.2 satisfy the inequalities of (6.27) to (6.28).

The system stability of the FMB control system is checked with the stability conditions in Theorem 6.2. The stability region is shown in Fig. 6.5 indicated by '∘' for the system parameters of $2 \le a \le 9$, $2 \le b \le 62$. Comparing Fig. 6.4 with Fig. 6.5, it can be seen that the RS FMB control system with perfectly matched premise membership functions has a larger stability region. However, the RS FMB control systems with imperfectly matched premise membership functions allow greater design flexibility on the membership functions of the fuzzy controller, which can be different from those of the fuzzy model.

Table 6.3 Feedback gains of the regional switching and traditional fuzzy controllers for Example 6.2 with $a = 9$ and $b = 58$.

RS Fuzzy Controller	Traditional Fuzzy Controller
$\mathbf{G}_{11} = [-2.315270 \quad 5.099520]$;	$\mathbf{G}_{11} = \mathbf{G}_{21} = \mathbf{G}_{31} = [-3.293223 \quad 3.700048]$;
$\mathbf{G}_{12} = [-0.905459 \quad -12.241911]$;	$\mathbf{G}_{12} = \mathbf{G}_{22} = \mathbf{G}_{32} = [-4.653888 \quad -10.749927]$;
$\mathbf{G}_{13} = [-0.930921 \quad -3.264101]$;	$\mathbf{G}_{13} = \mathbf{G}_{23} = \mathbf{G}_{33} = [36.253629 \quad 61.707869]$
$\mathbf{G}_{21} = [-0.718189 \quad 0.859555]$;	
$\mathbf{G}_{22} = [-0.135919 \quad -0.087910]$;	
$\mathbf{G}_{23} = [-0.137327 \quad 0.398319]$;	
$\mathbf{G}_{31} = [-0.004901 \quad 0.156980]$;	
$\mathbf{G}_{32} = [-0.020936 \quad 0.437946]$;	
$\mathbf{G}_{33} = [0.240529 \quad 0.766671]$	

Table 6.4 Minimum and maximum amplitudes of control signal $u(t)$ for RS and traditional fuzzy controllers for Example 6.2 with $a = 9$ and $b = 58$.

Initial Condition	Min/Max $u(t)$ for RS Fuzzy controller	Min/Max $u(t)$ for Traditional Fuzzy controller
$\mathbf{x}(0) = \begin{bmatrix} 10 & -10 \end{bmatrix}^T$	-5.2614/0.3051	-254.5424/0.1578
$\mathbf{x}(0) = \begin{bmatrix} 5 & -5 \end{bmatrix}^T$	-2.4906/0.3055	-61.5415/0.1578
$\mathbf{x}(0) = \begin{bmatrix} -5 & 5 \end{bmatrix}^T$	-1.9911/13.1114	-0.5542/9.4089
$\mathbf{x}(0) = \begin{bmatrix} -10 & 10 \end{bmatrix}^T$	-0.1608/110.0355	-0.5438/102.2530
$\mathbf{x}(0) = \begin{bmatrix} 10 & 0 \end{bmatrix}^T$	-0.5472/2.4053	-0.0000/362.5363
$\mathbf{x}(0) = \begin{bmatrix} 5 & 2.5 \end{bmatrix}^T$	-2.3552/5.7293	-5.6704/174.8370
$\mathbf{x}(0) = \begin{bmatrix} -5 & -2.5 \end{bmatrix}^T$	-1.1366/15.3709	-16.5867/25.1028
$\mathbf{x}(0) = \begin{bmatrix} -10 & 0 \end{bmatrix}^T$	-0.2869/23.1527	-0.8929/32.9322

For comparison purposes, we apply the traditional fuzzy controller (2.6) to the same system. It is stated in Remark 6.2 that the proposed RS fuzzy controller becomes the traditional one when all fuzzy controllers in different regions share the same sets of feedback gains and membership functions. The stability conditions in Theorem 6.2 are employed to check for the system stability. In this example, the traditional fuzzy controller produces exactly the same stability region as shown in Fig. 6.5. However, it can be shown in the following that the feedback gains given by the RS fuzzy controller are smaller in values than those of the traditional fuzzy controller. It is mainly because the regional fuzzy controller only deals with the nonlinear plant operating in a small operating region; the effect of the nonlinearity is less significant than that under the full operating region. Hence, by employing the regional switching fuzzy controller, potentially less energy is required to perform the control process.

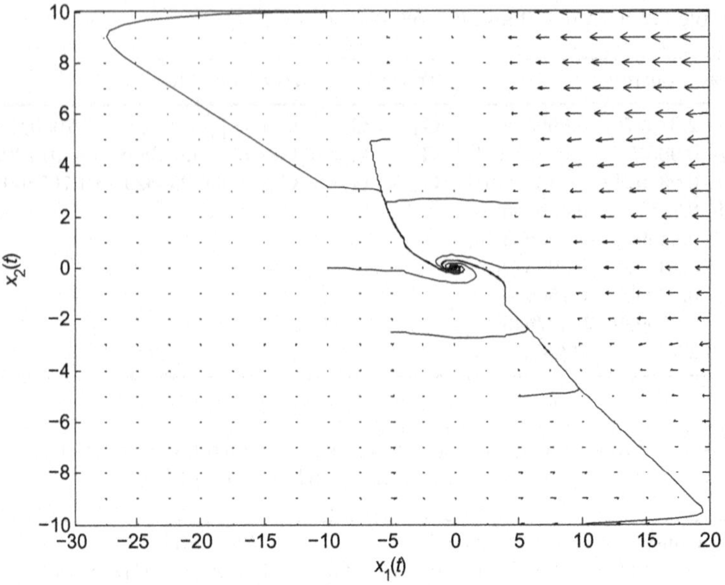

(a) Phase plot of $x_1(t)$ and $x_2(t)$ with RS fuzzy controller.

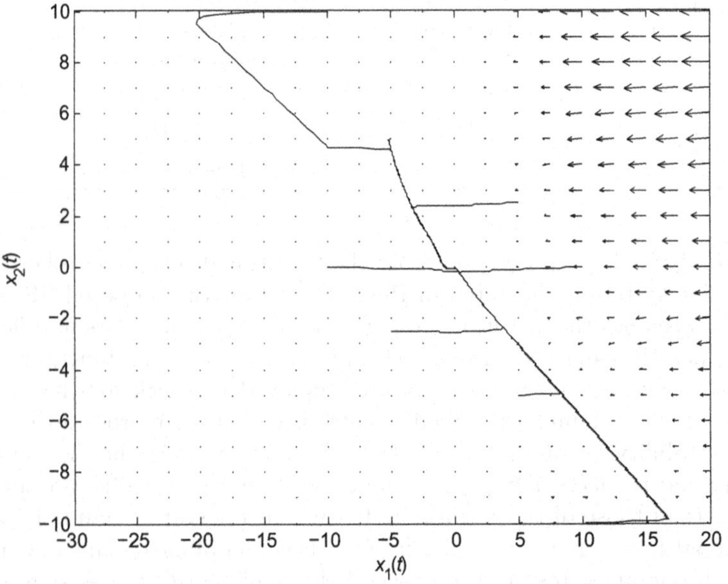

(b) Phase plot of $x_1(t)$ and $x_2(t)$ with traditional fuzzy controller.

Fig. 6.6 Phase portraits of the FMB control system with $a = 9$ and $b = 58$ for Example 6.2.

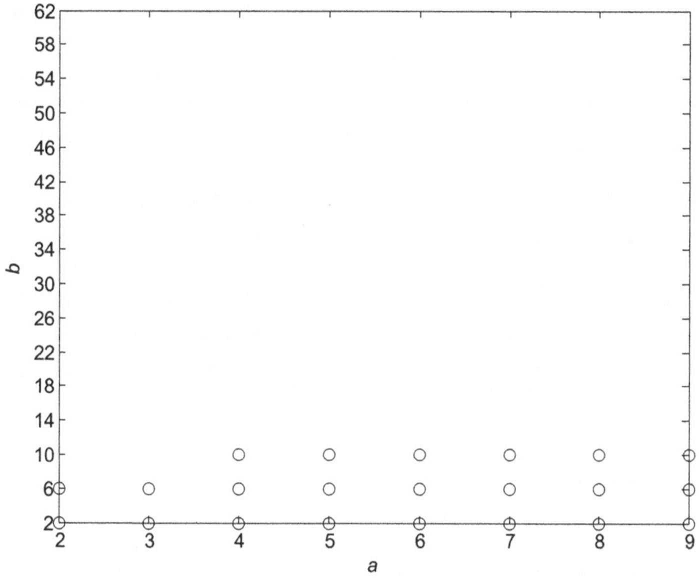

Fig. 6.7 Stability region indicated by 'o' given by stability conditions in [98] for Example 6.2.

Considering the system parameters of $a = 9$ and $b = 58$, the feedback gains of the regional switching and traditional fuzzy controllers are obtained with the MATLAB LMI toolbox based on Theorem 6.2, and listed in Table 6.3. It can be seen that the feedback gains of the traditional fuzzy controller are larger in values as compared with those of the regional switching fuzzy controller. The phase portraits of the FMB control system subject to different initial conditions are shown in Fig. 6.6. It can be seen that both the regional switching and traditional fuzzy controllers are able to stabilize the nonlinear plant successfully. The minimum and maximum amplitudes of the control signal $u(t)$ subject to different initial conditions are listed in Table 6.4. It can be seen that, in general, the minimum/maximum amplitudes of control signal $u(t)$ provided by the RS fuzzy controller are much lower than those of the traditional fuzzy controller.

We also consider the stability conditions in [98], which include the information of the membership function boundaries for the relaxation of stability conditions of FMB control systems subject to perfectly matched premise membership functions. The stability region given by the stability conditions in [98] for our example is shown in Fig. 6.7. It can be seen that the regional switching fuzzy controller outperforms the published one as it offers a larger stability region. Furthermore, comparing with the stability regions shown in Fig. 2.1 given by some published stability conditions, it can be seen that the stability region given by the proposed stability conditions in Theorem 6.2 is also larger. The effectiveness of the proposed stability conditions and the importance of the regional information are well demonstrated.

As mentioned early, the RS FMB control system with perfectly matched membership functions is able to produce a larger stability region. However, as the membership functions of the TS fuzzy model are required for the realization of the fuzzy controller, the plant parameter uncertainties cannot be handled simply by embedding them in the membership functions. Furthermore, a complex nonlinear plant may lead to complex membership functions, which will increase the implementation cost of the fuzzy controller. Hence, it is suggested to employ the fuzzy controller with imperfectly matched membership functions for the control process. When a feasible design cannot be achieved, the RS fuzzy control approach with perfectly matched premise membership functions can be employed to implement the fuzzy controller.

6.4 Conclusion

The stability of FMB control systems with imperfectly/perfectly matched premise membership functions and the RS fuzzy control technique has been investigated. A RS fuzzy controller, which consists of a number of regional fuzzy controllers, has been proposed to control the nonlinear plant. During the control process, one of the regional fuzzy controllers is employed at a time according to the current operating sub-region. Based on the regional information, the stability analysis and controller synthesis can be facilitated. LMI-based stability conditions have been derived to check for the system stability. Some simulation examples have been presented and the results show that the proposed stability conditions are able to produce more relaxed stability results.

Chapter 7
Fuzzy Combined Controller for Nonlinear Systems

7.1 Introduction

Lyapunov approach is the most common approach to investigate the system stability of fuzzy control systems based on the TS fuzzy model. The stability of the FMB control systems with imperfectly/perfectly matched premise membership functions are reviewed in Chapter 2. In general, the stability conditions for the FMB based control systems with perfectly matched premise membership functions are more relaxed. However, it is required that the TS fuzzy model in the form of (2.2) is uncertainty free. As discussed in Chapter 2, the fuzzy controller (2.6) demonstrates an inherent robustness property on dealing with nonlinear plant subject to a certain level of parameter uncertainties by presenting the parameter uncertainties to the membership functions of the fuzzy model. To deal with the parameter uncertainties, in this chapter, a TS fuzzy model subject to parameter uncertainties is proposed for doing the stability analysis. Stability conditions for this class of nonlinear systems were investigated in [74, 104].

In [62, 63, 65, 106], a switching fuzzy model was employed to describe the nonlinear plant with/without parameter uncertainties and to support the design of a switching fuzzy controller. The switching elements of the switching controller are able to approximate effectively the uncertain parameters to facilitate the control process. Hence, the switching control scheme is very robust to the parameter uncertainties and able to offer a consistent system performance. However, an undesirable chattering effect [100] occurs in the control signal caused by the high-frequency switching activities during the control process. Although the chattering effect can be alleviated by replacing the switching function with a saturation function[100], a steady-state error may be introduced in the system states.

In some published work, fuzzy logic was employed to combine various traditional controllers to merge their advantages together. In [83], a fuzzy sliding-mode controller used the sliding-surface function as the input of the

H.-K. Lam and F.H.F. Leung: FMB Control Systems, STUDFUZZ 264, pp. 123–150.
springerlink.com

fuzzy system, the number of fuzzy rules can be greatly reduced. In [37, 114], a fuzzy system was employed to estimate the values of the gains of the sliding-mode controller. Adaptive laws were derived to update the rules of the fuzzy systems. As the switching function of the sliding-mode controller is approximated by a continuous function, the chattering effect can be alleviated. In [37], an adaptive fuzzy controller was proposed to generate the control signals by estimating the values of the unknown parameters of the system. Based on these estimated parameter values, tracking control can be achieved by using the sliding-mode control approach. However, in these approaches, the way to determine the fuzzy rules is still an open question. Furthermore, the approximation error of the fuzzy systems will introduce steady-state errors to the system states or even cause the system to become unstable. In [123], switching elements were used in the controller for compensating the approximation error of the fuzzy system.

In this chapter, a fuzzy combined control scheme [52, 53, 122] is proposed to handle nonlinear plants subject to parameter uncertainties. The block diagram of the proposed fuzzy combined control scheme is shown in Fig. 7.1. A fuzzy combined model, which consists of the global and local fuzzy models, is proposed to represent a nonlinear system subject to parameter uncertainties. The global fuzzy model is employed to model the dynamics of the nonlinear plant in the full operating domain while the local fuzzy model is employed to model the dynamics in a small operating domain near the origin of the state space. Two fuzzy rules are proposed to combine the local and global fuzzy models to form the fuzzy combined model. Denote the grades of membership given by the two rules as $m_1(q(\mathbf{x}(t))) \geq 0$ and $m_2(q(\mathbf{x}(t))) \geq 0$ (, which will be explained in detail later on). The values of $m_1(q(\mathbf{x}(t)))$ and $m_2(q(\mathbf{x}(t)))$ indicates the contribution of the global and local fuzzy models respectively for the system modelling. Based on the fuzzy combined model, a fuzzy combined controller that integrates the advantages of a global switching controller and a local fuzzy controller (both of them are state-feedback controllers) is proposed. Stability conditions and switching laws are derived based on the Lyapunov stability theory [46, 100, 121].

The switching fuzzy controller is responsible for driving the system state towards the origin using its outstanding robustness property. When the system states are approaching the origin, the local fuzzy controller will gradually dominate the control process. The contribution of the switching fuzzy controller and local fuzzy controller to the control process is determined by the values of $m_1(q(\mathbf{x}(t)))$ and $m_2(q(\mathbf{x}(t)))$. By properly designing the membership functions of the fuzzy combined model and fuzzy combined controller, the chattering effect can be removed when the system operates near the origin. To alleviate the chattering effect in the transient period, a saturation function can be employed to replace the switching function of the global switching state-feedback controller. A simulation example will be presented to demonstrate the effectiveness of the proposed fuzzy combined control scheme.

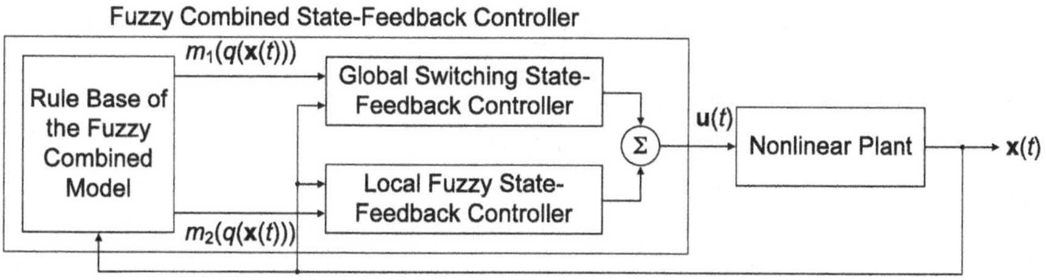

Fig. 7.1 Block diagram of the FMB Control system.

7.2 Fuzzy Combined Model

Consider a nonlinear plant operating in a domain characterized by the system states of $x_\gamma(t) \in \left[x_{\gamma_{min}}, x_{\gamma_{max}}\right]$, $\gamma = 1, 2, \cdots, n$, where $x_{\gamma_{min}}$ and $x_{\gamma_{max}}$ are scalars that representing the lower and upper bounds of the system state $x_\gamma(t)$. By using the TS fuzzy model, the nonlinear plant can be represented in the form of (2.2), which is named as the global TS fuzzy model in this chapter as it describes the dynamics of the nonlinear plant in the full operating domain.

We further consider a small operating domain of the nonlinear plant around the origin characterized by $x_\gamma(t) \in \left[\tilde{x}_{\gamma_{min}}, \tilde{x}_{\gamma_{max}}\right]$ where $\tilde{x}_{\gamma_{min}}$ and $\tilde{x}_{\gamma_{max}}$ are scalars representing the lower and upper bounds of the $x_\gamma(t)$ satisfying $x_{\gamma_{min}} \leq \tilde{x}_{\gamma_{min}} \leq \tilde{x}_{\gamma_{max}} \leq x_{\gamma_{max}}$, $\gamma = 1, 2, \cdots, n$. A local TS fuzzy model can be constructed for the small operating domain. Similar to the TS fuzzy model in Chapter 2, the rules of the local TS fuzzy model is described by p fuzzy rules of following format.

$$\begin{aligned} &\text{Rule } i\text{: IF } \tilde{f}_1(\mathbf{x}(t)) \text{ is } \tilde{M}_1^i \text{ AND } \cdots \text{ AND } \tilde{f}_\Psi(\mathbf{x}(t)) \text{ is } \tilde{M}_\Psi^i \\ &\qquad \text{THEN } \dot{\mathbf{x}}(t) = \tilde{\mathbf{A}}_i\mathbf{x}(t) + \tilde{\mathbf{B}}_i\mathbf{u}(t) \end{aligned} \tag{7.1}$$

where $\tilde{M}_\alpha^i$ is a fuzzy set of rule i corresponding to the function $\tilde{f}_\alpha(\mathbf{x}(t))$, $\alpha = 1, 2, \cdots, \Psi$; $i = 1, 2, \cdots, p$; Ψ is a positive integer; $\mathbf{x}(t) \in \Re^n$ is the system state vector; $\tilde{\mathbf{A}}_i \in \Re^{n\times n}$ and $\tilde{\mathbf{B}}_i \in \Re^{n\times m}$ are known system and input matrices, respectively; $\mathbf{u}(t) \in \Re^m$ is the input vector. The system dynamics is described by,

$$\dot{\mathbf{x}}(t) = \sum_{i=1}^{p} \tilde{w}_i(\mathbf{x}(t))(\tilde{\mathbf{A}}_i\mathbf{x}(t) + \tilde{\mathbf{B}}_i\mathbf{u}(t)) \tag{7.2}$$

where

$$\tilde{w}_i(\mathbf{x}(t)) \geq 0 \;\forall\, i, \sum_{i=1}^{p} \tilde{w}_i(\mathbf{x}(t)) = 1, \tag{7.3}$$

$$\tilde{w}_i(\mathbf{x}(t)) = \frac{\prod_{l=1}^{\Psi} \mu_{\tilde{M}_l^i}(\tilde{f}_l(\mathbf{x}(t)))}{\sum_{k=1}^{p}\prod_{l=1}^{\Psi} \mu_{\tilde{M}_l^k}(\tilde{f}_l(\mathbf{x}(t)))} \; \forall \, i, \tag{7.4}$$

$\tilde{w}_i(\mathbf{x}(t))$, $i = 1, 2, \cdots, p$, are the normalized grades of membership, $\mu_{\tilde{M}_l^i}(\tilde{f}_l(\mathbf{x}(t)))$, $l = 1, 2, \cdots, \Psi$, are the membership functions corresponding to the fuzzy set $\tilde{M}_l^i$.

The proposed fuzzy combined model, which is a fuzzy combination of the global and local TS fuzzy models, has two rules in the following format.

Rule 1: IF $q(\mathbf{x}(t))$ is ZE

$$\text{THEN } \dot{\mathbf{x}}(t) = \sum_{i=1}^{p} \tilde{w}_i(\mathbf{x}(t))(\tilde{\mathbf{A}}_i\mathbf{x}(t) + \tilde{\mathbf{B}}_i\mathbf{u}(t)) \tag{7.5}$$

Rule 2: IF $q(\mathbf{x}(t))$ is NZ

$$\text{THEN } \dot{\mathbf{x}}(t) = \sum_{i=1}^{p} w_i(\mathbf{x}(t))(\mathbf{A}_i\mathbf{x}(t) + \mathbf{B}_i\mathbf{u}(t)) \tag{7.6}$$

where rule 1 is for the local TS fuzzy model and rule 2 is for the global TS fuzzy model in the form of (2.2); ZE (zero) and NZ (non-zero) are fuzzy sets.

Remark 7.1. It should be noted that the global and local fuzzy models are of the same form with the same number of rules in this chapter. By considering the same operating domain, i.e., $x_{\gamma_{min}} = \tilde{x}_{\gamma_{min}}$ and $\tilde{x}_{\gamma_{max}} = x_{\gamma_{max}}$ for all γ, the two models will become one. However, in general, the global and local fuzzy models are not limited by these constraints and can be constructed using sector nonlinearity technique [122].

The function $q(\mathbf{x}(t))$, which determines if the system is operating inside the small operating region, is defined as,

$$q(\mathbf{x}(t)) = \max\left(\left|\frac{x_1(t) - \tilde{x}_1}{\overline{x}_1(t)}\right|, \left|\frac{x_2(t) - \tilde{x}_2}{\overline{x}_2(t)}\right|, \cdots, \left|\frac{x_n(t) - \tilde{x}_n}{\overline{x}_n(t)}\right|\right) \geq 0 \tag{7.7}$$

where $\tilde{x}_i = \frac{\tilde{x}_{i_{max}} + \tilde{x}_{i_{min}}}{2}$, $\overline{x}_i = \frac{\tilde{x}_{i_{max}} - \tilde{x}_{i_{min}}}{2}$, $i = 1, 2, \cdots, n$; $|\cdot|$ and $\max(\cdot)$ denote absolute-value and maximum operators, respectively. $q(\mathbf{x}(t)) \leq 1$ implies that the system is working inside the small operating domain characterized by $x_{\gamma_{min}} \leq \tilde{x}_{\gamma_{min}} \leq x_\gamma(t) \leq \tilde{x}_{\gamma_{max}} \leq x_{\gamma_{max}}$ for all γ; otherwise, it is outside the small operating domain. The inferred fuzzy combined model is defined as,

$$\dot{\mathbf{x}}(t) = m_1(\mathbf{x}(t)) \sum_{i=1}^{p} \tilde{w}_i(\mathbf{x}(t))(\tilde{\mathbf{A}}_i \mathbf{x}(t) + \tilde{\mathbf{B}}_i \mathbf{u}(t)) + m_2(\mathbf{x}(t)) \sum_{i=1}^{p} w_i(\mathbf{x}(t))(\mathbf{A}_i \mathbf{x}(t) + \mathbf{B}_i \mathbf{u}(t)) \tag{7.8}$$

where

$$m_k(\mathbf{x}(t)) \geq 0 \; \forall \, i, \sum_{k=1}^{2} m_k(\mathbf{x}(t)) = 1, \tag{7.9}$$

$m_1(\mathbf{x}(t)) = \frac{\mu_{ZE}(q(\mathbf{x}(t)))}{\mu_{ZE}(q(\mathbf{x}(t)))+\mu_{NZ}(q(\mathbf{x}(t)))}$ and $m_2(\mathbf{x}(t)) = \frac{\mu_{NZ}(q(\mathbf{x}(t)))}{\mu_{ZE}(q(\mathbf{x}(t)))+\mu_{NZ}(q(\mathbf{x}(t)))}$; $\mu_{ZE}(q(\mathbf{x}(t)))$ and $\mu_{NZ}(q(\mathbf{x}(t)))$ are the membership functions corresponding to the fuzzy terms of ZE and NZ, respectively. The membership function $\mu_{ZE}(q(\mathbf{x}(t)))$ is designed to cover the operating region of $0 \leq q(\mathbf{x}(t)) \leq 1$, while the membership function $\mu_{NZ}(q(\mathbf{x}(t)))$ covers the operating region of $q(\mathbf{x}(t)) > 1$.

It should be noted that the fuzzy combined model (7.8) is equivalent to the TS fuzzy model (2.2) for any values of $m_1(\mathbf{x}(t))$ and $m_2(\mathbf{x}(t))$. The fuzzy combined model has the property that the local TS fuzzy model is equivalent to the global TS fuzzy model for both $m_1(\mathbf{x}(t)) \neq 0$ and $m_2(\mathbf{x}(t)) \neq 0$, i.e.,

$$\dot{\mathbf{x}}(t) = \sum_{i=1}^{p} \tilde{w}_i(\mathbf{x}(t))(\tilde{\mathbf{A}}_i \mathbf{x}(t) + \tilde{\mathbf{B}}_i \mathbf{u}(t)) = \sum_{i=1}^{p} w_i(\mathbf{x}(t))(\mathbf{A}_i \mathbf{x}(t) + \mathbf{B}_i \mathbf{u}(t)) \text{ if } m_1(\mathbf{x}(t)) \neq 0 \text{ and } m_2(\mathbf{x}(t)) \neq 0. \tag{7.10}$$

The proof of (7.10) is given in the following.

Proof. The proof is simple and straightforward. It is assumed that the global and local fuzzy models are the exact models for the nonlinear plant. For example, based on the sector nonlinearity technique [122], an exact fuzzy model can be constructed and it is equivalent to the mathematical model. The local model is valid for the local operating domain while the global fuzzy model is for the full operating domain. Considering the local operating domain, both global and local fuzzy models are valid and equivalent to the mathematical model, i.e., (7.10) is true. It is obvious that the global and local fuzzy models are equivalent in the local operating domain characterized by $0 \leq m_1(\mathbf{x}(t)) \leq 1$ for all $m_2(\mathbf{x}(t))$. Consequently, we can draw the same conclusion that (7.10) is true for a tighter condition $m_1(\mathbf{x}(t)) \neq 0$ and $m_2(\mathbf{x}(t)) \neq 0$. This completes the proof.

7.3 Local, Global and Fuzzy Combined Controllers

A fuzzy combined controller, formed by a local fuzzy controller and a global switching controller, is proposed to control the nonlinear plant. The local fuzzy controller is a state-feedback fuzzy controller and the global one is a switching fuzzy controller. Their contribution to the control process depends on the membership functions of $\mu_{ZE}(q(\mathbf{x}(t)))$ and $\mu_{NZ}(q(\mathbf{x}(t)))$. The global switching controller has a stronger robustness property to handle parameter uncertainties by using the switching control technique. However, the switching activity will produce an undesired chattering effect, which can be removed thanks to the local state-feedback fuzzy controller. Based on the property of the local and global controllers, the membership functions $\mu_{ZE}(q(\mathbf{x}(t)))$ and $\mu_{NZ}(q(\mathbf{x}(t)))$ are designed in a way that when the system is working in the defined small operating region, the local fuzzy controller is employed; otherwise, the global switching controller is employed to drive the system states towards the small operating region. As a result, the local and global controllers are combined by the membership functions to form a fuzzy combined controller that integrates their advantages for the control process.

7.3.1 Local Fuzzy Controller

The local fuzzy controller has the same form of (2.6) and is described by c fuzzy rules of the following format:

$$\begin{aligned}&\text{Rule } j\text{: IF } g_1(\mathbf{x}(t)) \text{ is } N_1^j \text{ AND } \cdots \text{ AND } g_\Omega(\mathbf{x}(t)) \text{ is } N_\Omega^j\\ &\qquad \text{THEN } \mathbf{u}(t) = \tilde{\mathbf{G}}_j\mathbf{x}(t)\end{aligned} \tag{7.11}$$

where $\tilde{\mathbf{G}}_j \in \Re^{m\times n}$, $j = 1, 2, \cdots, c$, are constant feedback gains to be determined. The rest variables are defined in Section 2.3. The fuzzy controller is defined as follows,

$$\mathbf{u}(t) = \sum_{j=1}^{c} v_j(\mathbf{x}(t))\tilde{\mathbf{G}}_j\mathbf{x}(t) \tag{7.12}$$

where

$$v_j(\mathbf{x}(t)) \geq 0 \; \forall \; j, \sum_{j=1}^{c} v_j(\mathbf{x}(t)) = 1, \tag{7.13}$$

$$v_j(\mathbf{x}(t)) = \frac{\prod_{l=1}^{\Omega} \mu_{N_l^j}(g_l(\mathbf{x}(t)))}{\sum_{k=1}^{c}\prod_{l=1}^{\Omega} \mu_{N_l^k}(g_l(\mathbf{x}(t)))} \; \forall \; j, \tag{7.14}$$

$v_j(\mathbf{x}(t))$, $j = 1, 2, \cdots, c$, are the normalized grades of membership, $\mu_{N_l^j}(g_l(\mathbf{x}(t)))$, $l = 1, 2, \cdots, \Omega$, are the membership functions corresponding to the fuzzy set N_l^j.

Remark 7.2. The local fuzzy controller is a state-feedback fuzzy controller. Various LMIs based stability conditions have been given in Chapter 2 for FMB control systems with this class of fuzzy controller. It has the advantage that the structure of the fuzzy controller is simple, particularly when it has some simple form of membership functions. However, it is not effective as compared with other fuzzy control techniques, such as fuzzy sliding-mode control [83] and adaptive fuzzy control [5, 95, 123, 132], to deal with nonlinear plants with parameter uncertainties.

7.3.2 Global Switching Controller

The global switching state-feedback controller is defined as follows.

$$\mathbf{u}(t) = \sum_{j=1}^{p} n_j(\mathbf{x}(t))\mathbf{G}_j\mathbf{x}(t) \tag{7.15}$$

where $\mathbf{G}_j \in \Re^{m\times n}$, $j = 1, 2, \cdots, p$, are constant feedback gains to be determined; $n_j(\mathbf{x}(t))$ takes the value of either $-\frac{K}{\alpha_{min}}$ or $\frac{K}{\alpha_{min}}$ according to a switching scheme to be derived later; $\alpha_{min} > 0$ and $K \geq 1$ are scalars.

Remark 7.3. The global fuzzy controller is a switching controller that the switching components offer a strong robustness property to handle the parameter uncertainties. However, due to the switching activity, an undesirable chattering effect is introduced to the output of the system. Moreover, comparing with the local fuzzy controller, it is more complicated to implement the global fuzzy controller owing to the switching components.

7.3.3 Fuzzy Combined Controller

As discussed above, the global and local fuzzy controllers have their own advantages and disadvantages. In this section, we apply the fuzzy combination technique to combine these two types of fuzzy controllers such that the overall fuzzy controller, named fuzzy combined controller, integrates the advantages of the two types of fuzzy controllers but with their weaknesses removed. The proposed fuzzy combined controller is shown in Fig. 7.1 and has the following two rules to combine the global and local fuzzy controllers.

Rule 1: IF $q(\mathbf{x}(t))$ is ZE

$$\text{THEN } \mathbf{u}(t) = \sum_{j=1}^{c} v_j(\mathbf{x}(t))\tilde{\mathbf{G}}_j\mathbf{x}(t) \tag{7.16}$$

Rule 2: IF $q(\mathbf{x}(t))$ is NZ

$$\text{THEN } \mathbf{u}(t) = \sum_{j=1}^{p} n_j(\mathbf{x}(t))\mathbf{G}_j\mathbf{x}(t) \tag{7.17}$$

where rule 1 is for the local fuzzy controller and rule 2 is for the global switching state-feedback controller. The inferred fuzzy combined controller is given by,

$$\dot{\mathbf{x}}(t) = m_1(\mathbf{x}(t))\sum_{j=1}^{c} v_j(\mathbf{x}(t))\tilde{\mathbf{G}}_j\mathbf{x}(t) + m_2(\mathbf{x}(t))\sum_{j=1}^{p} n_j(\mathbf{x}(t))\mathbf{G}_j\mathbf{x}(t). \tag{7.18}$$

The fuzzy combined controller (7.18) is employed to control the nonlinear plant.

Remark 7.4. It can be seen from (7.18) that the contribution of each controller is determined by the normalized membership functions $m_1(\mathbf{x}(t))$ and $m_2(\mathbf{x}(t))$. The global switching controller (7.15) dominates the control process when $m_2(\mathbf{x}(t)) \gg m_1(\mathbf{x}(t))$ to drive the system states towards the origin. The local fuzzy controller (7.12) becomes dominant when the values of $m_1(\mathbf{x}(t)) \gg m_2(\mathbf{x}(t))$, which implies that the system is working in the operating domain near the origin. In this small operating domain, the chattering effect introduced by the global switching controller is reduced and then vanishes when $m_2(\mathbf{x}(t)) = \mathbf{0}$. When $m_2(\mathbf{x}(t)) = \mathbf{0}$, only the local fuzzy controller is responsible for the control process.

7.4 Stability Analysis

The stability of the FMB control system formed by the nonlinear plant represented in the form of (7.8) and the fuzzy combined controller (7.18) is investigated. The block diagram of the FMB control system is shown in Fig. 7.1. For brevity, $w_i(\mathbf{x}(t))$, $\tilde{w}_i(\mathbf{x}(t))$, $v_j(\mathbf{x}(t))$, $n_j(\mathbf{x}(t))$ and $m_k(\mathbf{x}(t))$ are denoted as w_i, $\tilde{w}_i$, v_j, n_j and m_k, respectively. It follows from (7.8) and (7.18), with the equality of $\sum_{i=1}^{p} w_i = \sum_{i=1}^{p} \tilde{w}_i = \sum_{j=1}^{c} v_j = \sum_{k=1}^{2} m_l = \sum_{i=1}^{p}\sum_{j=1}^{p}\sum_{k=1}^{c}\sum_{l=1}^{2} w_i\tilde{w}_j v_k m_l = 1$ given by the property of the membership functions, the FMB control system can be obtained as follows.

$$
\begin{aligned}
\dot{\mathbf{x}}(t) &= m_1 \sum_{i=1}^{p} \tilde{w}_i\big(\tilde{\mathbf{A}}_i + \tilde{\mathbf{B}}_i(m_1 \sum_{j=1}^{c} v_j \tilde{\mathbf{G}}_j + m_2 \sum_{j=1}^{p} n_j \mathbf{G}_j)\big)\mathbf{x}(t) \\
&\quad + m_2 \sum_{i=1}^{p} w_i\big(\mathbf{A}_i + \mathbf{B}_i(m_1 \sum_{j=1}^{c} v_j \tilde{\mathbf{G}}_j + m_2 \sum_{j=1}^{p} n_j \mathbf{G}_j)\big)\mathbf{x}(t) \\
&= m_1 m_1 \sum_{i=1}^{p} \tilde{w}_i(\tilde{\mathbf{A}}_i + \tilde{\mathbf{B}}_i \sum_{j=1}^{c} v_j \tilde{\mathbf{G}}_j)\mathbf{x}(t) \\
&\quad + m_1 m_2 \sum_{i=1}^{p} \tilde{w}_i(\tilde{\mathbf{A}}_i + \tilde{\mathbf{B}}_i \sum_{j=1}^{p} n_j \mathbf{G}_j)\mathbf{x}(t) \\
&\quad + m_2 m_1 \sum_{i=1}^{p} w_i(\mathbf{A}_i + \mathbf{B}_i \sum_{j=1}^{c} v_j \tilde{\mathbf{G}}_j)\mathbf{x}(t) \\
&\quad + m_2 m_2 \sum_{i=1}^{p} w_i(\mathbf{A}_i + \mathbf{B}_i \sum_{j=1}^{p} n_j \mathbf{G}_j)\mathbf{x}(t)
\end{aligned} \tag{7.19}
$$

It can be seen from (7.19) that the second and the third terms vanish for either $m_1 = 0$ or $m_2 = 0$. Under the case of $m_1 \neq 0$ or $m_2 \neq 0$, from the property of (7.10) which states that the local and global fuzzy model are equivalent in the overlapping region of m_1 and m_2, we have,

$$
\sum_{i=1}^{p} \tilde{w}_i(\tilde{\mathbf{A}}_i + \tilde{\mathbf{B}}_i \sum_{j=1}^{p} n_j \mathbf{G}_j)\mathbf{x}(t) = \sum_{i=1}^{p} w_i(\mathbf{A}_i + \mathbf{B}_i \sum_{j=1}^{p} n_j \mathbf{G}_j)\mathbf{x}(t) \tag{7.20}
$$

$$
\sum_{i=1}^{p} w_i(\mathbf{A}_i + \mathbf{B}_i \sum_{j=1}^{c} v_j \tilde{\mathbf{G}}_j)\mathbf{x}(t) = \sum_{i=1}^{p} \tilde{w}_i(\tilde{\mathbf{A}}_i + \tilde{\mathbf{B}}_i \sum_{j=1}^{c} v_j \tilde{\mathbf{G}}_j)\mathbf{x}(t) \tag{7.21}
$$

It follows from (7.19) to (7.21) that the FMB control system (7.17) can be written as follows.

$$\begin{aligned}\dot{\mathbf{x}}(t) &= m_1 m_1 \sum_{i=1}^{p} \tilde{w}_i(\tilde{\mathbf{A}}_i + \tilde{\mathbf{B}}_i \sum_{j=1}^{c} v_j \tilde{\mathbf{G}}_j)\mathbf{x}(t) \\ &\quad + m_1 m_2 \sum_{i=1}^{p} w_i(\mathbf{A}_i + \mathbf{B}_i \sum_{j=1}^{p} n_j \mathbf{G}_j)\mathbf{x}(t) \\ &\quad + m_2 m_1 \sum_{i=1}^{p} \tilde{w}_i(\tilde{\mathbf{A}}_i + \tilde{\mathbf{B}}_i \sum_{j=1}^{c} v_j \tilde{\mathbf{G}}_j)\mathbf{x}(t) \\ &\quad + m_2 m_2 \sum_{i=1}^{p} w_i(\mathbf{A}_i + \mathbf{B}_i \sum_{j=1}^{p} n_j \mathbf{G}_j)\mathbf{x}(t) \\ &= m_1 \sum_{i=1}^{p} \tilde{w}_i(\tilde{\mathbf{A}}_i + \tilde{\mathbf{B}}_i \sum_{j=1}^{c} v_j \tilde{\mathbf{G}}_j)\mathbf{x}(t) \\ &\quad + m_2 \sum_{i=1}^{p} w_i(\mathbf{A}_i + \mathbf{B}_i \sum_{j=1}^{p} n_j \mathbf{G}_j)\mathbf{x}(t) \end{aligned} \tag{7.22}$$

Remark 7.5. By using the property of (7.10), the cross terms in (7.19) can be removed to facilitate the stability analysis. It can be seen from (7.22) that the system consists of only the global and local FMB control systems with the contribution determined by m_1 and m_2.

Assumed that the input matrix exhibits the following property:

$$\mathbf{B}(\mathbf{x}(t)) = \sum_{i=1}^{p} w_i(\mathbf{x}(t))\mathbf{B}_i = \alpha(\mathbf{x}(t))\mathbf{B}_m \tag{7.23}$$

where $\mathbf{B}_m \in \Re^{n \times m}$ is a constant matrix.

Remark 7.6. It is assumed that $\alpha(\mathbf{x}(t))$ is a scalar nonlinear function that is bounded, unknown (because $w_i(\mathbf{x}(t))$ is unknown), non-zero and single-signed but with a known form. As the form of $\alpha(\mathbf{x}(t))$ is assumed to be known and it is in terms of the system states and parameters, the sign of $\alpha(\mathbf{x}(t))$ can be determined. Its bounds ($|\alpha(\mathbf{x}(t))| \in [\alpha_{min}, \alpha_{max}]$ where $\alpha_{min} \leq \alpha_{max}$) can be estimated analytically or numerically. Furthermore, it should be noted that $\alpha(\mathbf{x}(t)) \neq 0$ is required (otherwise it leads to an uncontrollable system with $\mathbf{B}(\mathbf{x}(t)) = \mathbf{0}$) for the stability analysis below.

From (7.22) and (7.23), we have,

$$\begin{aligned}
\dot{\mathbf{x}}(t) &= m_1 \sum_{i=1}^{p}\sum_{j=1}^{c} \tilde{w}_i v_j (\tilde{\mathbf{A}}_i + \tilde{\mathbf{B}}_i \tilde{\mathbf{G}}_j)\mathbf{x}(t) \\
&\quad + m_2\Big(\sum_{i=1}^{p} w_i(\mathbf{A}_i + \mathbf{B}_m\mathbf{G}_i) + \sum_{j=1}^{p}(\alpha(\mathbf{x}(t))n_j - w_j)\mathbf{B}_m\mathbf{G}_j\Big)\mathbf{x}(t) \\
&= m_1 \sum_{i=1}^{p}\sum_{j=1}^{c} \tilde{w}_i v_j \tilde{\mathbf{H}}_{ij}\mathbf{x}(t) \\
&\quad + m_2\Big(\sum_{i=1}^{p} w_i\mathbf{H}_i + \sum_{j=1}^{p}(\alpha(\mathbf{x}(t))n_j - w_j)\mathbf{B}_m\mathbf{G}_j\Big)\mathbf{x}(t).
\end{aligned} \tag{7.24}$$

where

$$\tilde{\mathbf{H}}_{ij} = \tilde{\mathbf{A}}_i + \tilde{\mathbf{B}}_i\tilde{\mathbf{G}}_j \;\forall\, i,\; j, \tag{7.25}$$

$$\mathbf{H}_i = \mathbf{A}_i + \mathbf{B}_m\mathbf{G}_i \;\forall\, i. \tag{7.26}$$

Consider the quadratic Lyapunov function (3.1) to investigate the stability of the FMB control system (7.24). From (3.1) and (7.24), we have

$$\begin{aligned}
\dot{V}(t) &= \dot{\mathbf{x}}(t)^T\mathbf{P}\mathbf{x}(t) + \mathbf{x}(t)^T\mathbf{P}\dot{\mathbf{x}}(t) \\
&= m_1 \sum_{i=1}^{p}\sum_{j=1}^{c} \tilde{w}_i v_j \mathbf{x}(t)^T(\tilde{\mathbf{H}}_{ij}^T\mathbf{P} + \mathbf{P}\tilde{\mathbf{H}}_{ij})\mathbf{x}(t) \\
&\quad + m_2 \sum_{i=1}^{p} w_i \mathbf{x}(t)^T(\mathbf{H}_i^T\mathbf{P} + \mathbf{P}\mathbf{H}_i)\mathbf{x}(t) \\
&\quad + 2m_2 \sum_{j=1}^{p}(\alpha(\mathbf{x}(t))n_j - w_j)\mathbf{x}(t)^T\mathbf{P}\mathbf{B}_m\mathbf{G}_j\mathbf{x}(t).
\end{aligned} \tag{7.27}$$

We choose the switching law as

$$\eta_j(\mathbf{x}(t)) = -\frac{K\mathrm{sgn}(\alpha(\mathbf{x}(t)))\mathrm{sgn}(\mathbf{x}(t)^T\mathbf{P}\mathbf{B}_m\mathbf{G}_j\mathbf{x}(t))}{\alpha_{min}} \;\forall\, j. \tag{7.28}$$

where $K \geq 1$.

From (7.27) to (7.28), we have

$$\dot{V}(t) \le m_1 \sum_{i=1}^{p} \sum_{j=1}^{c} \tilde{w}_i v_j \mathbf{x}(t)^T (\tilde{\mathbf{H}}_{ij}^T \mathbf{P} + \mathbf{P}\tilde{\mathbf{H}}_{ij}) \mathbf{x}(t)$$
$$+ m_2 \sum_{i=1}^{p} w_i \mathbf{x}(t)^T (\mathbf{H}_i^T \mathbf{P} + \mathbf{P}\mathbf{H}_i) \mathbf{x}(t)$$
$$+ 2m_2 \sum_{j=1}^{p} \Big(- \frac{K|\alpha(\mathbf{x}(t))|}{\alpha_{min}} + w_j \Big) |\mathbf{x}(t)^T \mathbf{P}\mathbf{B}_m \mathbf{G}_j \mathbf{x}(t)|. \tag{7.29}$$

Considering the last term in (7.29), as $K \ge 1$ and thus $\frac{K|\alpha(\mathbf{x}(t))|}{\alpha_{min}} \ge 1$, it is obvious that $-\frac{K|\alpha(\mathbf{x}(t))|}{\alpha_{min}} \le -1$. Based on the property of the membership functions, i.e., $w_j \in \big[0, 1\big]$ for all j, we have $-\frac{K|\alpha(\mathbf{x}(t))|}{\alpha_{min}} + w_j \le 0$. Consequently, it follows from (7.29) that we have,

$$\dot{V}(t) \le m_1 \sum_{i=1}^{p} \sum_{j=1}^{c} \tilde{w}_i v_j \mathbf{x}(t)^T (\tilde{\mathbf{H}}_{ij}^T \mathbf{P} + \mathbf{P}\tilde{\mathbf{H}}_{ij}) \mathbf{x}(t)$$
$$+ m_2 \sum_{i=1}^{p} w_i \mathbf{x}(t)^T (\mathbf{H}_i^T \mathbf{P} + \mathbf{P}\mathbf{H}_i) \mathbf{x}(t). \tag{7.30}$$

From (3.1) and (7.30), with the support of Remark 7.6, based on the the Lyapunov stability theory, $V(t) > 0$ and $\dot{V}(t) < 0$ for $\mathbf{x}(t) \neq \mathbf{0}$ implying the asymptotic stability of the FMB control system (7.19), i.e., $\mathbf{x}(t) \to 0$ when time $t \to \infty$, can be achieved if the stability conditions summarized in the following theorem are satisfied.

Theorem 7.1. *The FMB control system (7.19), formed by the nonlinear plant represented by the fuzzy combined model (7.8) and the fuzzy combined controller (7.18) connected in a closed loop, is asymptotically stable if there exist matrix* $\mathbf{P} = \mathbf{P}^T \in \Re^{n \times n}$ *such that the following LMIs are satisfied.*

$$\mathbf{P} > 0;$$

$$\tilde{\mathbf{H}}_{ij}^T \mathbf{P} + \mathbf{P}\tilde{\mathbf{H}}_{ij} < 0 \, \forall \, i, \, j;$$

$$\mathbf{H}_i^T \mathbf{P} + \mathbf{P}\mathbf{H}_i < 0 \, \forall \, i;$$

and the feedback gains $\tilde{\mathbf{G}}_j$ *and* $\mathbf{G}_j$ *for all* i *and* j *are predefined; and the switching law of the global switching controller is chosen as* $\eta_j(\mathbf{x}(t)) = -\frac{K sgn(\alpha(\mathbf{x}(t))) sgn(\mathbf{x}(t)^T \mathbf{P}\mathbf{B}_m \mathbf{G}_j \mathbf{x}(t))}{\alpha_{min}}$ *for all* j; $|\alpha(\mathbf{x}(t))| \in \big[\alpha_{min}, \alpha_{max}\big]$ *where* $\alpha_{min} \le \alpha_{max}$. *The membership function of the fuzzy combined controller corresponding to* ZE *is designed to cover the region of* $0 \le q(\mathbf{x}(t)) \le 1$, *while that corresponding to* NZ *is required to cover the region* $q(\mathbf{x}(t)) > 0$, *with* $q(\mathbf{x}(t))$ *defined in (7.7).*

Remark 7.7. It can be seen from (7.30) that $\dot{V}(t) < 0$ for $\mathbf{x}(t) \neq 0$, which implies the the asymptotic stability of the FMB control system (7.19) if the inequalities $\tilde{\mathbf{H}}_{ij}^T\mathbf{P} + \mathbf{P}\tilde{\mathbf{H}}_{ij} < 0$ and $\mathbf{H}_i^T\mathbf{P} + \mathbf{P}\mathbf{H}_i < 0$ are satisfied simultaneously. Considering $\mathbf{P}$ as the decision variables, these two inequalities are not linear in the feedback gains $\tilde{\mathbf{G}}_j$ and $\mathbf{G}_j$. Thus, convex programming techniques cannot be applied to find numerically the solution. By following the approach in the previous chapters, by changing of matrix variables, both inequalities can be transformed to LMIs. Denote $\mathbf{X} = \mathbf{P}^{-1}$ and design the feedback gains as $\tilde{\mathbf{G}}_j = \tilde{\mathbf{N}}_j\mathbf{X}^{-1}$ and $\mathbf{G}_j = \mathbf{N}_j\mathbf{X}^{-1}$ where $\tilde{\mathbf{N}}_j \in \Re^{m\times n}$ and $\mathbf{N}_j \in \Re^{m\times n}$ for all j. Pre- and post-multiplying $\mathbf{X}$ to $\tilde{\mathbf{H}}_{ij}^T\mathbf{P} + \mathbf{P}\tilde{\mathbf{H}}_{ij} < 0$, we have $\mathbf{X}(\tilde{\mathbf{H}}_{ij}^T\mathbf{P}+\mathbf{P}\tilde{\mathbf{H}}_{ij})\mathbf{X} = \mathbf{X}(\tilde{\mathbf{H}}_{ij}^T\mathbf{X}^{-1}+\mathbf{X}^{-1}\tilde{\mathbf{H}}_{ij})\mathbf{X} = \mathbf{X}\tilde{\mathbf{A}}_i^T+\tilde{\mathbf{A}}_i\mathbf{X}+\tilde{\mathbf{N}}_j^T\tilde{\mathbf{B}}_i^T+\tilde{\mathbf{B}}_i\tilde{\mathbf{N}}_j < 0$. Similarly, from $\mathbf{H}_i^T\mathbf{P} + \mathbf{P}\mathbf{H}_i < 0$, we have $\mathbf{X}(\mathbf{H}_i^T\mathbf{P} + \mathbf{P}\mathbf{H}_i)\mathbf{X} = \mathbf{X}(\mathbf{H}_i^T\mathbf{X}^{-1} + \mathbf{X}^{-1}\mathbf{H}_i)\mathbf{X} = \mathbf{X}\mathbf{A}_i^T + \mathbf{A}\mathbf{X} + \mathbf{N}_i^T\mathbf{B}_m^T + \mathbf{B}_m\mathbf{N}_i < 0$. Consequently, these inequalities becomes LMI conditions and that can be summarized in the following theory.

Theorem 7.2. *The FMB control system (7.19), formed by the nonlinear plant represented by the fuzzy combined model (7.8) and the fuzzy combined controller (7.18) connected in a closed loop, is asymptotically stable if there exist matrices* $\tilde{\mathbf{N}}_j \in \Re^{m\times n}$, $j = 1, 2, \cdots, c$, $\mathbf{N}_i \in \Re^{m\times n}$, $i = 1, 2, \cdots, p$ *and* $\mathbf{X} = \mathbf{X}^T \in \Re^{n\times n}$ *such that the following LMIs are satisfied.*

$$\mathbf{X} > 0;$$

$$\mathbf{X}\tilde{\mathbf{A}}_i^T + \tilde{\mathbf{A}}_i\mathbf{X} + \tilde{\mathbf{N}}_j^T\tilde{\mathbf{B}}_i^T + \tilde{\mathbf{B}}_i\tilde{\mathbf{N}}_j < 0 \;\forall\; i,\, j;$$

$$\mathbf{X}\mathbf{A}_i^T + \mathbf{A}\mathbf{X} + \mathbf{N}_i^T\tilde{\mathbf{B}}_m^T + \tilde{\mathbf{B}}_m\mathbf{N}_i < 0 \;\forall\; i;$$

and the feedback gains are chosen as $\tilde{\mathbf{G}}_j = \tilde{\mathbf{N}}_j\mathbf{X}^{-1}$ *and* $\mathbf{G}_j = \mathbf{N}_j\mathbf{X}^{-1}$ *for all i and j; and the switching law of the global switching controller is chosen as* $\eta_j(\mathbf{x}(t)) = -\frac{K sgn(\alpha(\mathbf{x}(t))) sgn(\mathbf{x}(t)^T\mathbf{P}\mathbf{B}_m\mathbf{G}_j\mathbf{x}(t))}{\alpha_{min}}$ *for all j;* $|\alpha(\mathbf{x}(t))| \in [\alpha_{min}, \alpha_{max}]$ *where* $\alpha_{min} \leq \alpha_{max}$*. The membership function of the fuzzy combined controller corresponding to ZE is designed to cover the region of* $0 \leq q(\mathbf{x}(t)) \leq 1$*, while corresponding to NZ is required to cover the region* $q(\mathbf{x}(t)) > 0$*, with* $q(\mathbf{x}(t))$ *defined in (7.7).*

Remark 7.8. It can be seen from the switching law (7.28) that the switching element will cause undesired chattering effect. In order to alleviate the chattering effect, the switching function can be replaced by a saturation function [100]. Then, (7.28) is replaced by

$$\eta_j(\mathbf{x}(t)) = -\frac{K\text{sgn}(\alpha(\mathbf{x}(t)))\text{sat}(\mathbf{x}(t)^T\mathbf{P}\mathbf{B}_m\mathbf{G}_j\mathbf{x}(t))}{\alpha_{min}} \;\forall\; i. \tag{7.31}$$

where

$$\mathrm{sat}(z) = \begin{cases} 1 & \text{for } \frac{z}{T} \geq 1 \\ -1 & \text{for } \frac{z}{T} \leq -1 \\ \frac{z}{T} & \text{otherwise} \end{cases} \tag{7.32}$$

and T is a non-zero positive scalar to be designed.

Remark 7.9. It should be noted that $\alpha(\mathbf{x}(t))$ is a single-signed scalar function; hence, $\mathrm{sgn}(\alpha(\mathbf{x}(t)))$ is not a switching signal. Consequently, it is not necessary to replace it by a saturation function.

Remark 7.10. The saturation function employed to replace the switching function for the alleviation of chattering effect may introduce steady-state error. The magnitude of the steady-state error is related to the value of T. A higher value of T will lead to a larger steady-state error. Hence, when the saturation function is employed, it is possible that the steady-state error is so large that the system states cannot be driven to the pre-defined operating domain. Consequently, the local fuzzy controller will never be activated to drive the system states to the origin. Hence, the value of T should be designed such that the system states can be driven into the valid operating domain of the local fuzzy model. Once the system states are inside the valid operating domain of the local fuzzy model, the local fuzzy controller will gradually replace the global switching controller. As a result, the chattering effect and the steady-state error will be eliminated eventually when the local fuzzy controller completely dominates the control process.

Remark 7.11. To achieve the stability conditions in Theorem 7.1, we employ the MFSI stability analysis approach. The stability conditions can be relaxed by considering the MFSD analysis approach. For example, the boundary and/or regional information of the membership functions can be considered in the stability analysis.

Example 7.1. A cart-pole typed inverted pendulum [65] subject to parameter uncertainties is considered in this example. The dynamic equation for the inverted pendulum is given by,

$$\ddot{\theta}(t) = \frac{g\sin(\theta(t)) - am_p l\dot{\theta}(t)^2\sin(2\theta(t))/2 - a\cos(\theta(t))u(t)}{4l/3 - am_p l\cos^2(\theta(t))} \tag{7.33}$$

where $\theta(t)$ is the angular displacement of the pendulum, $g = 9.8m/s^2$ is the acceleration due to gravity, $m_p \in [m_{p_{min}} \quad m_{p_{max}}] = [2 \quad 5]kg$ is the mass of the pendulum, $M_c \in [M_{min} \quad M_{max}] = [8 \quad 10]kg$ is the mass of the cart, $a = 1/(m_p+M_c)$, $2l = 1m$ is the length of the pendulum, and $u(t)$ is the force (N) applied to the cart. The proposed fuzzy combined controller is employed to balance the pole, i.e., $\theta(t) \to 0$ as time $t \to \infty$.

First, the fuzzy combined model is constructed to facilitate the design of the fuzzy combined controller. The inverted pendulum can be represented by the global fuzzy model [56] with the following four fuzzy rules:

$$\text{Rule } i\text{: IF } f_1(\mathbf{x}(t)) \text{ is } M_1^i \text{ AND } f_2(\mathbf{x}(t)) \text{ is } M_2^i$$
$$\text{THEN } \dot{\mathbf{x}}(t) = \mathbf{A}_i\mathbf{x}(t) + \mathbf{B}_i u(t), i = 1, 2, 3, 4 \tag{7.34}$$

The global fuzzy model is defined as follows.

$$\begin{aligned}\dot{\mathbf{x}}(t) &= \sum_{i=1}^{4} w_i(\mathbf{x}(t))(\mathbf{A}_i\mathbf{x}(t) + \mathbf{B}_i u(t)) \\ &= \sum_{i=1}^{4} w_i(\mathbf{x}(t))(\mathbf{A}_i\mathbf{x}(t) + \alpha(x_1(t))\mathbf{B}_m u(t))\end{aligned} \tag{7.35}$$

where $\mathbf{x}(t) = \begin{bmatrix} x_1(t) \ x_2(t) \end{bmatrix}^T = \begin{bmatrix} \theta(t) \ \dot{\theta}(t) \end{bmatrix}^T$; $f_1(\mathbf{x}(t)) = \frac{g - am_p l x_2(t)^2 \cos(x_1(t))}{4l/3 - am_p l \cos^2(x_1(t))} \times \frac{\sin(x_1(t))}{x_1(t)}$ and $f_2(x_1(t)) = \alpha(x_1(t)) = -\frac{a\cos(x_1(t))}{4l/3 - am_p l \cos^2(x_1(t))}\frac{\sin(x_1(t))}{x_1(t)}$; $\mathbf{A}_1 = \mathbf{A}_2 = \begin{bmatrix} 0 & 1 \\ f_{1_{min}} & 0 \end{bmatrix}$ and $\mathbf{A}_3 = \mathbf{A}_4 = \begin{bmatrix} 0 & 1 \\ f_{1_{max}} & 0 \end{bmatrix}$; $\mathbf{B}_1 = \mathbf{B}_3 = \begin{bmatrix} 0 \\ f_{2_{min}} \end{bmatrix}$, $\mathbf{B}_2 = \mathbf{B}_4 = \begin{bmatrix} 0 \\ f_{2_{max}} \end{bmatrix}$ and $\mathbf{B}_m = \begin{bmatrix} 0 \\ 1 \end{bmatrix}$. The inverted pendulum is considered working in the operating domain characterized by $x_1(t) = \theta(t) \in \left[-\frac{22\pi}{45}, \frac{22\pi}{45}\right]$ and $x_2(t) = \dot{\theta}(t) \in \left[-5, 5\right]$. Consequently, we have $f_{1_{min}} = 9.4047$ and $f_{1_{max}} = 20.6595$, $f_{2_{min}} = -0.1765$ and $f_{2_{max}} = -0.0034$. The normalized membership functions are defined as $w_i(\mathbf{x}(t)) = \frac{\mu_{M_1^i}(f_1(\mathbf{x}(t))) \times \mu_{M_2^i}(f_2(x_1(t)))}{\sum_{l=1}^{4}(\mu_{M_1^l}(f_1(\mathbf{x}(t))) \times \mu_{M_2^l}(f_2(x_1(t))))}$ for all i with $\mu_{M_1^\beta}(f_1(\mathbf{x}(t))) = \frac{-f_1(\mathbf{x}(t)) + f_{1_{max}}}{f_{1_{max}} - f_{1_{min}}}$ for $\beta = 1,\ 2$; $\mu_{M_1^\delta}(f_1(\mathbf{x}(t))) = 1 - \mu_{M_1^1}(f_1(\mathbf{x}(t)))$ for $\delta = 3,\ 4$; $\mu_{M_2^\kappa}(f_2(x_1(t))) = \frac{-f_2(x_1(t)) + f_{2_{max}}}{f_{2_{max}} - f_{2_{min}}}$, $\kappa = 1,\ 3$; $\mu_{M_2^\phi}(f_2(x_1(t))) = 1 - \mu_{M_2^1}(f_2(x_1(t)))$ for $\phi = 2,\ 4$.

Consider a small operating domain near the origin that is characterized by $x_1(t) \in \left[\tilde{x}_{1_{min}}, \tilde{x}_{1_{max}}\right] = \left[-0.5, 0.5\right]$ and $x_2(t) \in \left[\tilde{x}_{2_{min}}, \tilde{x}_{2_{max}}\right] = \left[-2.5, 2.5\right]$. The dynamical behaviour of the inverted pendulum in this small operating domain can be described by a local fuzzy model with the following four rules:

$$\text{Rule } i\text{: IF } \tilde{f}_1(\mathbf{x}(t)) \text{ is } \tilde{M}_1^i \text{ AND } \tilde{f}_2(x_1(t)) \text{ is } \tilde{M}_2^i$$
$$\text{THEN } \dot{\mathbf{x}}(t) = \tilde{\mathbf{A}}_i\mathbf{x}(t) + \tilde{\mathbf{B}}_i u(t), i = 1, 2, 3, 4 \tag{7.36}$$

The local fuzzy model is given as follows.

$$\dot{\mathbf{x}}(t) = \sum_{i=1}^{4} \tilde{w}_i(\mathbf{x}(t))(\tilde{\mathbf{A}}_i\mathbf{x}(t) + \tilde{\mathbf{B}}_i u(t)) \tag{7.37}$$

where $\tilde{f}_1(\mathbf{x}(t)) = f_1(\mathbf{x}(t))$ and $\tilde{f}_2(x_1(t)) = f_2(x_1(t))$; $\tilde{\mathbf{A}}_1 = \tilde{\mathbf{A}}_2 = \begin{bmatrix} 0 & 1 \\ \tilde{f}_{1_{min}} & 0 \end{bmatrix}$ and $\tilde{\mathbf{A}}_3 = \tilde{\mathbf{A}}_4 = \begin{bmatrix} 0 & 1 \\ \tilde{f}_{1_{max}} & 0 \end{bmatrix}$; $\tilde{\mathbf{B}}_1 = \tilde{\mathbf{B}}_3 = \begin{bmatrix} 0 \\ \tilde{f}_{2_{min}} \end{bmatrix}$ and $\tilde{\mathbf{B}}_2 = \tilde{\mathbf{B}}_4 = \begin{bmatrix} 0 \\ \tilde{f}_{2_{max}} \end{bmatrix}$; $\tilde{f}_{1_{min}} = 14.8691$ and $\tilde{f}_{1_{max}} = 20.6595$, $\tilde{f}_{2_{min}} = -0.1765$ and $\tilde{f}_{2_{max}} = -0.1086$. The normalized membership functions are defined as $\tilde{w}_i(\mathbf{x}(t)) = \frac{\mu_{\tilde{M}_1^i}(\tilde{f}_1(\mathbf{x}(t)))\times\mu_{\tilde{M}_2^i}(\tilde{f}_2(x_1(t)))}{\sum_{l=1}^{4}(\mu_{\tilde{M}_1^l}(\tilde{f}_1(\mathbf{x}(t)))\times\mu_{\tilde{M}_2^l}(\tilde{f}_2(x_1(t))))}$ for all i with $\mu_{\tilde{M}_1^\beta}(\tilde{f}_1(\mathbf{x}(t))) = \frac{-\tilde{f}_1(\mathbf{x}(t))+\tilde{f}_{1_{max}}}{\tilde{f}_{1_{max}}-\tilde{f}_{1_{min}}}$ for $\beta = 1, 2$; $\mu_{\tilde{M}_1^\delta}(\tilde{f}_1(\mathbf{x}(t))) = 1 - \mu_{\tilde{M}_1^1}(\tilde{f}_1(\mathbf{x}(t)))$ for $\delta = 3, 4$; $\mu_{\tilde{M}_2^\kappa}(\tilde{f}_2(x_1(t))) = \frac{-\tilde{f}_2(\mathbf{x}(t))+\tilde{f}_{2_{max}}}{\tilde{f}_{2_{max}}-\tilde{f}_{2_{min}}}$, $\kappa = 1, 3$; $\mu_{\tilde{M}_2^\phi}(\tilde{f}_2(x_1(t))) = 1 - \mu_{\tilde{M}_2^1}(\tilde{f}_2(x_1(t)))$ for $\phi = 2, 4$.

Based on the global and local fuzzy models, the fuzzy combined model with the following two rules is employed to represent the inverted pendulum (7.33).

Rule 1: IF $q(\mathbf{x}(t))$ is ZE

$$\text{THEN } \dot{\mathbf{x}}(t) = \sum_{i=1}^{4} \tilde{w}_i(\mathbf{x}(t))(\tilde{\mathbf{A}}_i\mathbf{x}(t) + \tilde{\mathbf{B}}_i u(t)) \tag{7.38}$$

Rule 2: IF $q(\mathbf{x}(t))$ is NZ

$$\text{THEN } \dot{\mathbf{x}}(t) = \sum_{i=1}^{4} w_i(\mathbf{x}(t))(\mathbf{A}_i\mathbf{x}(t) + \mathbf{B}_i\mathbf{u}(t)) \tag{7.39}$$

The nonlinear function $q(\mathbf{x}(t))$ is defined as,

$$q(\mathbf{x}(t)) = \max\left(\left|\frac{x_1(t) - \tilde{x}_1}{\overline{x}_1(t)}\right|, \left|\frac{x_2(t) - \tilde{x}_2}{\overline{x}_2(t)}\right|\right) \geq 0 \tag{7.40}$$

where $\tilde{x}_i = \frac{\tilde{x}_{i_{max}}+\tilde{x}_{i_{min}}}{2}$, $\overline{x}_i = \frac{\tilde{x}_{i_{max}}-\tilde{x}_{i_{min}}}{2}$, $i = 1, 2$.

The fuzzy combined model is given by,

$$\begin{aligned}\dot{\mathbf{x}}(t) = {} & m_1(\mathbf{x}(t)) \sum_{i=1}^{4} \tilde{w}_i(\mathbf{x}(t))(\tilde{\mathbf{A}}_i\mathbf{x}(t) + \tilde{\mathbf{B}}_i u(t)) \\ & + m_2(\mathbf{x}(t)) \sum_{i=1}^{4} w_i(\mathbf{x}(t))(\mathbf{A}_i\mathbf{x}(t) + \mathbf{B}_i u(t)).\end{aligned} \tag{7.41}$$

The membership functions corresponding to the fuzzy terms ZE and NZ are defined as follows and shown in Fig. 7.2.

$$\mu_{ZE}(q(\mathbf{x}(t))) = \begin{cases} q(\mathbf{x}(t)) & \text{for } 0 \le q(\mathbf{x}(t)) \le 1 \\ 0 & \text{for } q(\mathbf{x}(t)) > 1 \end{cases} \tag{7.42}$$

$$\mu_{NZ}(q(\mathbf{x}(t))) = \begin{cases} 0 & \text{for } q(\mathbf{x}(t)) < 0.5 \\ q(\mathbf{x}(t)) - 0.5 & \text{for } 0.5 \le q(\mathbf{x}(t)) \le 1.5 \\ 1 & \text{for } q(\mathbf{x}(t)) > 1.5 \end{cases} \tag{7.43}$$

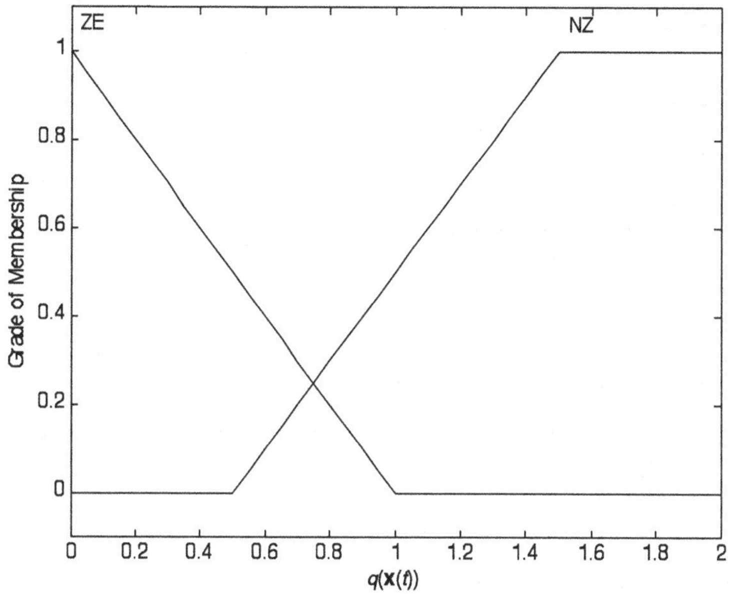

Fig. 7.2 Membership functions of $\mu_{ZE}(q(\mathbf{x}(t)))$ and $\mu_{NZ}(q(\mathbf{x}(t)))$.

A fuzzy combined controller is designed to balance the inverted pendulum. Based on the local fuzzy model (7.37), a local fuzzy controller is designed with four rules of the following format.

$$\begin{aligned} &\text{Rule } j\text{: IF } x_1(t) \text{ is } N_1^j \text{ AND } x_2(t) \text{ is } N_2^j \\ &\qquad \text{THEN } u(t) = \tilde{\mathbf{G}}_j\mathbf{x}(t), j = 1, 2, 3, 4 \end{aligned} \tag{7.44}$$

The local fuzzy controller is given as follows.

$$u(t) = \sum_{j=1}^{4} v_j(\mathbf{x}(t))\tilde{\mathbf{G}}_j\mathbf{x}(t) \tag{7.45}$$

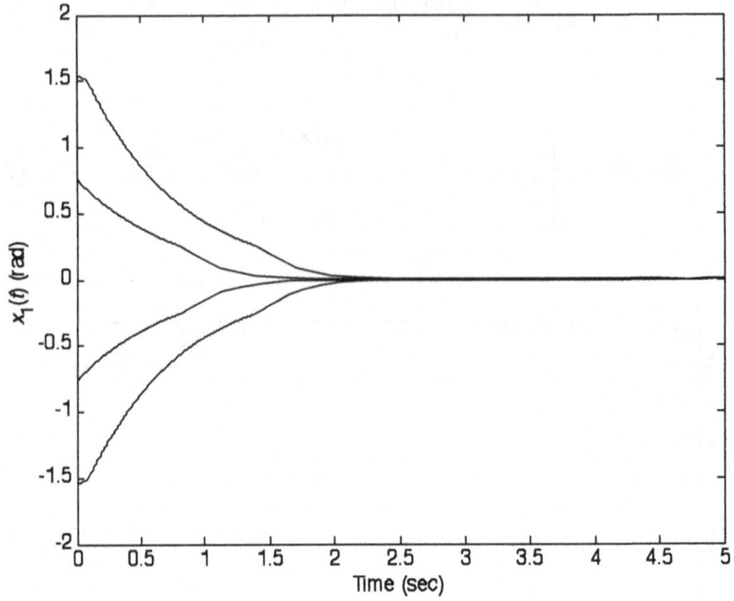

(a) $x_1(t)$.

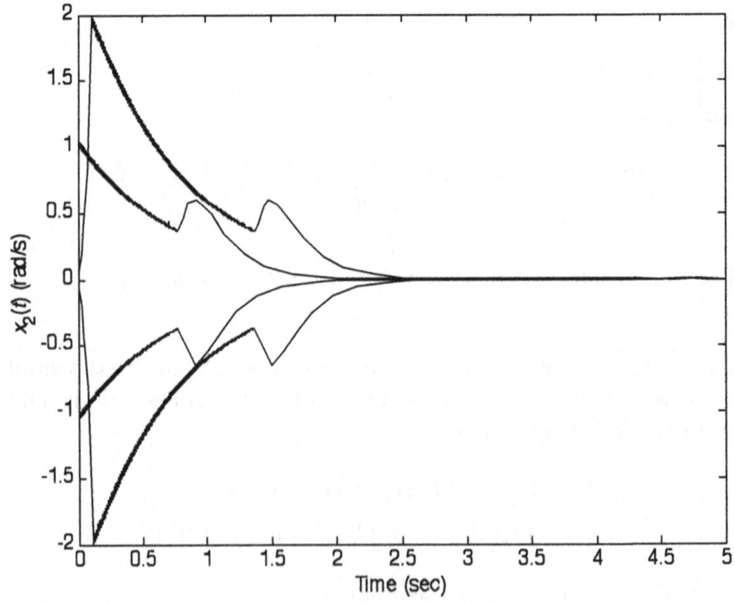

(b) $x_2(t)$.

Fig. 7.3 System responses with the fuzzy combined controller (7.49) for $m_p = 2kg$ and $M_c = 8kg$.

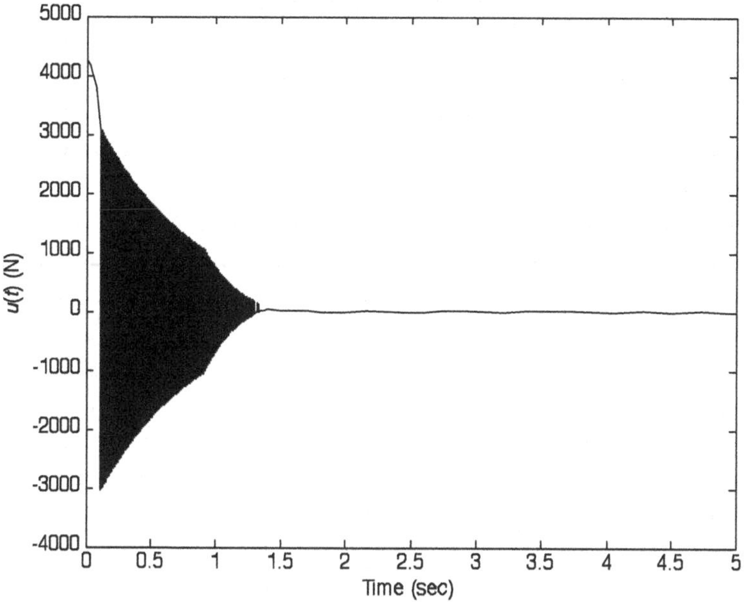

Fig. 7.4 Control signal of the fuzzy combined controller (7.49) with $\mathbf{x}(0) = \begin{bmatrix}\frac{22\pi}{45} & 0\end{bmatrix}^T$ for $m_p = 2kg$ and $M_c = 8kg$.

where $\mu_{N_1^\beta}(x_1(t)) = \frac{-|x_1(t)|+\tilde{x}_{1_{max}}}{\tilde{x}_{1_{max}}-\tilde{x}_{1_{min}}}$ for $\beta = 1, 2$; $\mu_{N_1^\delta}(x_1(t)) = 1 - \mu_{N_1^1}(x_1(t))$ for $\delta = 3, 4$; $\mu_{N_2^\kappa}(x_2(t)) = \frac{-|x_2(t)|+\tilde{x}_{2_{max}}}{\tilde{x}_{2_{max}}-\tilde{x}_{2_{min}}}$, $\kappa = 1, 3$; $\mu_{N_2^\phi}(x_2(t)) = 1 - \mu_{N_2^1}(x_2(t))$ for $\phi = 2, 4$; The normalized membership functions are defined as $v_j(\mathbf{x}(t)) = \frac{\mu_{N_1^j}(x_1(t))\times\mu_{N_2^j}(x_2(t))}{\sum_{l=1}^{4}(\mu_{N_1^l}(x_1(2))\times\mu_{N_2^l}(x_2(t)))}$ for all j. The feedback gains are chosen as $\tilde{\mathbf{G}}_1 = \begin{bmatrix}265.5473 \; 67.9887\end{bmatrix}$, $\tilde{\mathbf{G}}_2 = \begin{bmatrix}431.5755 \; 110.4972\end{bmatrix}$, $\tilde{\mathbf{G}}_3 = \begin{bmatrix}298.3541 \; 67.9887\end{bmatrix}$ and $\tilde{\mathbf{G}}_4 = \begin{bmatrix}484.8941 \; 110.4972\end{bmatrix}$ such that the eigenvalues of $\tilde{\mathbf{H}}_{ii}$, $i = 1, 2, 3$ and 4 are -4 and -8.

Based on the global fuzzy model of (7.41), the global switching controller is designed as follows.

$$u(t) = \sum_{j=1}^{4} n_j(\mathbf{x}(t))\mathbf{G}_j\mathbf{x}(t) \tag{7.46}$$

where the switching function $n_j(\mathbf{x}(t))$ is given in (7.28) and $K = 1.5$, $\alpha_{min} = |f_{2_{max}}| = 0.0034$. The feedback gains are chosen as $\mathbf{G}_1 = \mathbf{G}_2 = \begin{bmatrix}-18.8691 \; -4.0000\end{bmatrix}$ and $\mathbf{G}_3 = \mathbf{G}_4 = \begin{bmatrix}-24.6595 \; -4.0000\end{bmatrix}$ such that the eigenvalues of $\mathbf{H}_i$, $i = 1, 2, 3$ and 4 are all -2.

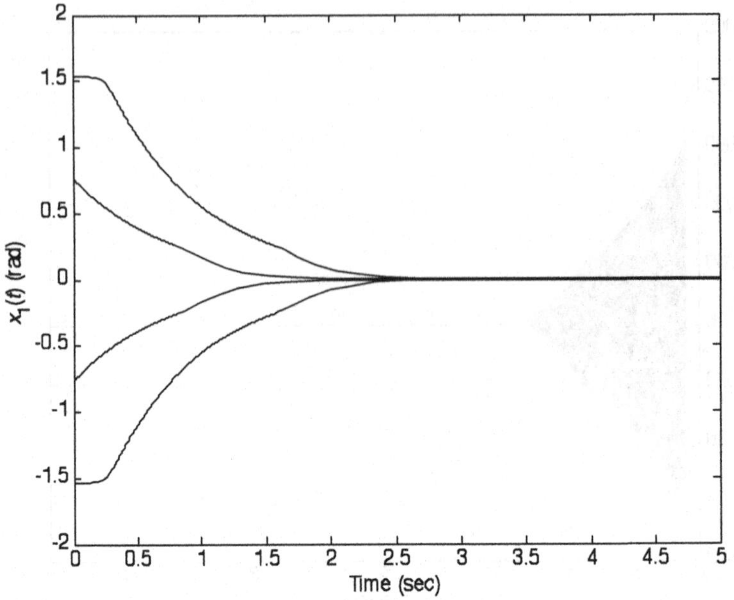

(a) $x_1(t)$.

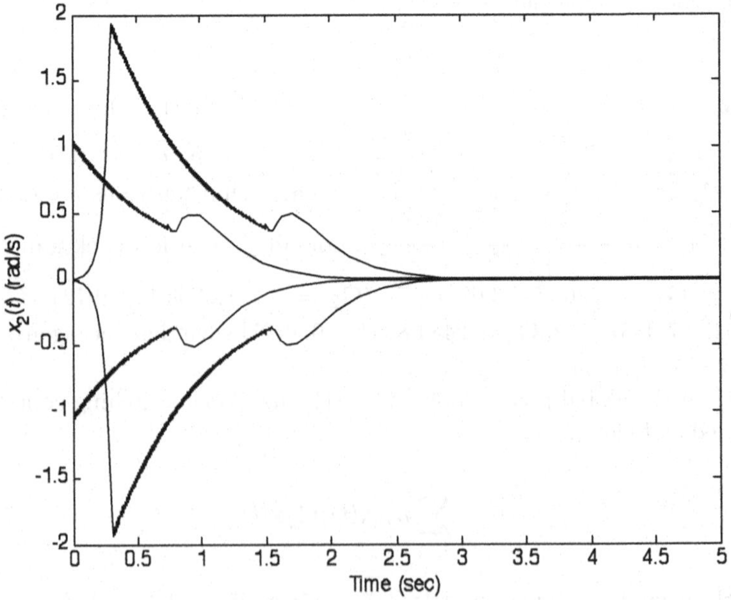

(b) $x_2(t)$.

Fig. 7.5 System responses with the fuzzy combined controller (7.49) for $m_p = 5kg$ and $M_c = 10kg$.

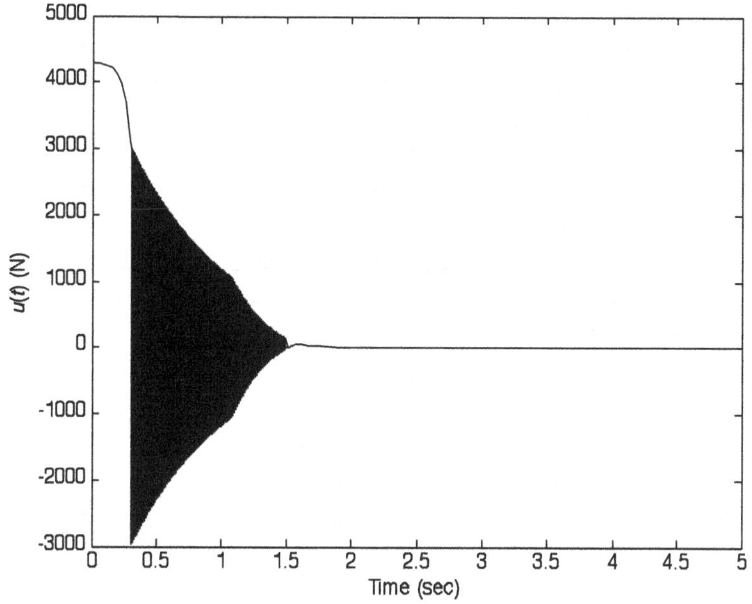

Fig. 7.6 Control signal of the fuzzy combined controller (7.49) with $\mathbf{x}(0) = \begin{bmatrix} \frac{22\pi}{45} & 0 \end{bmatrix}^T$ for $m_p = 5kg$ and $M_c = 10kg$.

Based on the fuzzy combined model (7.41), the fuzzy combined controller is designed with the following two rules:

$$\text{Rule 1: IF } q(\mathbf{x}(t)) \text{ is } ZE$$

$$\text{THEN } u(t) = \sum_{j=1}^{4} v_j(\mathbf{x}(t))\tilde{\mathbf{G}}_j\mathbf{x}(t) \tag{7.47}$$

$$\text{Rule 2: IF } q(\mathbf{x}(t)) \text{ is } NZ$$

$$\text{THEN } u(t) = \sum_{j=1}^{4} n_j(\mathbf{x}(t))\mathbf{G}_j\mathbf{x}(t) \tag{7.48}$$

The fuzzy combined controller is given as follows.

$$\dot{\mathbf{x}}(t) = m_1(\mathbf{x}(t))\sum_{j=1}^{4} v_j(\mathbf{x}(t))\tilde{\mathbf{G}}_j\mathbf{x}(t) + m_2(\mathbf{x}(t))\sum_{j=1}^{4} n_j(\mathbf{x}(t))\mathbf{G}_j\mathbf{x}(t) \tag{7.49}$$

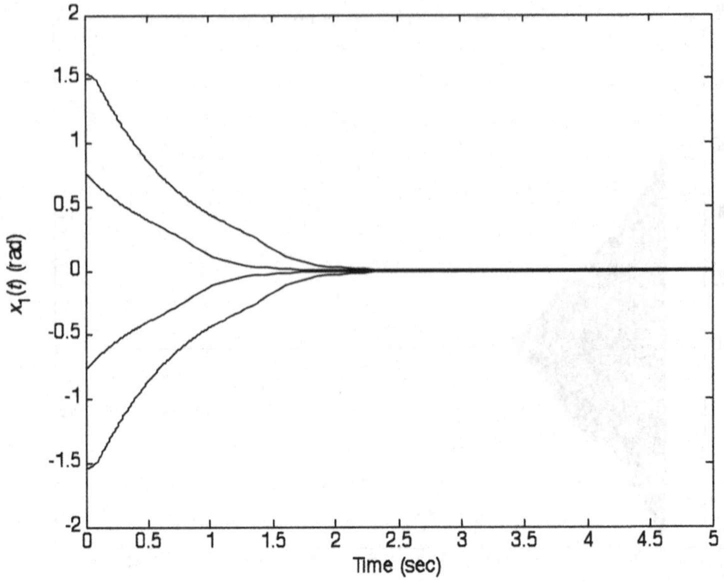

(a) $x_1(t)$.

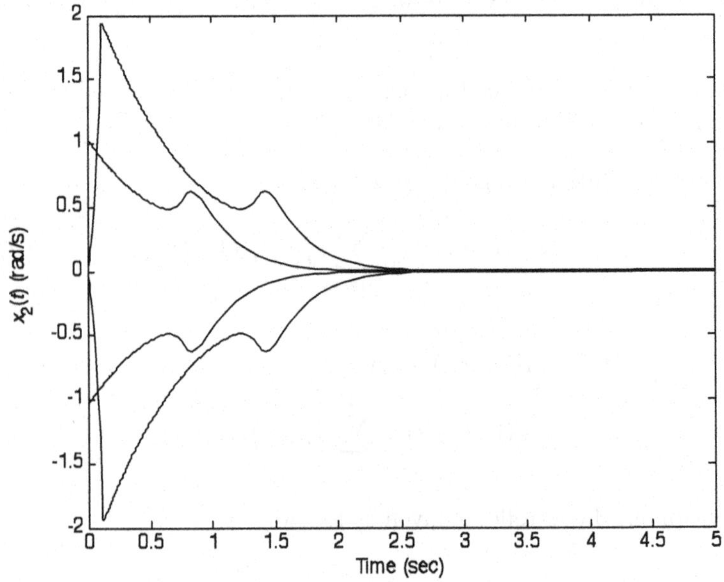

(b) $x_2(t)$.

Fig. 7.7 System responses and control signals with the fuzzy combined controller using the saturation function for $m_p = 2kg$ and $M_c = 8kg$.

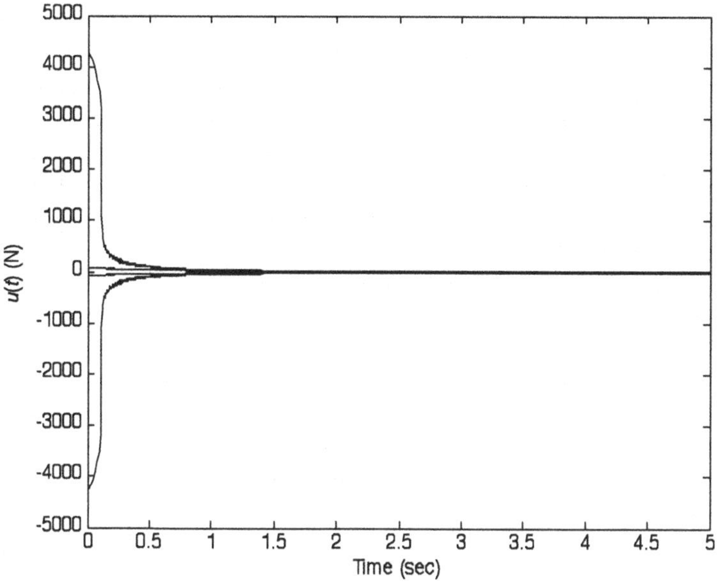

(c) $u(t)$.

Fig. 7.7 *(continued)*

To check for the system stability, the stability conditions in Theorem 7.1 are employed. With the MATLAB LMI toolbox, it is found that $\mathbf{P} = \begin{bmatrix} 5.1158 & 0.4197 \\ 0.4197 & 0.3082 \end{bmatrix}$ satisfies all LMIs in Theorem 7.1. Hence, the FMB control system is guaranteed to be asymptotically stable.

The fuzzy combined controller (7.49) is employed to control the inverted pendulum (7.33) subject to parameter uncertainties. The system responses of the FMB control system with the initial state conditions of $\mathbf{x}(0) = \left[\frac{22\pi}{45}\ 0\right]^T$, $\mathbf{x}(0) = \left[\frac{11\pi}{45}\ 0\right]^T$, $\mathbf{x}(0) = \left[-\frac{11\pi}{45}\ 0\right]^T$ and $\mathbf{x}(0) = \left[-\frac{22\pi}{45}\ 0\right]^T$ for $m_p = 2kg$ and $M_c = 8kg$ are shown in Fig. 7.3. The control signal for $\mathbf{x}(0) = \left[\frac{22\pi}{45}\ 0\right]^T$ is shown in Fig. 7.4.

We change the system parameters to $m_p = 5kg$ and $M_c = 10kg$ to test the robustness property of the fuzzy combined controller. The same fuzzy combined controller (7.49) is employed to control the inverted pendulum with the same initial state conditions. The system responses of the FMB control system are shown in Fig. 7.5. The control signal for the fuzzy combined controller (7.49) with $\mathbf{x}(0) = \left[\frac{22\pi}{45}\ 0\right]^T$ is shown in Fig. 7.6. It can be seen that the proposed fuzzy combined controller is able to stabilize successfully the inverted pendulum subject to parameter uncertainties. From the figures,

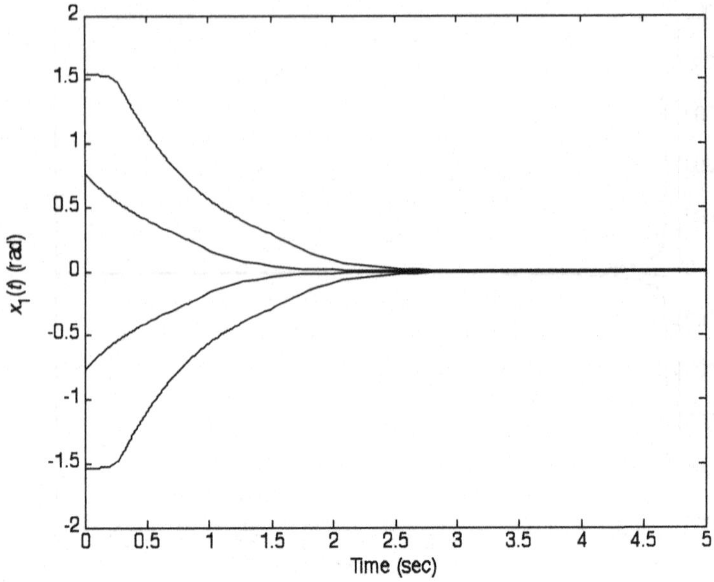

(a) $x_1(t)$.

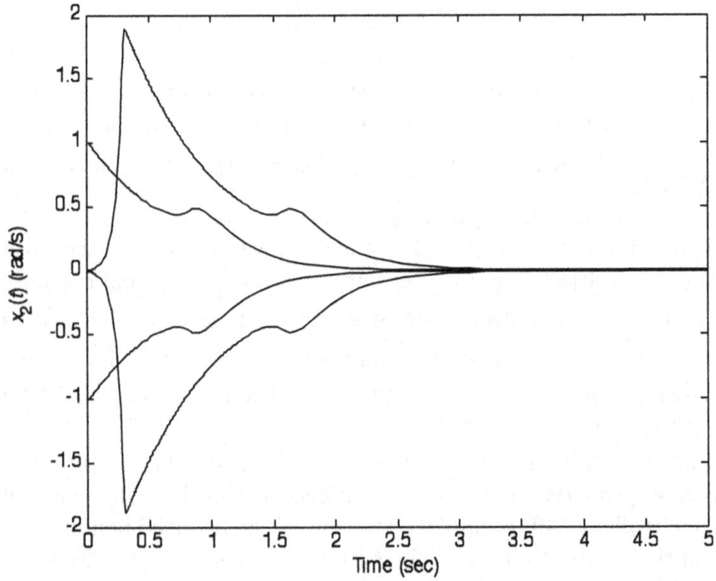

(b) $x_2(t)$.

Fig. 7.8 System responses and control signals with the fuzzy combined controller using the saturation function for $m_p = 5kg$ and $M_c = 10kg$.

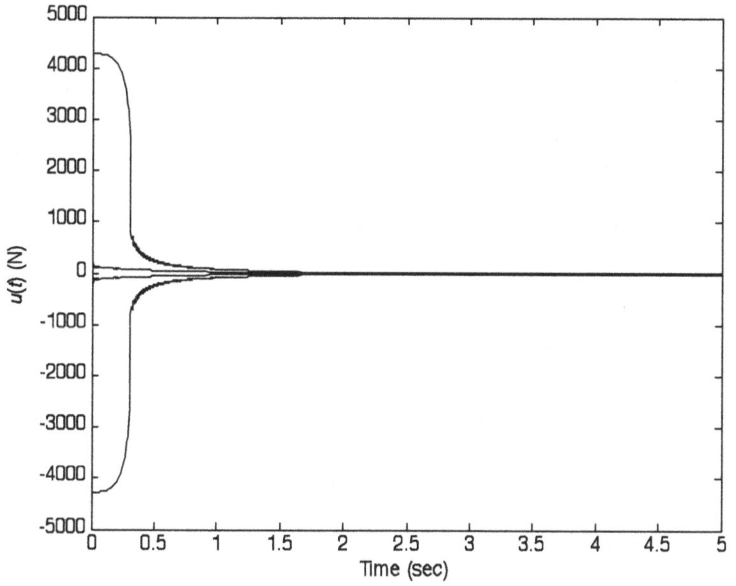

(c) $u(t)$.

Fig. 7.8 (*continued*)

it can be seen that the chattering effect happens during the transient state when the global switching controller dominates the control process. When the system states are near the origin, the chattering effect reduces and eventually vanishes as the local fuzzy controller dominates the control process.

As discussed in Remark 7.8, the saturated function can be employed to replace the switching function (7.31) to alleviate the chattering effect. By choosing $T = 1$, the fuzzy combined controller with the switching law (7.31) using the saturation function is employed to stabilize the inverted pendulum. Under this setup, the system responses and control signal of the FMB control system for $m_p = 2kg$ and $M_c = 8kg$, and $m_p = 5kg$ and $M_c = 10kg$, under various initial state conditions are shown in Fig. 7.7 and Fig.7.8, respectively. It can be seen from these figures that the fuzzy combined controller with the saturation function can also stabilize the inverted pendulum successfully. Moreover, the chattering effect is significantly reduced and no steady-state error is found.

For comparison purposes, the traditional fuzzy controller (2.6) rewritten as follows is employed to control the inverted pendulum.

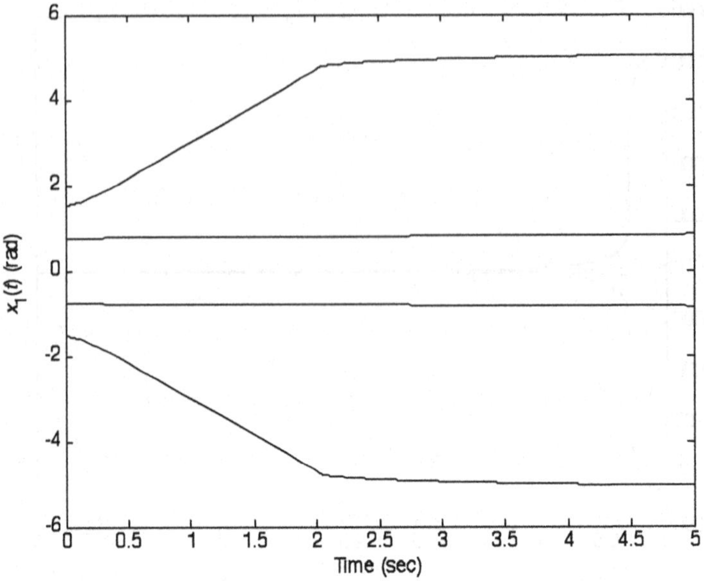

(a) $x_1(t)$.

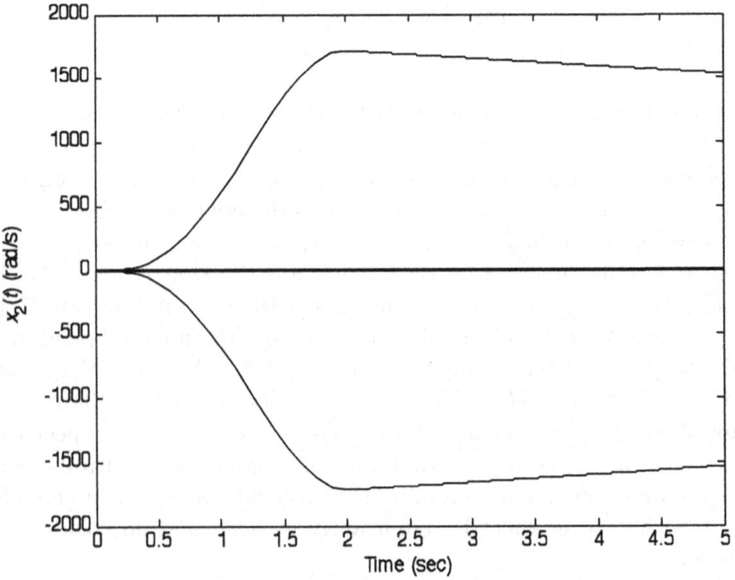

(b) $x_2(t)$.

Fig. 7.9 System responses with the traditional fuzzy controller (7.50) for $m_p = 2kg$ and $M_c = 8kg$.

$$u(t) = \sum_{j=1}^{4} \overline{m}_j(\mathbf{x}(t))\overline{\mathbf{G}}_j\mathbf{x}(t) \tag{7.50}$$

where the normalized membership functions are defined as $\overline{m}_i(\mathbf{x}(t)) = \frac{\mu_{V_1^i}(x_1(t))\times\mu_{V_2^i}(x_2(t))}{\sum_{l=1}^{4}(\mu_{V_1^l}(x_1(2))\times\mu_{V_2^l}(x_2(t)))}$ for all i with $\mu_{V_1^\beta}(x_1(t)) = \frac{-|x_1(t)|+\tilde{x}_{1_{max}}}{\tilde{x}_{1_{max}}-\tilde{x}_{1_{min}}}$ for $\beta = 1, 2$; $\mu_{V_1^\delta}(x_1(t)) = 1-\mu_{V_1^1}(x_1(t))$ for $\delta = 3, 4$; $\mu_{V_2^\kappa}(x_2(t)) = \frac{-|x_2(t)|+\tilde{x}_{2_{max}}}{\tilde{x}_{2_{max}}-\tilde{x}_{2_{min}}}$, $\kappa = 1, 3$; $\mu_{V_2^\phi}(x_2(t)) = 1-\mu_{V_2^1}(x_2(t))$ for $\phi = 2, 4$; $x_1(t) \in \left[x_{1_{min}}, x_{1_{max}}\right] = \left[-\frac{22\pi}{45}, \frac{22\pi}{45}\right]$ and $x_2(t) \in \left[x_{2_{min}}, x_{2_{max}}\right] = \left[-5, 5\right]$.

By applying the PDC design technique, and designing the feedback gains $\overline{\mathbf{G}}_j$ under the same design criterion as that of the global switching state-feedback controller, i.e. the eigenvalues of $\overline{\mathbf{H}}_{ii}$ are both assigned to be -2 for all i, we have $\overline{\mathbf{G}}_1 = \left[75.9474\ 22.6629\right]$, $\overline{\mathbf{G}}_2 = \left[3942.5647\ 1176.4735\right]$, $\overline{\mathbf{G}}_3 = \left[139.7140\ 22.6629\right]$ and $\overline{\mathbf{G}}_4 = \left[7252.8000\ 1176.4735\right]$. From the stability conditions in [122], the FMB control system is guaranteed to be asymptotically stable if there exists a solution $\overline{\mathbf{P}} > 0$ to the LMIs of $\overline{\mathbf{H}}_{ij}^T\overline{\mathbf{P}} + \overline{\mathbf{P}}\ \overline{\mathbf{H}}_{ij} < 0$, $i = 1, 2, 3, 4$, where $\overline{\mathbf{H}}_{ij} = \mathbf{A}_i + \mathbf{B}_i\overline{\mathbf{G}}_j$ for all i and j. By using the MATLAB LMI toolbox, it can be shown that no feasible solution of $\overline{\mathbf{P}}$ can be found. The system responses and control signal given by the traditional fuzzy controller (7.50) for $m_p = 2kg$ and $M_c = 8kg$ under various initial state conditions are shown in Fig. 7.9. Referring to this figure, it can be seen that the inverted pendulum cannot be stabilized by the traditional fuzzy controller (7.50).

It should be noted that in this example, the feedback gains are determined by the pole placement technique before applying the stability conditions in Theorem 7.1. By using the stability conditions in Theorem 7.2, the feedback gains can be obtained numerically with convex programming techniques. It can be shown that the fuzzy controller (2.6) with the feedback gains obtained by the stability conditions in Theorem 2.1 is able to stabilize the inverted pendulum.

7.5 Conclusion

A fuzzy combined model, which consists of the global and local fuzzy models, has been proposed to represent nonlinear plants subject to parameter uncertainties. A fuzzy combined controller, which consists of a global switching controller and a local fuzzy controller, has been designed based on the fuzzy combined model. The global switching controller is responsible for driving the system states towards the origin. When the FMB control system is operating near the origin, the local fuzzy controller takes over gradually to control the process. The contribution of both controllers to the control process is determined by the fuzzy combination rules. As the nonlinearity of the local fuzzy model is not so strong as compared with that of the global fuzzy model, the local fuzzy controller is easier to be obtained. Stability conditions in terms of LMIs have been derived based on the Lyapunov stability theory. Owing to the properties of the proposed fuzzy combined model and fuzzy combined controller, the cross terms of the FMB control system vanish such that the stability conditions contain only the system matrices of the local and global FMB systems individually. Furthermore, the chattering effect (because of the switching function) and the steady-state error introduced by the global switching controller (because of the saturation function) will gradually vanish when the system states are inside the valid operating domain of the local fuzzy model. A simulation example has been given to illustrate the design procedure and the merits of the proposed fuzzy combined control scheme.

Chapter 8
Time-Delay FMB Control Systems

8.1 Introduction

Time-delay nonlinear systems can be found in many real-life engineering processes. As the time delay is one of the sources to cause system instability, it is important to extend the FMB control techniques to this class of nonlinear systems to put the fuzzy controllers into practice.

To deal with the time-delay nonlinear systems using fuzzy control techniques, in general, two stability analysis approaches can be found in the literature, namely delay-independent and delay-dependent stability analysis approaches. Delay-independent stability conditions for time-delay FMB control systems were derived in [7, 8, 124] based on Lyapunov-Krasovskii or Lyapunov-Razumikhin based approaches.

In the delay-independent stability analysis approach, the time delay is not considered in the stability analysis and thus the stability conditions are not related to the time-delay information. As a result, once the time-delay FMB control system is guaranteed to be stable, it is stable for any value of time delay. Hence, delay-independent stability conditions are particularly useful for nonlinear systems subject to unknown or inestimable value of time delay.

In [10–12, 34, 75, 79, 113, 126, 129], delay-dependent stability conditions were derived based on the Lyapunov-Krasovskii based approach with the time-delay information is considered in the stability analysis. To facilitate the stability analysis, some inequalities have been proposed to approximate the upper bound of some terms related to the time delay. Other forms of inequalities were proposed in [31, 133] to serve the same purpose to reduce the conservativeness of the stability analysis. These inequalities have been employed in [10–12, 34, 75, 79, 113, 126, 129] to investigate the stability of time-delay FMB control systems. It was shown in [12, 75, 79, 126, 129] that some inequalities are able to achieve relaxed stability analysis results. Furthermore, by introducing some slack matrices, the stability conditions can be further relaxed. In the delay-dependent stability conditions, the time-delay information is one of the elements to determine the system stability. As

H.-K. Lam and F.H.F. Leung: FMB Control Systems, STUDFUZZ 264, pp. 151–171.
springerlink.com

a result, less conservative stability conditions can be achieved with time-delay information being considered. The delay-dependent stability conditions are good for time-delay FMB control systems with known or estimable values of time delays.

In general, both the delay-independent and delay-dependent stability analysis results have their own advantages for different classes of time-delay nonlinear systems.

8.2 Time-Delay Fuzzy Model and Fuzzy Controller

Let p be the number of fuzzy rules describing the time-delay nonlinear plant. The i-th rule is of the following format.

$$\begin{aligned}&\text{Rule } i\text{: IF } f_1(\mathbf{x}(t)) \text{ is } M_1^i \text{ AND } \cdots \text{ AND } f_\Psi(\mathbf{x}(t)) \text{ is } M_\Psi^i \\ &\quad \text{THEN } \dot{\mathbf{x}}(t) = \mathbf{A}_i\mathbf{x}(t) + \mathbf{A}_{di}\mathbf{x}(t-\tau_d) + \mathbf{B}_i\mathbf{u}(t)\end{aligned} \tag{8.1}$$

where $\mathbf{x}(t-\tau_d) \in \Re^n$ is the delayed system state vector; $\mathbf{A}_{di} \in \Re^{n\times n}$ are constant system matrices for all i and $\tau_d \geq 0$ denotes a constant time delay. The rest variables are defined in Section 2.2.

The system dynamics of the time-delay nonlinear plant is described as follows.

$$\begin{aligned}\dot{\mathbf{x}}(t) &= \sum_{i=1}^{p} w_i(\mathbf{x}(t))(\mathbf{A}_i\mathbf{x}(t) + \mathbf{A}_{di}\mathbf{x}(t-\tau_d) + \mathbf{B}_i\mathbf{u}(t)) \\ &= \sum_{i=1}^{p} w_i(\mathbf{x}(t)) \begin{bmatrix}\mathbf{A}_i & \mathbf{A}_{di} & \mathbf{B}_i\end{bmatrix} \begin{bmatrix}\mathbf{x}(t) \\ \mathbf{x}(t-\tau_d) \\ \mathbf{u}(t)\end{bmatrix}\end{aligned} \tag{8.2}$$

It is assumed that $\mathbf{x}(t) = \varphi(t)$ for $t \in \begin{bmatrix}-\tau_d & 0\end{bmatrix}$ where $\varphi(t)$ denotes the initial condition of $\mathbf{x}(t)$.

Based on the time-delay fuzzy model (8.2), the fuzzy controller in the form of (2.6) sharing the same premise membership functions of the time-delay fuzzy model is employed to perform the control process, i.e. drive the system states $\mathbf{x}(t)$ to the origin as time t tends to infinity.

8.2.1 Stability Analysis and Performance Design

In this section, the system stability of the time-delay FMB control systems formed by (8.2) and (2.6) connected in a closed loop is investigated. Two stability analysis approaches, namely delay-independent and delay-dependent approaches, are employed for the investigation of system stability. Based on the Lyapunov stability theory [46, 100, 121], LMI-based stability conditions are derived to guarantee the system stability.

For brevity, the normalized membership functions of $w_i(\mathbf{x}(t))$ is denoted as w_i in the following analysis. Furthermore, the property of $\sum_{i=1}^{p} w_i = \sum_{i=1}^{p}\sum_{j=1}^{p} w_i w_j = 1$ given by the membership functions is used.

8.2.2 Delay-Independent Approach

The system stability of the time-delay FMB system based on the time-delay independent approach is investigated. As its name tells, the stability analysis is carried out without considering the time delay. Hence, once the time-delay FMB control system is guaranteed to be stable, it is stable for any value of time delay. To investigate the system stability of the time-delay FMB control system, the following Lyapunov functional candidate is considered.

$$V_1(t) = \mathbf{x}(t)^T \mathbf{P}_1 \mathbf{x}(t) + \int_{t-\tau_d}^{t} \mathbf{x}(\varphi)^T \mathbf{S} \mathbf{x}(\varphi) d\varphi \tag{8.3}$$

where $0 < \mathbf{P}_1 = \mathbf{P}_1^T \in \Re^{n \times n}$ and $0 < \mathbf{S} = \mathbf{S}^T \in \Re^{n \times n}$. It can be seen that $V_1(t) > 0$ for all $\mathbf{x}(t)$.

In the following, it will be shown that $\dot{V}_1(t) < 0$ (for both $\mathbf{x}(t) \neq \mathbf{0}$ and $\mathbf{x}(t-\tau_d) \neq \mathbf{0}$) which implies the asymptotic stability of the time-delay FMB control system.

It follows from (8.2) and (8.3) that we have

$$\begin{aligned} \dot{V}(t) &= \mathbf{x}(t)^T \mathbf{P}_1 \dot{\mathbf{x}}(t) + \dot{\mathbf{x}}(t)^T \mathbf{P}_1 \mathbf{x}(t) + \mathbf{x}(t)^T \mathbf{S} \mathbf{x}(t) - \mathbf{x}(t-\tau_d)^T \mathbf{S} \mathbf{x}(t-\tau_d) \\ &= \sum_{i=1}^{p} w_i \begin{bmatrix} \mathbf{x}(t) \\ \mathbf{x}(t-\tau_d) \\ \mathbf{u}(t) \end{bmatrix}^T \left(\mathbf{P}^T \begin{bmatrix} \mathbf{A}_i & \mathbf{A}_{di} & \mathbf{B}_i \\ \mathbf{0} & \mathbf{0} & \mathbf{0} \\ \mathbf{0} & \mathbf{0} & \mathbf{0} \end{bmatrix} \right. \\ &\quad \left. + \begin{bmatrix} \mathbf{A}_i & \mathbf{A}_{di} & \mathbf{B}_i \\ \mathbf{0} & \mathbf{0} & \mathbf{0} \\ \mathbf{0} & \mathbf{0} & \mathbf{0} \end{bmatrix}^T \mathbf{P} + \begin{bmatrix} \mathbf{S} & \mathbf{0} & \mathbf{0} \\ \mathbf{0} & -\mathbf{S} & \mathbf{0} \\ \mathbf{0} & \mathbf{0} & \mathbf{0} \end{bmatrix} \right) \begin{bmatrix} \mathbf{x}(t) \\ \mathbf{x}(t-\tau_d) \\ \mathbf{u}(t) \end{bmatrix} \end{aligned} \tag{8.4}$$

where $\mathbf{P} = \begin{bmatrix} \mathbf{P}_1 & \mathbf{0} & \mathbf{0} \\ \mathbf{P}_2 & \mathbf{P}_3 & \mathbf{0} \\ \mathbf{P}_4 & \mathbf{P}_5 & \mathbf{P}_6 \end{bmatrix} \in \Re^{(2n+m)\times(2n+m)}$, $\mathbf{P}_2 \in \Re^{n \times n}$, $\mathbf{P}_3 \in \Re^{n \times n}$, $\mathbf{P}_4 \in \Re^{m \times n}$, $\mathbf{P}_5 \in \Re^{m \times n}$ and $\mathbf{P}_6 \in \Re^{m \times m}$.

From the fuzzy controller (2.6), it is obvious that we have $\sum_{i=1}^{p} w_i \mathbf{G}_j \mathbf{x}(t) - \mathbf{u}(t) = \mathbf{0}$. Thus, we have $\sum_{i=1}^{p} w_i \begin{bmatrix} \mathbf{0} & \mathbf{0} & \mathbf{0} \\ \mathbf{0} & \mathbf{0} & \mathbf{0} \\ \mathbf{G}_i & \mathbf{0} & -\mathbf{I} \end{bmatrix} \begin{bmatrix} \mathbf{x}(t) \\ \mathbf{x}(t-\tau_d) \\ \mathbf{u}(t) \end{bmatrix} = \begin{bmatrix} \mathbf{0} \\ \mathbf{0} \\ \mathbf{0} \end{bmatrix}$ which leads to the following property to facilitate the stability analysis.

$$\sum_{i=1}^{p} w_i \begin{bmatrix} \mathbf{x}(t) \\ \mathbf{x}(t-\tau_d) \\ \mathbf{u}(t) \end{bmatrix}^T \left(\mathbf{P}^T \begin{bmatrix} \mathbf{0} & \mathbf{0} & \mathbf{0} \\ \mathbf{0} & \mathbf{0} & \mathbf{0} \\ \mathbf{G}_i & \mathbf{0} & -\mathbf{I} \end{bmatrix} + \begin{bmatrix} \mathbf{0} & \mathbf{0} & \mathbf{0} \\ \mathbf{0} & \mathbf{0} & \mathbf{0} \\ \mathbf{G}_i & \mathbf{0} & -\mathbf{I} \end{bmatrix}^T \mathbf{P} \right) \begin{bmatrix} \mathbf{x}(t) \\ \mathbf{x}(t-\tau_d) \\ \mathbf{u}(t) \end{bmatrix} = \begin{bmatrix} 0 \\ 0 \\ 0 \end{bmatrix} \tag{8.5}$$

Adding (8.5) to (8.4), we have

$$\dot{V}(t) = \sum_{i=1}^{p} w_i \begin{bmatrix} \mathbf{x}(t) \\ \mathbf{x}(t-\tau_d) \\ \mathbf{u}(t) \end{bmatrix}^T \left(\mathbf{P}^T \begin{bmatrix} \mathbf{A}_i & \mathbf{A}_{di} & \mathbf{B}_i \\ \mathbf{0} & \mathbf{0} & \mathbf{0} \\ \mathbf{G}_i & \mathbf{0} & -\mathbf{I} \end{bmatrix} + \begin{bmatrix} \mathbf{A}_i & \mathbf{A}_{di} & \mathbf{B}_i \\ \mathbf{0} & \mathbf{0} & \mathbf{0} \\ \mathbf{G}_i & \mathbf{0} & -\mathbf{I} \end{bmatrix}^T \mathbf{P} + \begin{bmatrix} \mathbf{S} & \mathbf{0} & \mathbf{0} \\ \mathbf{0} & -\mathbf{S} & \mathbf{0} \\ \mathbf{0} & \mathbf{0} & \mathbf{0} \end{bmatrix} \right) \begin{bmatrix} \mathbf{x}(t) \\ \mathbf{x}(t-\tau_d) \\ \mathbf{u}(t) \end{bmatrix}. \tag{8.6}$$

We choose a particular form of $\mathbf{P}^{-1}$ for mathematical development of the LMI-based stability conditions in the following analysis. Denote $\mathbf{X} = \mathbf{P}^{-1} = \begin{bmatrix} \mathbf{X}_1 & \mathbf{0} & \mathbf{0} \\ \mathbf{X}_1 & \mathbf{X}_1 & \mathbf{0} \\ \mathbf{X}_2 & \mathbf{X}_3 & \mathbf{X}_4 \end{bmatrix}$, $\mathbf{X}_1 = \mathbf{X}_1^T = \mathbf{P}_1^{-1} \in \Re^{n \times n}$, $\mathbf{X}_2 \in \Re^{m \times n}$, $\mathbf{X}_3 \in \Re^{m \times n}$ and $\mathbf{X}_4 \in \Re^{m \times m}$.

Denote the feedback gains as $\mathbf{G}_i = \mathbf{N}_i \mathbf{X}_1^{-1}$ where $\mathbf{N}_i \in \Re^{m \times n}$ for all i, $\mathbf{Y} = \mathbf{X}_1 \mathbf{S} \mathbf{X}_1$ and $\mathbf{z}(t) = \mathbf{X}^{-1} \begin{bmatrix} \mathbf{x}(t) \\ \mathbf{x}(t-\tau_d) \\ \mathbf{u}(t) \end{bmatrix}$. From (8.6), we have

$$\begin{aligned} \dot{V}(t) &= \sum_{i=1}^{p} w_i \mathbf{z}(t)^T \mathbf{X}^T \left(\mathbf{P}^T \begin{bmatrix} \mathbf{A}_i & \mathbf{A}_{di} & \mathbf{B}_i \\ \mathbf{0} & \mathbf{0} & \mathbf{0} \\ \mathbf{G}_i & \mathbf{0} & -\mathbf{I} \end{bmatrix} + \begin{bmatrix} \mathbf{A}_i & \mathbf{A}_{di} & \mathbf{B}_i \\ \mathbf{0} & \mathbf{0} & \mathbf{0} \\ \mathbf{G}_i & \mathbf{0} & -\mathbf{I} \end{bmatrix}^T \mathbf{P} + \begin{bmatrix} \mathbf{S} & \mathbf{0} & \mathbf{0} \\ \mathbf{0} & -\mathbf{S} & \mathbf{0} \\ \mathbf{0} & \mathbf{0} & \mathbf{0} \end{bmatrix} \right) \mathbf{X}\mathbf{z}(t) \\ &= \sum_{i=1}^{p} w_i \mathbf{z}(t)^T \mathbf{Q}_i \mathbf{z}(t) \end{aligned} \tag{8.7}$$

where $\mathbf{Q}_i = \begin{bmatrix} \mathbf{Q}_1^{(11)} & * & * \\ \mathbf{X}_1 \mathbf{A}_{di}^T + \mathbf{X}_3^T \mathbf{B}_i^T - \mathbf{Y} & -\mathbf{Y} & * \\ \mathbf{N}_i - \mathbf{X}_2 + \mathbf{X}_4^T \mathbf{B}_i^T & -\mathbf{X}_3 & -\mathbf{X}_4 - \mathbf{X}_4^T \end{bmatrix}$, $\mathbf{Q}_1^{(11)} = (\mathbf{A}_i + \mathbf{A}_{di})\mathbf{X}_1 + \mathbf{X}_1(\mathbf{A}_i + \mathbf{A}_{di})^T + \mathbf{B}_i \mathbf{X}_2 + \mathbf{X}_2^T \mathbf{B}_i^T$, the symbol "$*$" denotes the transposed element in the corresponding position of the matrix.

From (8.3) and (8.7), based on the the Lyapunov stability theory, $V(t) > 0$ and $\dot{V}(t) < 0$ for $\mathbf{z}(t) \neq \mathbf{0}$ ($\mathbf{x}(t) \neq \mathbf{0}$ and $\mathbf{x}(t-\tau_d) \neq \mathbf{0}$) implying the asymptotic stability of the time-delay FMB control system, i.e., $\mathbf{x}(t) \to 0$ and $\mathbf{x}(t-\tau_d) \to 0$ when time $t \to \infty$, can be achieved if the stability conditions summarized in the following theorem are satisfied.

Theorem 8.1. *(Delay-Independent Approach): The time-delay FMB control system, formed by the nonlinear plant represented by the time-delay fuzzy model (8.2) and the fuzzy controller (2.6) with* $m_j(\mathbf{x}(t)) = w_j(\mathbf{x}(t))$ *and* $c = p$ *connected in a closed loop is asymptotically stable if there exist matrices* $\mathbf{N}_i \in \Re^{m\times n}$, $i = 1, 2, \cdots, p$, $\mathbf{X}_1 = \mathbf{X}_1^T \in \Re^{n\times n}$, $\mathbf{X}_2 \in \Re^{m\times n}$, $\mathbf{X}_3 \in \Re^{m\times n}$, $\mathbf{X}_4 \in \Re^{m\times m}$, $\mathbf{Y} = \mathbf{Y}^T \in \Re^{n\times n}$ *such that the following LMIs are satisfied.*

$$\mathbf{X}_1 > 0;$$

$$\mathbf{Y} > 0;$$

$$\mathbf{Q}_i < 0 \ \forall \ i;$$

and the feedback gains are designed as $\mathbf{G}_i = \mathbf{N}_i\mathbf{X}_1^{-1}$ *for all* i.

Remark 8.1. The number of LMI stability conditions are reduced to $p+2$ compared with that of the stability conditions in [7, 8, 10–12, 34, 75, 79, 113, 124, 126, 129]. Consequently, the computational demand on searching for the solution to the stability conditions can be alleviated. The computational advantage is obvious when the time-delay FMB control system has a large number of rules.

Remark 8.2. The stability conditions can be relaxed by considering the following modifications.

1. Choose $\mathbf{X} = \begin{bmatrix} \mathbf{X}_1 & \mathbf{0} & \mathbf{0} \\ \varepsilon_1\mathbf{X}_1 & \varepsilon_2\mathbf{X}_1 & \mathbf{0} \\ \mathbf{X}_2 & \mathbf{X}_3 & \mathbf{X}_4 \end{bmatrix}$ where ε_1 is a scalar and ε_2 is a nonzero positive scalar. The scalars of ε_1 and ε_2 offer higher degrees of freedom on searching for the solution. However, the stability conditions become BMIs which cannot be solved directly using convex programming techniques. The GA-convex programming technique introduced in Chapter 4 can be employed in this case to find the solution numerically.
2. Choose $\mathbf{X}_k = \sum_{i=1}^{p} w_i\mathbf{X}_{ki}$, $k = 2, 3, 4$. Consequently, more slack matrices are introduced to the stability conditions, which increase the degrees of freedom on searching for the solution of the stability conditions. However, the number of stability conditions will be increased.

Remark 8.3. The system performance of the time-delay FMB control systems can be realized by considering the performance index in (3.14), Section 3.2.2. As the fuzzy controller considered in this chapter is the same as the one in Chapter 3. The LMI-based performance conditions $\overline{\mathbf{W}}_j < 0$ for all j in Theorem 3.2 given in Section 3.2.2 can be used together with the stability conditions in Theorem 8.1 for the realization of a stable and well-performed time-delay FMB control system subject to the weighting matrices $\mathbf{J}_1$ and $\mathbf{J}_2$.

Remark 8.4. The stability analysis is based on the assumption that $\mathbf{P}$ is invertible. If there exists a solution to the stability conditions in Theorem 8.1, it implies that $\mathbf{X}_1 > 0$ and $-\mathbf{X}_4 - \mathbf{X}_4^T < 0$. As $\mathbf{X}_1$ and $\mathbf{X}_4$ are both non-singular matrices and they are the diagonal elements in $\mathbf{X}$, these are the sufficient conditions to ensure that $\mathbf{X}$ is a non-singular matrix. Hence, the existence of $\mathbf{X} = \mathbf{P}^{-1}$ is ensured.

Example 8.1. Consider a time-delay TS fuzzy model with the following rules.

Rule i: IF $x_1(t)$ is M_1^i

$$\text{THEN } \dot{\mathbf{x}}(t) = \mathbf{A}_i\mathbf{x}(t) + \mathbf{A}_{di}\mathbf{x}(t - \tau_d) + \mathbf{B}_i u(t), i = 1, 2 \tag{8.8}$$

where $\mathbf{A}_1 = \begin{bmatrix} -2.1 & 0.1 \\ -0.2 & -0.9 \end{bmatrix}$, $\mathbf{A}_2 = \begin{bmatrix} -1.9 & 0 \\ -0.2 & -1.1 \end{bmatrix}$; $\mathbf{A}_{d1} = \begin{bmatrix} -1.1 & 0.1 \\ -0.8 & -0.9 \end{bmatrix}$, $\mathbf{A}_{d2} = \begin{bmatrix} -0.9 & 0 \\ -1.1 & -1.2 \end{bmatrix}$; $\mathbf{B}_1 = \begin{bmatrix} 1 \\ 1 \end{bmatrix}$, $\mathbf{B}_2 = \begin{bmatrix} -1 \\ 2 \end{bmatrix}$. The membership functions are chosen as $w_1(x_1(t)) = \mu_{M_1^1}(x_1(t)) = \frac{1}{1+e^{-2x_1(t)}}$ and $w_2(x_1(t)) = \mu_{N_1^2}(x_1(t)) = 1 - w_1(x_1(t))$.

Considering the open-loop fuzzy system (i.e., setting $\mathbf{B}_i = \mathbf{0}$), the delay-independent stability conditions in Theorem 8.1 is employed to check for the system stability. It can be found that a feasible solution can be obtained with the MATLAB LMI toolbox. However, no feasible solution can be found for the stability conditions in [7, 133].

To illustrate the effectiveness of the LMI-based performance conditions in Theorem 3.2 presented in Section 3.2.2, we consider the stability conditions in Theorem 8.1 with the performance conditions $\overline{\mathbf{W}}_j < 0$ for all j given in Section 3.2.2. Considering $\eta = 10^{-10}$ and $\mathbf{J}_2 = 1$, the feedback gains with different weighting matrices, $\mathbf{J}_1$, for the performance conditions are given in Table 8.1. It can be seen that different $\mathbf{J}_1$ put different weights on the system states. Take $\mathbf{J}_1 = \begin{bmatrix} 100 & 0 \\ 0 & 1 \end{bmatrix}$ as an example and refer to the performance index (3.14), the weight of 100 is put on $x_1(t)$. The physical meaning is to suppress the integral of energy contributed by the system state $x_1(t)$ 100 times more than those contributed by the rest, i.e., $x_2(t)$.

The system state responses with $\tau_d = 1s$ and the control signals of the time-delay FMB control system under the initial state condition of $\mathbf{x}(t) = \begin{bmatrix} 1 & 0 \end{bmatrix}^T$ are shown in Fig. 8.1. The initial system state function is defined as $\varphi(t) = \begin{bmatrix} 1 & 0 \end{bmatrix}$ for $t \in \begin{bmatrix} -\tau_d, 0 \end{bmatrix}$. Referring to this figure, it can be seen that the fuzzy controllers with different sets of feedback gains are able to stabilize the time-delay nonlinear system. The fuzzy controller with $\mathbf{J}_1 = \begin{bmatrix} 100 & 0 \\ 0 & 1 \end{bmatrix}$ offers better system response on $x_1(t)$ in terms of rising time and overshoot/undershoot magnitude as the heaviest weight is put on $x_1(t)$ in $\mathbf{J}_1$. For the feedback gains corresponding to $\mathbf{J}_1 = \begin{bmatrix} 1 & 0 \\ 0 & 100 \end{bmatrix}$, as the heaviest weight is put on $x_2(t)$ in $\mathbf{J}_1$,

Table 8.1 Feedback gains under delay-dependent stability approach with different $\mathbf{J}_1$ for Example 8.1.

$\mathbf{J}_1$	Feedback Gains
$\mathbf{J}_1 = \begin{bmatrix} 1 & 0 \\ 0 & 1 \end{bmatrix}$	$\mathbf{G}_1 = \begin{bmatrix} 1.0349 & -10.0025 \end{bmatrix}$, $\mathbf{G}_2 = \begin{bmatrix} 0.9584 & -8.9833 \end{bmatrix}$
$\mathbf{J}_1 = \begin{bmatrix} 100 & 0 \\ 0 & 1 \end{bmatrix}$	$\mathbf{G}_1 = \begin{bmatrix} 2.4915 & -38.8894 \end{bmatrix}$, $\mathbf{G}_2 = \begin{bmatrix} 9.2812 & -52.5048 \end{bmatrix}$
$\mathbf{J}_1 = \begin{bmatrix} 1 & 0 \\ 0 & 100 \end{bmatrix}$	$\mathbf{G}_1 = \begin{bmatrix} 0.4619 & -19.9519 \end{bmatrix}$, $\mathbf{G}_2 = \begin{bmatrix} 0.3537 & -18.7751 \end{bmatrix}$

this fuzzy controller offers better system response on $x_2(t)$ in terms of rising time and overshoot/undershoot magnitude.

To test the delay-independent stability conditions in Theorem 8.1, we choose a larger time delay, i.e., $\tau_d = 5s$ for the simulation example. The system state responses with $\tau_d = 5s$ and the control signals of the time-delay FMB control system under the same initial conditions are shown in Fig. 8.2. It can be seen from the figures that the time-delay FMB control system is stable. Based on the delay-independent stability conditions in Theorem 8.1, the time-delay FMB control system is asymptotically stable for any values of time delay.

8.3 Delay-Dependent Approach

In this section, the system stability of the time-delay FMB control systems based on the delay-dependent approach is investigated that the time delay is taken into consideration in stability analysis. As a result, the system stability is related to the value of the time delay, which is brought to the stability conditions. Compared to the delay-independent analysis approach in the previous section, the time delay provides additional information for the stability analysis. Consequently, a more relaxed stability analysis result can be achieved.

The following Lyapunov functional candidate [75] is considered to investigate the system stability.

$$V(t) = V_1(t) + V_2(t) \tag{8.9}$$

where $V_1(t)$ is defined in (8.3) and

$$V_2(t) = \int_{-h_d}^{0} \int_{t+\sigma}^{t} \dot{\mathbf{x}}(\varphi)^T \mathbf{R} \dot{\mathbf{x}}(\varphi) d\varphi d\sigma \tag{8.10}$$

Denote the upper bound of τ_d as h_d, i.e., $\tau_d \leq h_d$. From (8.2), (8.10), and using the property that $\mathbf{R} > 0$ and $(\mathbf{a}_i - \mathbf{a}_j)^T \mathbf{R} (\mathbf{a}_i - \mathbf{a}_j) \geq 0 \Longrightarrow \mathbf{a}_i^T \mathbf{R} \mathbf{a}_i + \mathbf{a}_j^T \mathbf{R} \mathbf{a}_j \geq$

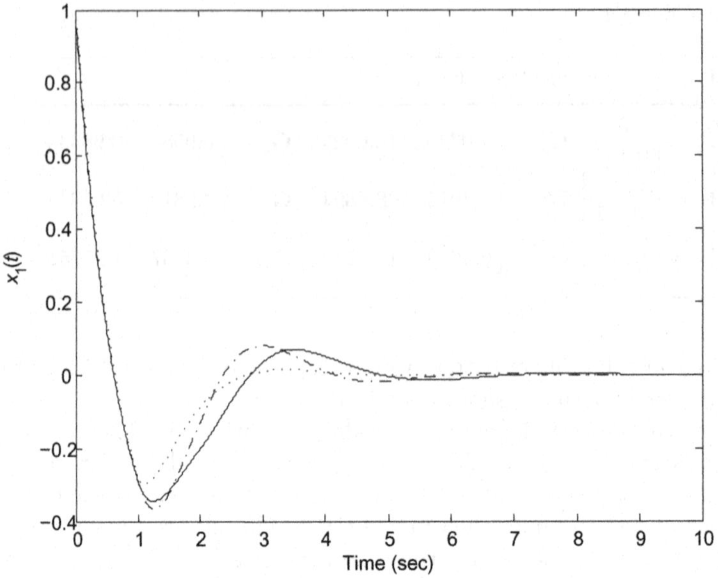

(a) $x_1(t)$.

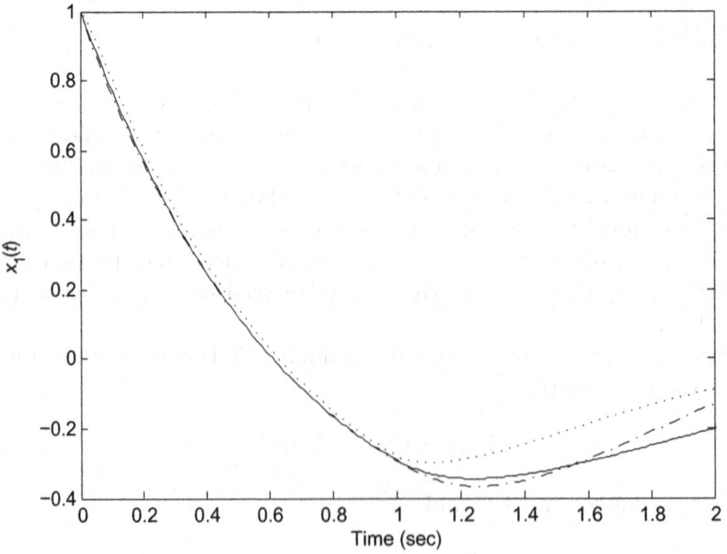

(b) $x_1(t)$ for $0s \leq t \leq 2s$.

Fig. 8.1 System state responses and control signals of Example 8.1 with $\tau_d = 1s$ under the fuzzy controller with $\mathbf{J}_1 = \begin{bmatrix} 1 & 0 \\ 0 & 1 \end{bmatrix}$ (solid lines), $\mathbf{J}_1 = \begin{bmatrix} 100 & 0 \\ 0 & 1 \end{bmatrix}$ (dotted lines) and $\mathbf{J}_1 = \begin{bmatrix} 1 & 0 \\ 0 & 100 \end{bmatrix}$ (dash-dot lines).

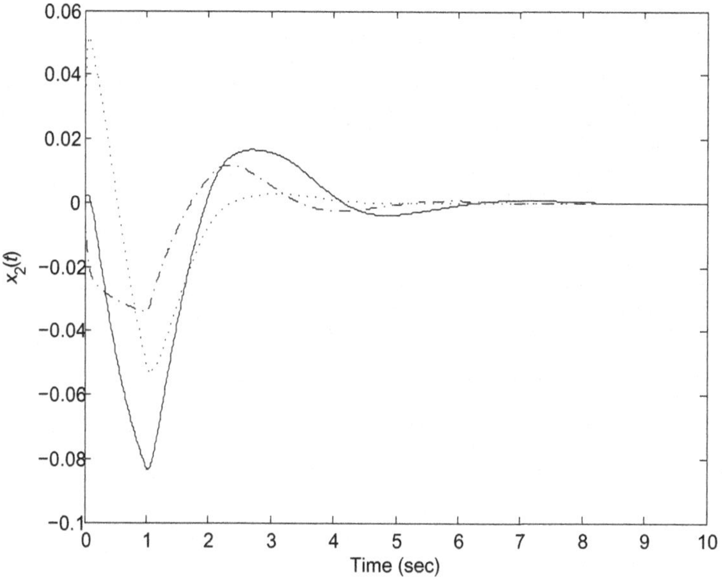

(c) $x_2(t)$.

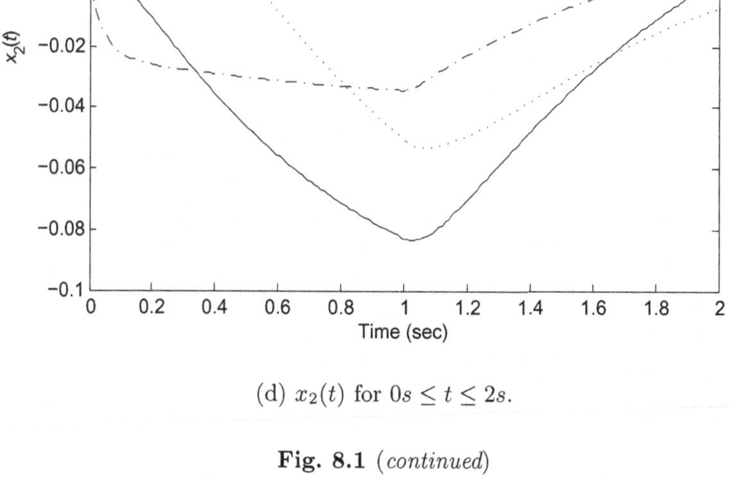

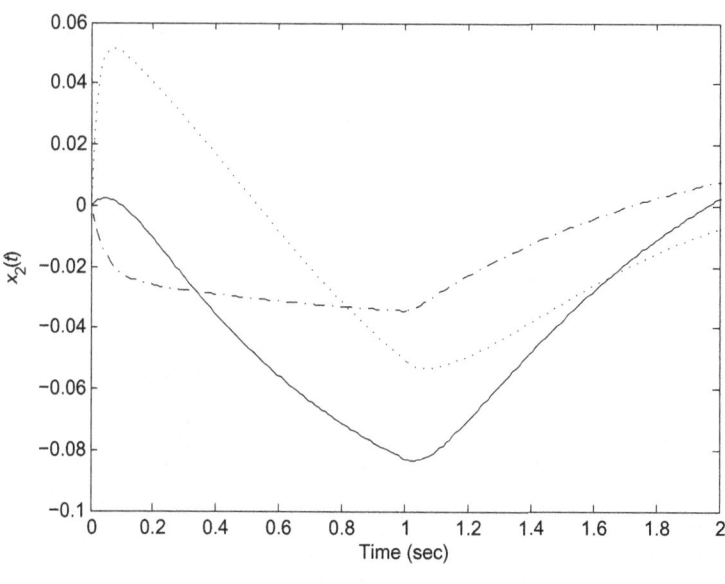

(d) $x_2(t)$ for $0s \leq t \leq 2s$.

Fig. 8.1 *(continued)*

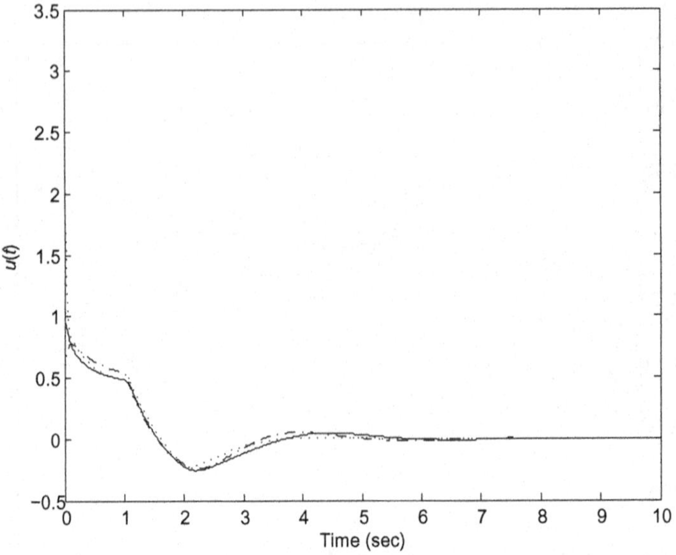

(e) $u(t)$.

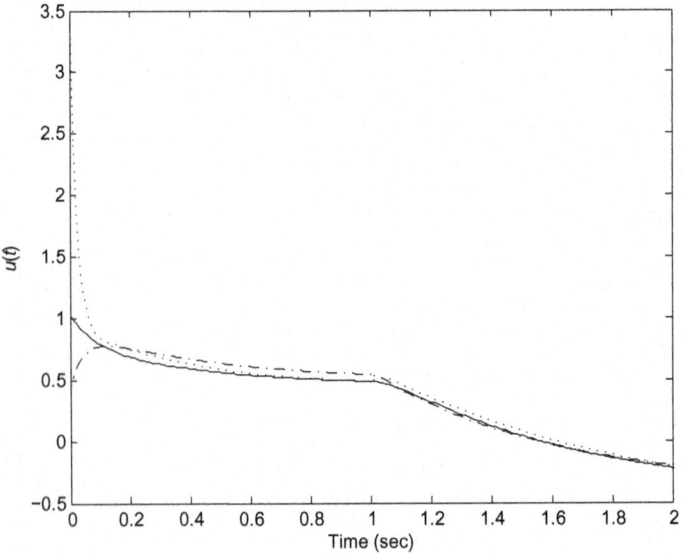

(f) $u(t)$ for $0s \leq t \leq 2s$.

Fig. 8.1 (*continued*)

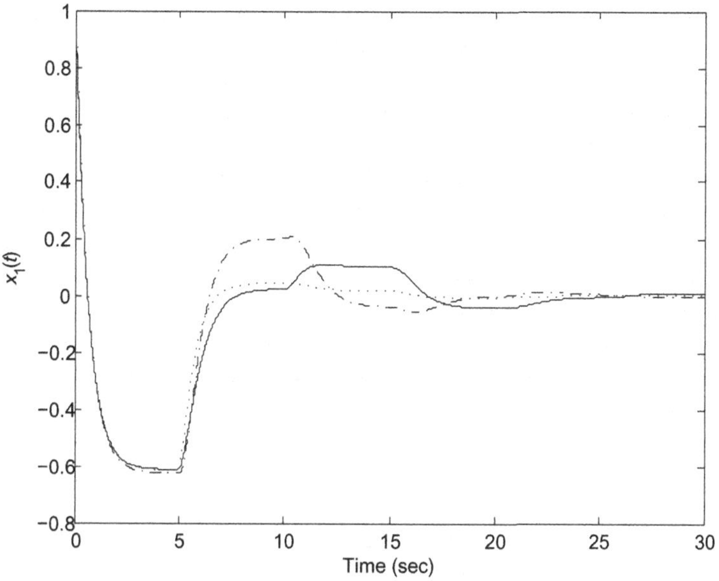

(a) $x_1(t)$.

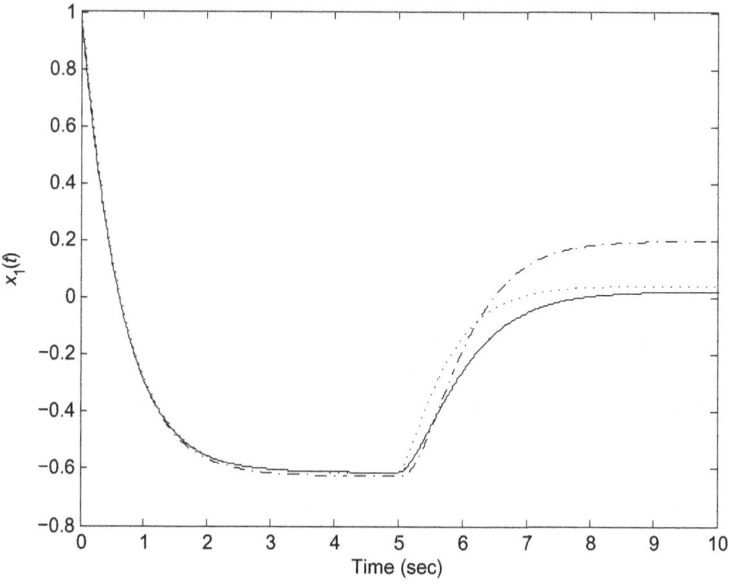

(b) $x_1(t)$ for $0s \le t \le 10s$.

Fig. 8.2 System state responses and control signals of Example 8.1 with $\tau_d = 5s$ under the fuzzy controller with $\mathbf{J}_1 = \begin{bmatrix} 1 & 0 \\ 0 & 1 \end{bmatrix}$ (solid lines), $\mathbf{J}_1 = \begin{bmatrix} 100 & 0 \\ 0 & 1 \end{bmatrix}$ (dotted lines) and $\mathbf{J}_1 = \begin{bmatrix} 1 & 0 \\ 0 & 100 \end{bmatrix}$ (dash-dot lines).

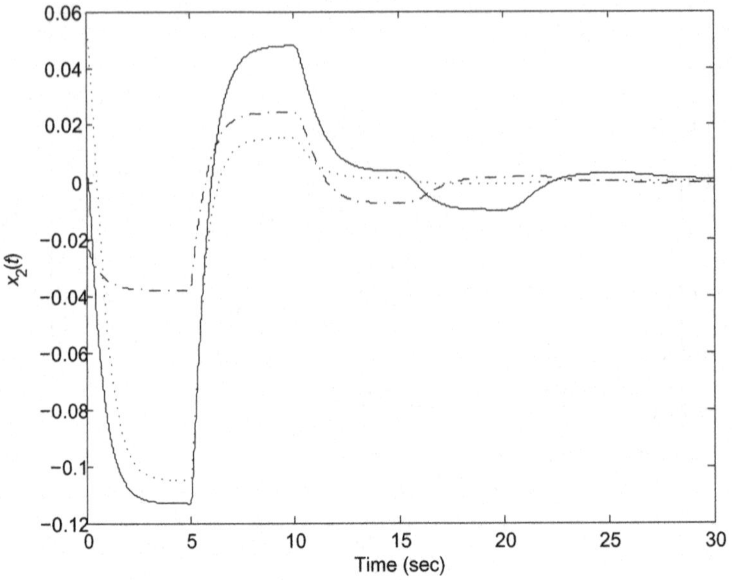

(c) $x_2(t)$.

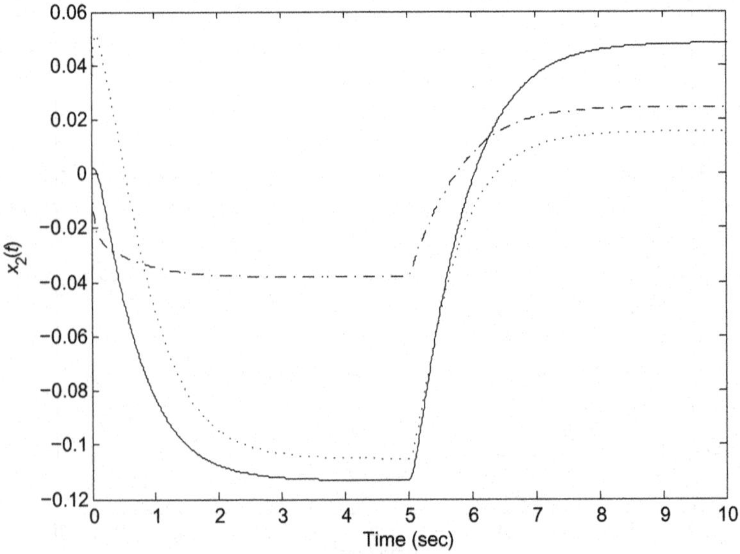

(d) $x_2(t)$ for $0s \leq t \leq 10s$.

Fig. 8.2 (*continued*)

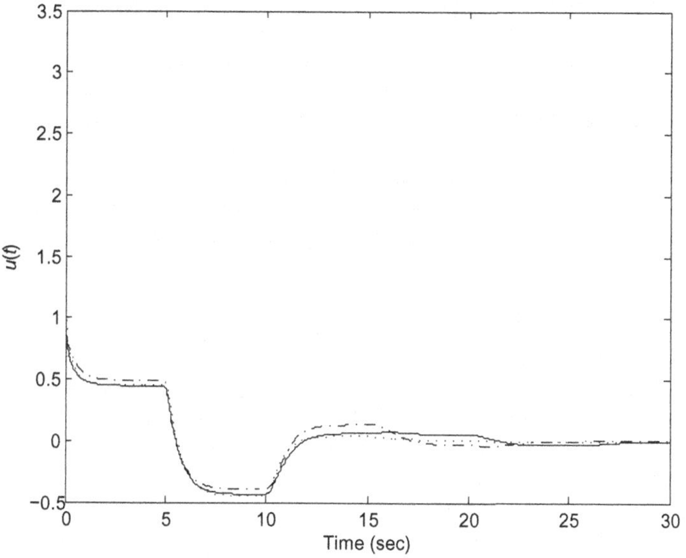

(e) $u(t)$.

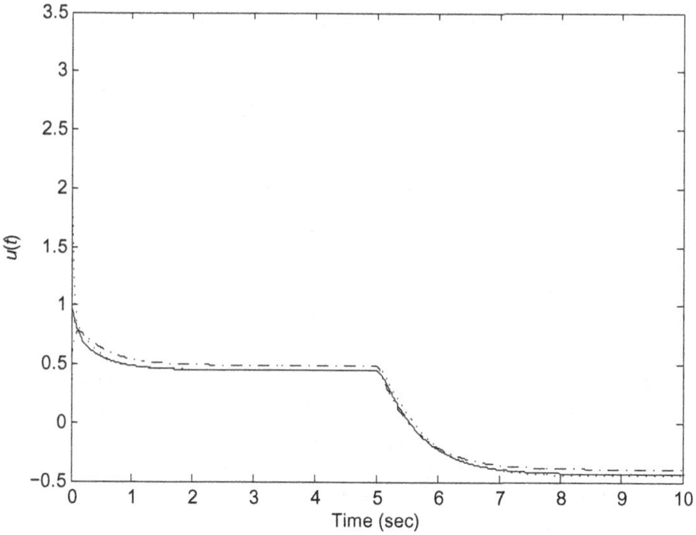

(f) $u(t)$ for $0s \leq t \leq 10s$.

Fig. 8.2 (*continued*)

$\mathbf{a}_i^T\mathbf{R}\mathbf{a}_j + \mathbf{a}_j^T\mathbf{R}\mathbf{a}_i$ where $\mathbf{a}_i \in \Re^n$, $i, j = 1, 2, \cdots, p$, are arbitrary vectors, the time derivative of $V_2(t)$ is obtained as follows.

$$\begin{aligned}
\dot{V}_2(t) &= h_d\dot{\mathbf{x}}(t)^T\mathbf{R}\dot{\mathbf{x}}(t) - \int_{t-h_d}^{t} \dot{\mathbf{x}}(\varphi)^T\mathbf{R}\dot{\mathbf{x}}(\varphi)d\varphi \\
&= h_d \sum_{i=1}^{p}\sum_{j=1}^{p} w_i w_j \begin{bmatrix} \mathbf{x}(t) \\ \mathbf{x}(t-\tau_d) \\ \mathbf{u}(t) \end{bmatrix}^T \begin{bmatrix} \mathbf{A}_i^T \\ \mathbf{A}_{di}^T \\ \mathbf{B}_i^T \end{bmatrix} \mathbf{R} \begin{bmatrix} \mathbf{A}_j^T \\ \mathbf{A}_{dj}^T \\ \mathbf{B}_j^T \end{bmatrix}^T \begin{bmatrix} \mathbf{x}(t) \\ \mathbf{x}(t-\tau_d) \\ \mathbf{u}(t) \end{bmatrix} \\
&\quad - \int_{t-h_d}^{t} \dot{\mathbf{x}}(\varphi)^T\mathbf{R}\dot{\mathbf{x}}(\varphi)d\varphi \\
&\le h_d \sum_{i=1}^{p} w_i \begin{bmatrix} \mathbf{x}(t) \\ \mathbf{x}(t-\tau_d) \\ \mathbf{u}(t) \end{bmatrix}^T \begin{bmatrix} \mathbf{A}_i^T \\ \mathbf{A}_{di}^T \\ \mathbf{B}_i^T \end{bmatrix} \mathbf{R} \begin{bmatrix} \mathbf{A}_i^T \\ \mathbf{A}_{di}^T \\ \mathbf{B}_i^T \end{bmatrix}^T \begin{bmatrix} \mathbf{x}(t) \\ \mathbf{x}(t-\tau_d) \\ \mathbf{u}(t) \end{bmatrix} \\
&\quad - \int_{t-h_d}^{t} \dot{\mathbf{x}}(\varphi)^T\mathbf{R}\dot{\mathbf{x}}(\varphi)d\varphi
\end{aligned} \tag{8.11}$$

Before proceeding further, the following Lemma is introduced to facilitate the stability analysis.

Lemma 8.1. *[133] The following integral inequality holds for any arbitrary matrices $\mathbf{T}_1 \in \Re^{n\times n}$, $\mathbf{T}_2 \in \Re^{n\times n}$, $0 < \mathbf{R} = \mathbf{R}^T \in \Re^{n\times n}$ and a scalar $h_d \ge 0$.*

$$\begin{aligned}
- \int_{t-h_d}^{t} \dot{\mathbf{x}}(\varphi)^T\mathbf{R}\dot{\mathbf{x}}(\varphi)d\varphi &\le \begin{bmatrix} \mathbf{x}(t) \\ \mathbf{x}(t-h_d) \end{bmatrix}^T \begin{bmatrix} \mathbf{T}_1 + \mathbf{T}_1^T & -\mathbf{T}_1^T + \mathbf{T}_2 \\ -\mathbf{T}_1 + \mathbf{T}_2^T & -\mathbf{T}_2 - \mathbf{T}_2^T \end{bmatrix} \begin{bmatrix} \mathbf{x}(t) \\ \mathbf{x}(t-h_d) \end{bmatrix} \\
&\quad + h_d \begin{bmatrix} \mathbf{x}(t) \\ \mathbf{x}(t-h_d) \end{bmatrix}^T \begin{bmatrix} \mathbf{T}_1^T \\ \mathbf{T}_2^T \end{bmatrix} \mathbf{R}^{-1} \begin{bmatrix} \mathbf{T}_1^T \\ \mathbf{T}_2^T \end{bmatrix}^T \begin{bmatrix} \mathbf{x}(t) \\ \mathbf{x}(t-h_d) \end{bmatrix}
\end{aligned}$$

where $\mathbf{x}(t) \in \Re^n$ with continuous first derivative.

From (8.11) and Lemma 8.1, we have

$$\begin{aligned}
\dot{V}_2(t) &\le \sum_{i=1}^{p} w_i \begin{bmatrix} \mathbf{x}(t) \\ \mathbf{x}(t-\tau_d) \\ \mathbf{u}(t) \end{bmatrix}^T \left(h_d \begin{bmatrix} \mathbf{A}_i^T \\ \mathbf{A}_{di}^T \\ \mathbf{B}_i^T \end{bmatrix} \mathbf{R} \begin{bmatrix} \mathbf{A}_i^T \\ \mathbf{A}_{di}^T \\ \mathbf{B}_i^T \end{bmatrix}^T + h_d \begin{bmatrix} \mathbf{T}_{1i}^T \\ \mathbf{T}_{2i}^T \\ \mathbf{0} \end{bmatrix} \mathbf{R}^{-1} \begin{bmatrix} \mathbf{T}_{1i}^T \\ \mathbf{T}_{2i}^T \\ \mathbf{0} \end{bmatrix}^T \right. \\
&\quad \left. + \begin{bmatrix} \mathbf{T}_{1i} + \mathbf{T}_{1i}^T & -\mathbf{T}_{1i}^T + \mathbf{T}_{2i} & \mathbf{0} \\ -\mathbf{T}_{1i} + \mathbf{T}_{2i}^T & -\mathbf{T}_{2i} - \mathbf{T}_{2i}^T & \mathbf{0} \\ \mathbf{0} & \mathbf{0} & \mathbf{0} \end{bmatrix} \right) \begin{bmatrix} \mathbf{x}(t) \\ \mathbf{x}(t-\tau_d) \\ \mathbf{u}(t) \end{bmatrix}
\end{aligned} \tag{8.12}$$

where $\mathbf{T}_{1i} \in \Re^{n\times n}$ and $\mathbf{T}_{2i} \in \Re^{n\times n}$ for all i.

From (8.6) and (8.12), recalling that $\mathbf{z}(t) = \mathbf{X}^{-1} \begin{bmatrix} \mathbf{x}(t) \\ \mathbf{x}(t-\tau_d) \\ \mathbf{u}(t) \end{bmatrix}$, (8.9) can be written as the following compact form.

$$\dot{V}(t) \le \sum_{i=1}^{p} w_i \begin{bmatrix} \mathbf{x}(t) \\ \mathbf{x}(t-\tau_d) \\ \mathbf{u}(t) \end{bmatrix}^T \overline{\mathbf{Q}}_i \begin{bmatrix} \mathbf{x}(t) \\ \mathbf{x}(t-\tau_d) \\ \mathbf{u}(t) \end{bmatrix}$$
$$= \sum_{i=1}^{p} w_i \mathbf{z}(t)^T \mathbf{X}^T \overline{\mathbf{Q}}_i \mathbf{X}\mathbf{z}(t) \tag{8.13}$$

where

$$\overline{\mathbf{Q}}_i = \mathbf{P}^T \begin{bmatrix} \mathbf{A}_i & \mathbf{A}_{di} & \mathbf{B}_i \\ \mathbf{0} & \mathbf{0} & \mathbf{0} \\ \mathbf{G}_i & \mathbf{0} & -\mathbf{I} \end{bmatrix} + \begin{bmatrix} \mathbf{A}_i & \mathbf{A}_{di} & \mathbf{B}_i \\ \mathbf{0} & \mathbf{0} & \mathbf{0} \\ \mathbf{G}_i & \mathbf{0} & -\mathbf{I} \end{bmatrix}^T \mathbf{P} + \begin{bmatrix} \mathbf{S} & \mathbf{0} & \mathbf{0} \\ \mathbf{0} & -\mathbf{S} & \mathbf{0} \\ \mathbf{0} & \mathbf{0} & \mathbf{0} \end{bmatrix}$$
$$+ h_d \begin{bmatrix} \mathbf{A}_i^T \\ \mathbf{A}_{di}^T \\ \mathbf{B}_i^T \end{bmatrix} \mathbf{R} \begin{bmatrix} \mathbf{A}_i^T \\ \mathbf{A}_{di}^T \\ \mathbf{B}_i^T \end{bmatrix}^T + h_d \begin{bmatrix} \mathbf{T}_{1i}^T \\ \mathbf{T}_{2i}^T \\ \mathbf{0} \end{bmatrix} \mathbf{R}^{-1} \begin{bmatrix} \mathbf{T}_{1i}^T \\ \mathbf{T}_{2i}^T \\ \mathbf{0} \end{bmatrix}^T$$
$$+ \begin{bmatrix} \mathbf{T}_{1i} + \mathbf{T}_{1i}^T & -\mathbf{T}_{1i}^T + \mathbf{T}_{2i} & \mathbf{0} \\ -\mathbf{T}_{1i} + \mathbf{T}_{2i}^T & -\mathbf{T}_{2i} - \mathbf{T}_{2i}^T & \mathbf{0} \\ \mathbf{0} & \mathbf{0} & \mathbf{0} \end{bmatrix}. \tag{8.14}$$

It can be seen that $\dot{V}(t) < 0$ when the stability conditions $\mathbf{X}^T\overline{\mathbf{Q}}_i\mathbf{X}^T < 0$ for all i, which imply the asymptotic stability of the time-delay FMB control system. In order to turn the stability conditions into LMIs, we consider $\mathbf{M} = \mathbf{R}^{-1} \in \Re^{n\times n}$, $\mathbf{G}_i = \mathbf{N}_i\mathbf{X}_1^{-1}$, $\mathbf{N}_i \in \Re^{m\times n}$, $\mathbf{U}_{1i} = \mathbf{X}_1^T\mathbf{T}_{1i}\mathbf{X}_1 \in \Re^{n\times n}$, $\mathbf{U}_{2i} = \mathbf{X}_1^T\mathbf{T}_{2i}\mathbf{X}_1 \in \Re^{n\times n}$ and $\mathbf{Y} = \mathbf{X}_1\mathbf{S}\mathbf{X}_1$. Furthermore, with the property $\mathbf{R} = \mathbf{R}^T > 0$, we consider the inequality $(\mathbf{X}_1 - \zeta\mathbf{R}^{-1})^T\mathbf{R}(\mathbf{X}_1 - \zeta\mathbf{R}^{-1}) = \mathbf{X}_1^T\mathbf{R}\mathbf{X}_1 - \zeta\mathbf{X}_1^T - \zeta\mathbf{X}_1 + \zeta^2\mathbf{R}^{-1} \ge 0$ that leads to

$$\mathbf{X}_1^T\mathbf{R}\mathbf{X}_1 \ge \zeta\mathbf{X}_1^T + \zeta\mathbf{X}_1 - \zeta^2\mathbf{R}^{-1} = 2\zeta\mathbf{X}_1 - \zeta^2\mathbf{R}^{-1} \tag{8.15}$$

where ζ is an arbitrary scalar.

By Schur complement and with (8.15), $\mathbf{X}^T\overline{\mathbf{Q}}_i\mathbf{X} < 0$ for all i are implied by the following inequalities.

$$\mathbf{\Theta}_i = \begin{bmatrix} \mathbf{\Theta}_i^{(11)} & * & * & * & * \\ \mathbf{\Theta}_i^{(21)} & -\mathbf{Y} - \mathbf{U}_{2i} - \mathbf{U}_{2i}^T & * & * & * \\ \mathbf{\Theta}_i^{(31)} & -\mathbf{X}_3 & -\mathbf{X}_4 - \mathbf{X}_4^T & * & * \\ \mathbf{\Theta}_i^{(41)} & h_d(\mathbf{A}_{di}\mathbf{X}_1 + \mathbf{B}_i\mathbf{X}_3) & h_d\mathbf{B}_i\mathbf{X}_4 & -h_d\mathbf{M} & * \\ h_d(\mathbf{U}_{1i} + \mathbf{U}_{2i}) & h_d\mathbf{U}_{2i} & \mathbf{0} & \mathbf{0} & \mathbf{\Theta}_i^{(55)} \end{bmatrix} < 0 \tag{8.16}$$

where $\mathbf{\Theta}_i^{(11)} = (\mathbf{A}_i + \mathbf{A}_{di})\mathbf{X}_1 + \mathbf{X}_1(\mathbf{A}_i + \mathbf{A}_{di})^T + \mathbf{B}_i\mathbf{X}_2 + \mathbf{X}_2^T\mathbf{B}_i^T$, $\mathbf{\Theta}_i^{(21)} = \mathbf{X}_1\mathbf{A}_{di}^T + \mathbf{X}_3^T\mathbf{B}_i - \mathbf{Y} - \mathbf{U}_{1i} - \mathbf{U}_{2i}$, $\mathbf{\Theta}_i^{(31)} = \mathbf{N}_i - \mathbf{X}_2 + \mathbf{X}_4^T\mathbf{B}_i^T$, $\mathbf{\Theta}_i^{(41)} = h_d((\mathbf{A}_i + \mathbf{A}_{di})\mathbf{X}_1 + \mathbf{B}_i\mathbf{X}_2)$, $\mathbf{\Theta}_i^{(55)} = -h_d(2\zeta\mathbf{X}_1 - \zeta^2\mathbf{M})$.

Table 8.2 Upper bounds of time delay given by various stability conditions for Example 8.2.

Stability Conditions	Maximum time delay h_d (Sec.)
Theorem 8.2	2.3767
[10]	1.2246
[34]	1.0124
[11]	0.7171
[12]	0.0343

From (8.9) and (8.13), based on the the Lyapunov stability theory, $V(t) > 0$ and $\dot{V}(t) < 0$ for $\mathbf{z}(t) \neq \mathbf{0}$ ($\mathbf{x}(t) \neq \mathbf{0}$ and $\mathbf{x}(t-\tau_d) \neq \mathbf{0}$) implying the asymptotic stability of the time-delay FMB control system, i.e., $\mathbf{x}(t) \to 0$ and $\mathbf{x}(t-\tau_d) \to 0$ when time $t \to \infty$, can be achieved if the stability conditions summarized in the following theorem are satisfied.

Theorem 8.2. *(Delay-Dependent Approach): The time-delay FMB control system, formed by the nonlinear plant represented by the time-delay fuzzy model (8.2) and the fuzzy controller (2.6) with* $m_j(\mathbf{x}(t)) = w_j(\mathbf{x}(t))$ *and* $c = p$ *connected in a closed loop is asymptotically stable if there exist pre-defined scalars* $h_d > 0$ *and* ζ *and there exist matrices* $\mathbf{M} = \mathbf{M}^T \in \Re^{n\times n}$, $\mathbf{N}_i \in \Re^{m\times n}$, $\mathbf{U}_{1i} \in \Re^{n\times n}$, $\mathbf{U}_{2i} \in \Re^{n\times n}$, $i = 1, 2, \cdots, p$, $\mathbf{X}_1 = \mathbf{X}_1^T \in \Re^{n\times n}$, $\mathbf{X}_2 \in \Re^{m\times n}$, $\mathbf{X}_3 \in \Re^{m\times n}$, $\mathbf{X}_4 \in \Re^{m\times m}$, $\mathbf{Y} = \mathbf{Y}^T \in \Re^{n\times n}$ *such that the following LMIs are satisfied.*

$$\mathbf{X}_1 > 0;$$

$$\mathbf{M} > 0;$$

$$\mathbf{Y} > 0;$$

$$\mathbf{\Theta}_i < 0 \;\forall\, i;$$

and the feedback gains are designed as $\mathbf{G}_i = \mathbf{N}_j\mathbf{X}_1^{-1}$ *for all* i.

Table 8.3 Feedback gains under delay-dependent stability approach with different $\mathbf{J}_1$ for Example 8.2.

$\mathbf{J}_1$	Feedback Gains
$\mathbf{J}_1 = \begin{bmatrix} 1 & 0 \\ 0 & 1 \end{bmatrix}$	$\mathbf{G}_1 = \begin{bmatrix} -0.0728 & -0.9403 \end{bmatrix}$, $\mathbf{G}_2 = \begin{bmatrix} -0.0779 & -0.8023 \end{bmatrix}$
$\mathbf{J}_1 = \begin{bmatrix} 100 & 0 \\ 0 & 1 \end{bmatrix}$	$\mathbf{G}_1 = \begin{bmatrix} -0.4058 & -0.9449 \end{bmatrix}$, $\mathbf{G}_2 = \begin{bmatrix} -0.4883 & -0.8045 \end{bmatrix}$
$\mathbf{J}_1 = \begin{bmatrix} 1 & 0 \\ 0 & 100 \end{bmatrix}$	$\mathbf{G}_1 = \begin{bmatrix} -0.0392 & -9.9649 \end{bmatrix}$, $\mathbf{G}_2 = \begin{bmatrix} -0.0286 & -6.9614 \end{bmatrix}$

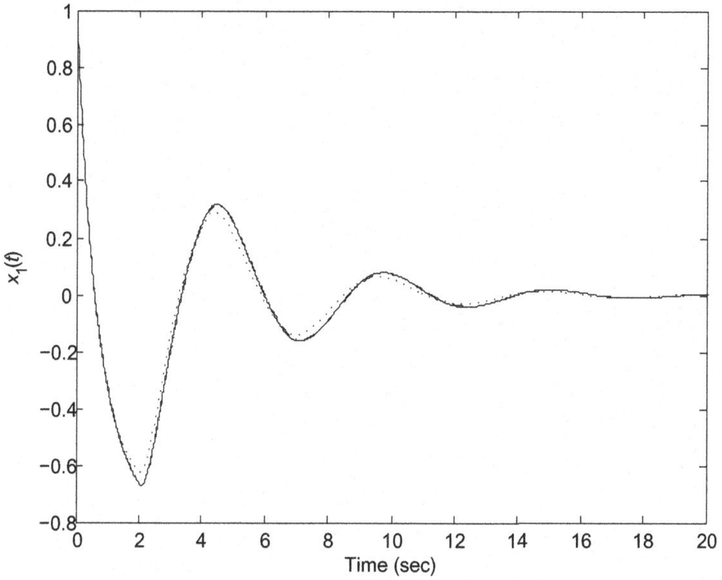

(a) $x_1(t)$.

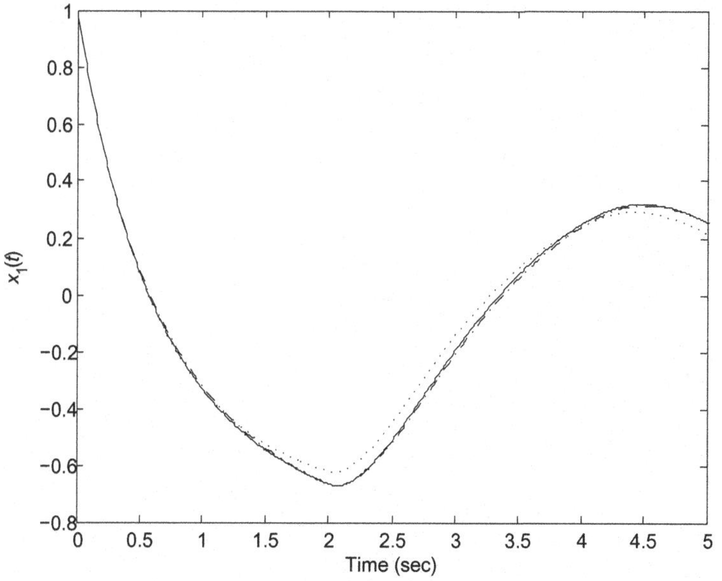

(b) $x_1(t)$ for $0s \leq t \leq 5s$.

Fig. 8.3 System state responses and control signals of Example 8.2 with $\tau_d = 2s$ under the fuzzy controller with $\mathbf{J}_1 = \begin{bmatrix} 1 & 0 \\ 0 & 1 \end{bmatrix}$ (solid lines), $\mathbf{J}_1 = \begin{bmatrix} 100 & 0 \\ 0 & 1 \end{bmatrix}$ (dotted lines) and $\mathbf{J}_1 = \begin{bmatrix} 1 & 0 \\ 0 & 100 \end{bmatrix}$ (dash-dot lines).

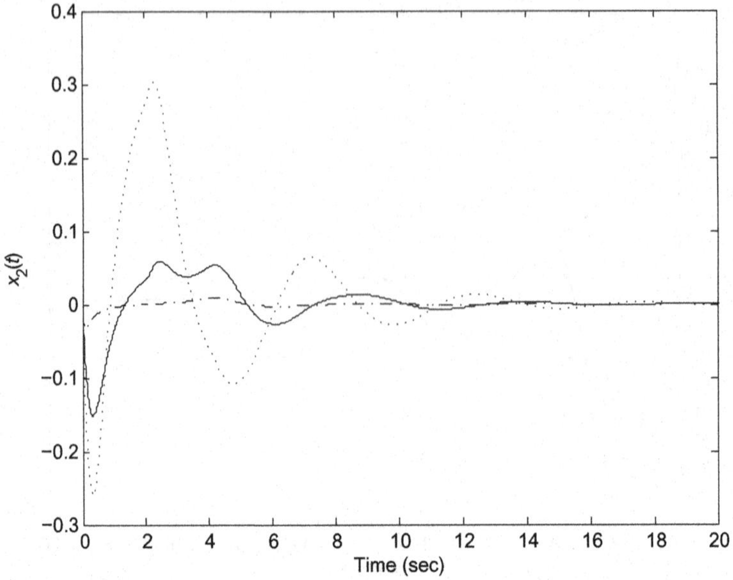

(c) $x_2(t)$.

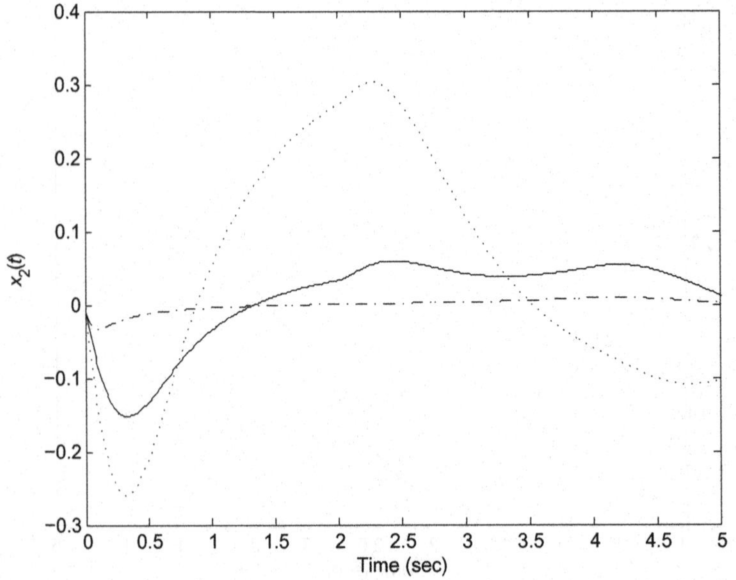

(d) $x_2(t)$ for $0s \leq t \leq 5s$.

Fig. 8.3 (*continued*)

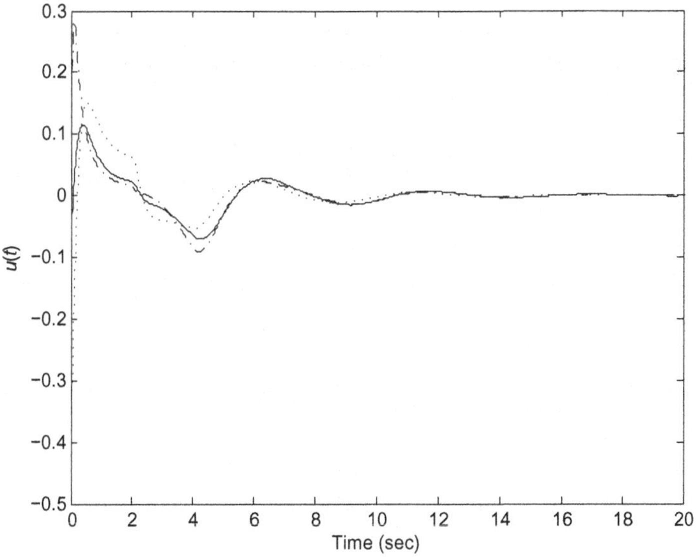

(e) $u(t)$.

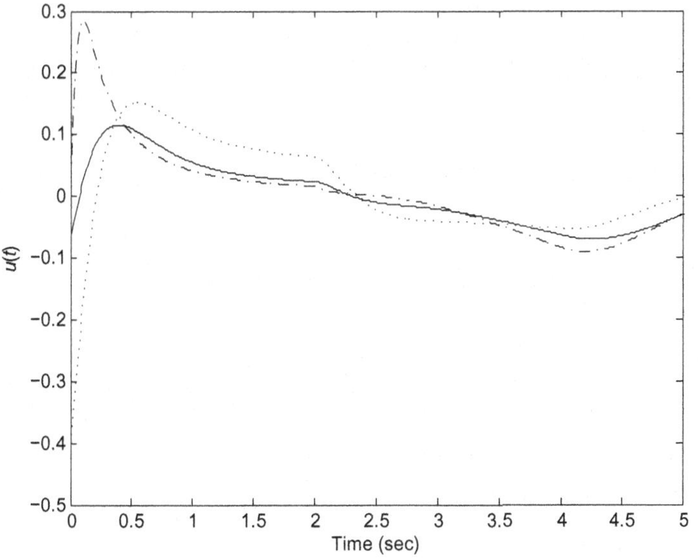

(f) $u(t)$ for $0s \leq t \leq 5s$.

Fig. 8.3 (*continued*)

Remark 8.5. The MFSD stability analysis approach introduced in Chapter 3 to Chapter 5 can be employed to both delay-dependent/independent stability analysis approaches to relax the stability analysis results by using the information of membership functions.

Example 8.2. Consider a fuzzy model with the following rules [113].

$$\begin{aligned}&\text{Rule } i\text{: IF } x_1(t) \text{ is } M_1^i\\ &\qquad \text{THEN } \dot{\mathbf{x}}(t) = \mathbf{A}_i\mathbf{x}(t) + \mathbf{A}_{di}\mathbf{x}(t-\tau_d) + \mathbf{B}_i u(t), i = 1, 2\end{aligned} \tag{8.17}$$

where $\mathbf{A}_1 = \begin{bmatrix} -2 & 0 \\ 0 & -0.9 \end{bmatrix}$, $\mathbf{A}_2 = \begin{bmatrix} -1 & 0.5 \\ 0 & -1 \end{bmatrix}$; $\mathbf{A}_{d1} = \begin{bmatrix} -1 & 0 \\ -1 & -1 \end{bmatrix}$, $\mathbf{A}_{d2} = \begin{bmatrix} -1 & 0 \\ -0.1 & -1 \end{bmatrix}$; $\mathbf{B}_1 = \begin{bmatrix} 0 \\ 1 \end{bmatrix}$, $\mathbf{B}_2 = \begin{bmatrix} 0 \\ 10 \end{bmatrix}$. The membership functions are chosen as the ones in Example 8.1.

Considering an opening-loop system by setting $\mathbf{B}_i = \mathbf{0}$, the upper bounds of time delay given by some existing approaches and Theorem 8.2 are listed in Table 8.2. It can be seen that Theorem 8.2 offers the largest upper bound of time delay, i.e., 2.3767s with $\zeta = 0.5$.

LMI-based stability conditions in Theorem 8.2 and performance conditions $\overline{\mathbf{W}}_j < 0$ for all j in Theorem 3.2 given in Section 3.2.2 are employed to obtain the feedback gains of the fuzzy controller for the time-delay fuzzy system (8.17). With the MATLAB LMI toolbox, choosing $\tau_d = 2s$, $\zeta = 0.5$, $\eta = 10^{-10}$, and $\mathbf{J}_2 = 1$ and corresponding to various weighting matrices $\mathbf{J}_1$, the feedback gains are tabulated in Table 8.3. Fig. 8.3 shows the system state responses with $\tau_d = 2s$ and the control signals of the time-delay FMB control systems under the initial state condition of $\mathbf{x}(t) = \begin{bmatrix} 1 & 0 \end{bmatrix}^T$. The initial system state function is defined as $\varphi(t) = \begin{bmatrix} 1 & 0 \end{bmatrix}$ for $t \in \begin{bmatrix} -\tau_d, 0 \end{bmatrix}$. Referring to this figure, it can be seen that the fuzzy controllers with different sets of feedback gains can stabilize the time-delay fuzzy system. Furthermore, the fuzzy controller with $\mathbf{J}_1 = \begin{bmatrix} 100 & 0 \\ 0 & 1 \end{bmatrix}$ offers better system response on $x_1(t)$ in terms of rising time and overshoot/undershoot magnitude as more weight is put on $x_1(t)$ in $\mathbf{J}_1$. For the fuzzy controller with $\mathbf{J}_1 = \begin{bmatrix} 1 & 0 \\ 0 & 100 \end{bmatrix}$, it can be seen that the overshoot/undershoot of $x_2(t)$ is suppressed effectively.

8.4 Conclusion

The stability of the time-delay FMB control systems has been investigated under the delay-independent and delay-independent stability approaches. The time-delay FMB control system is represented as a descriptor system to facilitate the stability analysis and controller synthesis. Based on the Lyapunov stability theory, delay-independent and delay-dependent LMI-based stability conditions have been derived to guarantee the system stability. As the delay-dependent stability conditions do not require the information of the time delay, it is more suitable to nonlinear plants with unknown or inestimable time delay. However, if the time delay is known or estimable, the delay-dependent stability conditions are more favourable as it usually offers less conservative result. Simulation examples have been given to show the merits of the proposed approach. Furthermore, LMI-based performance conditions given in Chapter 3 have been employed to realize the system performance.

Chapter 9
Sampled-Data FMB Model Reference Control Systems

9.1 Introduction

Fuzzy-model-based control approach is a promising approach dealing with nonlinear systems. By taking advantage of the TS fuzzy model, which represents the dynamics of the nonlinear plant in a favourable form, the system analysis and controller synthesis can be facilitated. Flouring stability analysis results and LMI-based controller synthesis techniques for continuous-time FMB control systems have been achieved and briefly discussed in Chapter 2. In the previous chapters, the stabilization control problem that the system states are driven to the origin is considered. The tracking control problem is another one that has drawn a great deal of research interest in the control community. The objective of the tracking control problem is to drive the system states of the nonlinear plant to follow some pre-defined trajectories. Compared with the stabilization control problem, the tracking control problem is relatively difficult to be handled.

Tracking control problem for continuous-time and discrete-time FMB systems was investigated in [13, 116, 117, 119]. However, the study on the sampled-data FMB control system is seldom found. The sampled-data FMB control systems are mixed continuous-time and discrete-time systems. The nonlinear plant to be handled is a continuous-time system while the sampled-data controller is a discrete-time one. The control signals of the sampled-data controller will be held constant during the sampling period. As a result, due to the sampling activity, discontinuity is introduced to the closed-loop system that complicates the system dynamics and makes the stability analysis difficult. Although the sampling activity exhibits undesirable characteristics to the system analysis, the sampled-data controllers can be implemented by microcontrollers or digital computers to enhance the design flexibility and lower the implementation cost. Hence, it is worthwhile investigating the sampled-data FMB control systems to put the sampled-data FMB control approach into practice.

H.-K. Lam and F.H.F. Leung: FMB Control Systems, STUDFUZZ 264, pp. 173–190.
springerlink.com

As discontinuity is introduced by the sampling activity, the Lyapunov function candidates proposed for pure continuous-time or discrete-time FMB control systems cannot be applied. A descriptor transformation technique was proposed in [30] to investigate the stability of the linear sampled-data control systems. In [32, 56, 66], the linear analysis approach was employed and extended to analyze the stability of nonlinear sampled-data control system using the descriptor representation. The descriptor approach was adopted to investigate the system stability of the time-delay and/or sampled-data fuzzy control systems [60]. Furthermore, the tracking control problem handled by a sampled-data FMB controller can also be found in [71]. In [51], the effect of the analogue-to-digital and digital-to-analogue converters to the system stability were considered.

In this chapter, a sampled-data FMB model reference tracking control systems is considered. A sampled-data fuzzy controller is proposed to drive the system states of the continuous-time nonlinear system to follow those of a stable reference model. A Lyapunov function is employed to study the system stability. LMI-based stability conditions are derived to design the sampled-data fuzzy controllers, of which the tracking performance is governed by an H_∞ tracking performance function.

9.2 Reference Model and Sampled-Data Fuzzy Controller

A sampled-data FMB control system is formed by a continuous-time nonlinear plant represented by a TS fuzzy model (2.2) and a sampled-data fuzzy controller connected in a closed loop. The tracking control problem is considered in this chapter. A sampled-data fuzzy controller is employed to drive the system states to follow those of a stable linear reference model.

9.2.1 Reference Model

The stable linear reference model is defined as follows.

$$\dot{\hat{\mathbf{x}}}(t) = \mathbf{A}\hat{\mathbf{x}}(t) + \mathbf{B}\mathbf{r}(t) \tag{9.1}$$

where $\hat{\mathbf{x}}(t) \in \Re^n$ is the system state vector of the reference model; $\mathbf{A} \in \Re^{n\times n}$ and $\mathbf{B} \in \Re^{n\times m}$ are the constant system and input matrices, $\mathbf{r}(t) \in \Re^m$ is the external input vector.

Remark 9.1. The reference model can also be a stable nonlinear system. In this case, the system and input matrices become $\mathbf{A}(\mathbf{x}(t)) \in \Re^{n\times n}$ and $\mathbf{B}(\mathbf{x}(t)) \in \Re^{n\times m}$, respectively.

9.2.2 Sampled-Data Fuzzy Controller

A sampled-data fuzzy controller with p fuzzy rules is proposed based on the TS fuzzy model (2.2) representing the nonlinear plant. The j-th rule of the sampled-data fuzzy controller is of the following format.

$$\begin{aligned}&\text{Rule } j\text{: IF } g_1(\mathbf{x}(t_\gamma)) \text{ is } N_1^j \text{ AND } \cdots \text{ AND } g_\Omega(\mathbf{x}(t_\gamma)) \text{ is } N_\Omega^j\\ &\qquad \text{THEN } \mathbf{u}(t) = \mathbf{G}_j\mathbf{e}(t_\gamma), t_\gamma \leq t < t_{\gamma+1} \qquad (9.2)\end{aligned}$$

where N_β^j is a fuzzy set of rule j corresponding to the function $g_\beta(\mathbf{x}(t_\gamma))$, $\beta = 1, 2, \cdots, \Omega$; $j = 1, 2, \cdots, p$; Ω is a positive integer; $\mathbf{G}_j \in \Re^{m\times n}$, $j = 1, 2, \cdots, p$, are constant feedback gains to be determined; $\mathbf{x}(t) \in \Re^n$ is the system states of the nonlinear plant represented by the TS fuzzy model (2.2); $\mathbf{e}(t_\gamma) = \mathbf{x}(t_\gamma) - \hat{\mathbf{x}}(t_\gamma)$; $t_\gamma = \gamma h_s$, $\gamma = 0, 1, 2, \cdots, \infty$, denotes a sampling instant; $h_s = t_{\gamma+1} - t_\gamma$ denotes the constant sampling period. The inferred sampled-data fuzzy controller is defined as follows.

$$\begin{aligned}\mathbf{u}(t) &= \sum_{j=1}^{p} m_j(\mathbf{x}(t_\gamma))\mathbf{G}_j\mathbf{e}(t_\gamma)\\ &= \sum_{j=1}^{p} m_j(\mathbf{x}(t_\gamma))\mathbf{G}_j\mathbf{e}(t-\tau_s(t)), t_\gamma \leq t < t_{\gamma+1} \qquad (9.3)\end{aligned}$$

where $\tau_s(t) = t - t_\gamma < h_s$ for $t_\gamma \leq t < t_{\gamma+1}$,

$$m_j(\mathbf{x}(t_\gamma)) \geq 0 \; \forall \; j, \sum_{j=1}^{p} m_j(\mathbf{x}(t_\gamma)) = 1, \qquad (9.4)$$

$$m_j(\mathbf{x}(t_\gamma)) = \frac{\prod_{l=1}^{\Omega} \mu_{N_l^j}(g_l(\mathbf{x}(t_\gamma)))}{\sum_{k=1}^{p}\prod_{l=1}^{\Omega} \mu_{N_l^k}(g_l(\mathbf{x}(t_\gamma)))} \; \forall \; j, \qquad (9.5)$$

$m_j(\mathbf{x}(t_\gamma))$, $j = 1, 2, \cdots, p$, are the normalized grades of membership, $\mu_{N_\beta^j}(g_\beta(\mathbf{x}(t_\gamma)))$, $\beta = 1, 2, \cdots, \Omega$, are the membership functions corresponding to the fuzzy set N_β^j. It can be seen from (9.3) that $\mathbf{u}(t) = \mathbf{u}(t_\gamma)$, which holds constant value for $t_\gamma \leq t < t_{\gamma+1}$.

9.3 Stability Analysis of Sampled-Data FMB Model Reference Control Systems

In this section, the stability of the sampled-data FMB model reference tracking control system formed by the nonlinear plant represented by the TS fuzzy model (2.2), the reference model (9.1) and the sampled-data fuzzy controller of (9.3) is investigated. The block diagram of the model reference control system is shown in Fig. 9.1.

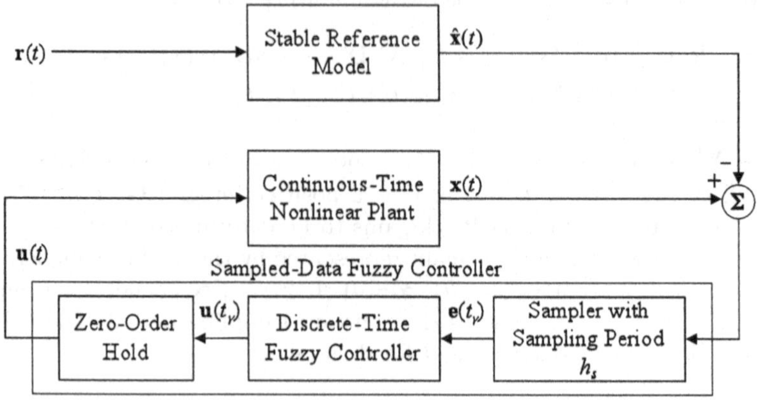

Fig. 9.1 Block diagram of the sampled-data FMB model reference tracking control system.

Referring to Fig. 9.1, the reference model and the continuous nonlinear plant offer the system state vector $\hat{\mathbf{x}}(t)$ and $\mathbf{x}(t)$, respectively. The sampled-data fuzzy controller will take $\hat{\mathbf{x}}(t)$ and $\mathbf{x}(t)$ at every h_s seconds (sampling period) as the input. Based on the sampled system state vectors, a control signal $\mathbf{u}(t)$ will be generated and kept constant by the zero order hold (ZOH) for every sampling period. The sampled-data fuzzy controller will be designed such that the the system state $\mathbf{x}(t)$ approaches $\hat{\mathbf{x}}(t)$ when time t tends to infinity.

9.3.1 Sampled-Data FMB Model Reference Control Systems

Consider the TS fuzzy model (2.2) and the reference model (9.1). With the inequality of $\sum_{i=1}^{p} w_i(\mathbf{x}(t)) = 1$, we have the error system defined as follows.

$$\dot{\mathbf{e}}(t) = \dot{\mathbf{x}}(t) - \dot{\hat{\mathbf{x}}}(t)$$
$$= \sum_{i=1}^{p} w_i(\mathbf{x}(t))(\mathbf{A}_i\mathbf{e}(t) + \mathbf{B}_i\mathbf{u}(t)) + \mathbf{m}_e(t) \qquad (9.6)$$

where $\mathbf{m}_e(t) = \sum_{i=1}^{p} w_i(\mathbf{x}(t))(\mathbf{A}_i - \mathbf{A})\hat{\mathbf{x}}(t) - \mathbf{B}\mathbf{r}(t)$.

Remark 9.2. It should be noted that $\mathbf{m}_e(t)$ is bounded due to the properties of the membership functions and the stable reference model, i.e., $w_i(\mathbf{x}(t))$, $\hat{\mathbf{x}}(t)$ and $\mathbf{r}(t)$ are bounded.

Consider the sampled-data fuzzy controller (9.3) and the error system (9.6), with the property of the membership functions, i.e., $\sum_{i=1}^{p} w_i(\mathbf{x}(t)) = \sum_{j=1}^{p} m_j(\mathbf{x}(t_\gamma)) = \sum_{i=1}^{p}\sum_{j=1}^{p} w_i(\mathbf{x}(t))m_j(\mathbf{x}(t_\gamma)) = 1$, the error system can be rewritten as follows.

$$
\begin{aligned}
\dot{\mathbf{e}}(t) &= \sum_{i=1}^{p} w_i(\mathbf{x}(t))(\mathbf{A}_i\mathbf{e}(t) + \mathbf{B}_i \sum_{j=1}^{p} m_j(\mathbf{x}(t_\gamma))\mathbf{G}_j\mathbf{e}(t-\tau_s(t))) + \mathbf{m}_e(t) \\
&= \sum_{i=1}^{p}\sum_{j=1}^{p} w_i(\mathbf{x}(t))m_j(\mathbf{x}(t_\gamma))(\mathbf{A}_i\mathbf{e}(t) + \mathbf{B}_i\mathbf{G}_j\mathbf{e}(t-\tau_s(t))) + \mathbf{m}_e(t) \quad (9.7)
\end{aligned}
$$

9.3.2 Stability Analysis

The stability of the error system (9.7) is investigated in this system. In the following analysis, for brevity, $w_i(\mathbf{x}(t))$ and $m_j(\mathbf{x}(t_\gamma))$ are denoted as w_i and m_j, respectively. Consider the following Lyapunov functional candidate.

$$
V(t) = \mathbf{e}(t)^T\mathbf{P}_1\mathbf{e}(t) + \int_{-h_s}^{0}\int_{t+\sigma}^{t} \dot{\mathbf{e}}(\varphi)^T\mathbf{R}\dot{\mathbf{e}}(\varphi)d\varphi d\sigma \quad (9.8)
$$

where $0 < \mathbf{P}_1 = \mathbf{P}_1^T \in \Re^{n\times n}$ and $0 < \mathbf{R} = \mathbf{R}^T \in \Re^{n\times n}$. From (9.7) and (9.8), we have,

$$
\begin{aligned}
\dot{V}(t) &= \dot{\mathbf{e}}(t)^T\mathbf{P}_1\mathbf{e}(t) + \mathbf{e}(t)^T\mathbf{P}_1\dot{\mathbf{e}}(t) \\
&\quad + h_s\dot{\mathbf{e}}(t)^T\mathbf{R}\dot{\mathbf{e}}(t) - \int_{t-h_s}^{t} \dot{\mathbf{e}}(\varphi)^T\mathbf{R}\dot{\mathbf{e}}(\varphi)d\varphi \\
&= \sum_{i=1}^{p}\sum_{j=1}^{p} w_i m_j \begin{bmatrix} \mathbf{e}(t) \\ \mathbf{e}(t-\tau_s(t)) \end{bmatrix}^T \left(\mathbf{P}^T \begin{bmatrix} \mathbf{A}_i & \mathbf{B}_i\mathbf{G}_j \\ \mathbf{0} & \mathbf{0} \end{bmatrix} \right. \\
&\quad \left. + \begin{bmatrix} \mathbf{A}_i & \mathbf{B}_i\mathbf{G}_j \\ \mathbf{0} & \mathbf{0} \end{bmatrix}^T \mathbf{P} \right) \begin{bmatrix} \mathbf{e}(t) \\ \mathbf{e}(t-\tau_s(t)) \end{bmatrix} + h_s\dot{\mathbf{e}}(t)^T\mathbf{R}\dot{\mathbf{e}}(t) \\
&\quad - \int_{t-h_s}^{t} \dot{\mathbf{e}}(\varphi)^T\mathbf{R}\dot{\mathbf{e}}(\varphi)d\varphi + \mathbf{e}(t)^T\mathbf{P}_1\mathbf{m}_e(t) + \mathbf{m}_e(t)^T\mathbf{P}_1\mathbf{e}(t) \quad (9.9)
\end{aligned}
$$

where $\mathbf{P} = \begin{bmatrix} \mathbf{P}_1 & \mathbf{0} \\ \mathbf{P}_2 & \mathbf{P}_3 \end{bmatrix} \in \Re^{2n\times 2n}$, $\mathbf{P}_2 \in \Re^{n\times n}$ and $\mathbf{P}_3 \in \Re^{n\times n}$.

Considering the fact that $\tau_s(t) = t - t_\gamma < h_s$, we have $-\int_{t-h_s}^{t} \dot{\mathbf{e}}(\varphi)^T\mathbf{R}\dot{\mathbf{e}}(\varphi)d\varphi \leq -\int_{t-\tau_s(t)}^{t} \dot{\mathbf{e}}(\varphi)^T\mathbf{R}\dot{\mathbf{e}}(\varphi)d\varphi$. Hence, it follows from (9.9) that we have

$$\dot{V}(t) \leq \sum_{i=1}^{p}\sum_{j=1}^{p} w_i m_j \begin{bmatrix} \mathbf{e}(t) \\ \mathbf{e}(t-\tau_s(t)) \end{bmatrix}^T \left(\mathbf{P}^T \begin{bmatrix} \mathbf{A}_i & \mathbf{B}_i\mathbf{G}_j \\ \mathbf{0} & \mathbf{0} \end{bmatrix} \right.$$
$$\left. + \begin{bmatrix} \mathbf{A}_i & \mathbf{B}_i\mathbf{G}_j \\ \mathbf{0} & \mathbf{0} \end{bmatrix}^T \mathbf{P} \right) \begin{bmatrix} \mathbf{e}(t) \\ \mathbf{e}(t-\tau_s(t)) \end{bmatrix} + h_s \sum_{i=1}^{p}\sum_{j=1}^{p}\sum_{k=1}^{p}\sum_{l=1}^{p} w_i m_j w_k m_l$$
$$\times \begin{bmatrix} \mathbf{e}(t) \\ \mathbf{e}(t-\tau_s(t)) \\ \mathbf{m}_e(t) \end{bmatrix}^T \begin{bmatrix} \mathbf{A}_i^T \\ \mathbf{G}_j^T\mathbf{B}_i^T \\ \mathbf{I} \end{bmatrix} \mathbf{R} \begin{bmatrix} \mathbf{A}_k^T \\ \mathbf{G}_l^T\mathbf{B}_k^T \\ \mathbf{I} \end{bmatrix}^T \begin{bmatrix} \mathbf{e}(t) \\ \mathbf{e}(t-\tau_s(t)) \\ \mathbf{m}_e(t) \end{bmatrix}$$
$$- \int_{t-\tau_s(t)}^{t} \dot{\mathbf{e}}(\varphi)^T \mathbf{R} \dot{\mathbf{e}}(\varphi) d\varphi + \mathbf{e}(t)^T \mathbf{P}_1 \mathbf{m}_e(t) + \mathbf{m}_e(t)^T \mathbf{P}_1 \mathbf{e}(t) \quad (9.10)$$

In order to deal with the integral terms in (9.10), we consider the Newton-Leibniz rule that we have $\int_{t-\tau_s(t)}^{t} \dot{\mathbf{e}}(\varphi)d\varphi = \mathbf{e}(t) - \mathbf{e}(t-\tau_s(t))$. As a result, the following equality is achieved to facilitate the stability analysis.

$$2\sum_{i=1}^{p}\sum_{j=1}^{p} w_i m_j \begin{bmatrix} \mathbf{e}(t) \\ \mathbf{e}(t-\tau_s(t)) \end{bmatrix}^T \begin{bmatrix} \mathbf{T}_{ij} \\ \mathbf{V}_{ij} \end{bmatrix}$$
$$\times \Big(- \int_{t-\tau_s(t)}^{t} \dot{\mathbf{e}}(\varphi)d\varphi + \mathbf{e}(t) - \mathbf{e}(t-\tau_s(t)) \Big) = 0 \quad (9.11)$$

where $\mathbf{T}_{ij} \in \Re^{n\times n}$ and $\mathbf{V}_{ij} \in \Re^{n\times n}$ for all i and j.

Denote $\boldsymbol{\Upsilon}_{ij} = \sum_{i=1}^{p}\sum_{j=1}^{p} w_i m_j \begin{bmatrix} \mathbf{T}_{ij} \\ \mathbf{V}_{ij} \end{bmatrix}^T \begin{bmatrix} \mathbf{e}(t) \\ \mathbf{e}(t-\tau_s(t)) \end{bmatrix}$. Considering the integral terms in (9.10), we have

$$\boldsymbol{\Lambda} \triangleq - \int_{t-\tau_s(t)}^{t} \dot{\mathbf{e}}(\varphi)^T \mathbf{R} \dot{\mathbf{e}}(\varphi) d\varphi$$
$$= - \int_{t-\tau_s(t)}^{t} \dot{\mathbf{e}}(\varphi)^T \mathbf{R} \dot{\mathbf{e}}(\varphi) d\varphi + 2\sum_{i=1}^{p}\sum_{j=1}^{p} w_i m_j \begin{bmatrix} \mathbf{e}(t) \\ \mathbf{e}(t-\tau_s(t)) \end{bmatrix}^T \begin{bmatrix} \mathbf{T}_{ij} \\ \mathbf{V}_{ij} \end{bmatrix}$$
$$\times \Big(- \int_{t-\tau_s(t)}^{t} \dot{\mathbf{e}}(\varphi)d\varphi + \mathbf{e}(t) - \mathbf{e}(t-\tau_s(t)) \Big)$$
$$= 2\sum_{i=1}^{p}\sum_{j=1}^{p} w_i m_j \begin{bmatrix} \mathbf{e}(t) \\ \mathbf{e}(t-\tau_s(t)) \end{bmatrix}^T \begin{bmatrix} \mathbf{T}_{ij} \\ \mathbf{V}_{ij} \end{bmatrix} (\mathbf{e}(t) - \mathbf{e}(t-\tau_s(t)))$$
$$+ \tau_s(t) \boldsymbol{\Upsilon}_{ij}^T \mathbf{R}^{-1} \boldsymbol{\Upsilon}_{kl} - \int_{t-\tau_s(t)}^{t} (\boldsymbol{\Upsilon}_{ij} + \mathbf{R}\dot{\mathbf{e}}(\varphi))^T \mathbf{R}^{-1} (\boldsymbol{\Upsilon}_{kl} + \mathbf{R}\dot{\mathbf{e}}(\varphi)) d\varphi. \quad (9.12)$$

It is obvious that the last term of (9.12) is negative definite. Based on fact that $\tau_s(t) = t - t_\gamma < h_s$, we have

$$\boldsymbol{\Lambda} \leq \sum_{i=1}^{p}\sum_{j=1}^{p} w_i m_j \begin{bmatrix} \mathbf{e}(t) \\ \mathbf{e}(t-\tau_s(t)) \end{bmatrix}^T \begin{bmatrix} \mathbf{T}_{ij} \\ \mathbf{V}_{ij} \end{bmatrix} \left(\begin{bmatrix} \mathbf{T}_{ij} \\ \mathbf{V}_{ij} \end{bmatrix} [\mathbf{I} \ -\mathbf{I}] \right.$$
$$\left. + \begin{bmatrix} \mathbf{I} \\ -\mathbf{I} \end{bmatrix} [\mathbf{T}_{ij}^T \ \mathbf{V}_{ij}^T] \right) \begin{bmatrix} \mathbf{e}(t) \\ \mathbf{e}(t-\tau_s(t)) \end{bmatrix} + h_s \boldsymbol{\Upsilon}_{ij}^T \mathbf{R}^{-1} \boldsymbol{\Upsilon}_{kl}. \qquad (9.13)$$

From (9.10) and (9.13), we have

$$\dot{V}(t) \leq \sum_{i=1}^{p}\sum_{j=1}^{p} w_i m_j \begin{bmatrix} \mathbf{e}(t) \\ \mathbf{e}(t-\tau_s(t)) \end{bmatrix}^T \mathbf{Q}_{ij} \begin{bmatrix} \mathbf{e}(t) \\ \mathbf{e}(t-\tau_s(t)) \end{bmatrix}$$
$$+ h_s \sum_{i=1}^{p}\sum_{j=1}^{p}\sum_{k=1}^{p}\sum_{l=1}^{p} w_i m_j w_k m_l \begin{bmatrix} \mathbf{e}(t) \\ \mathbf{e}(t-\tau_s(t)) \\ \mathbf{m}_e(t) \end{bmatrix}^T \begin{bmatrix} \mathbf{A}_i^T \\ \mathbf{G}_j^T \mathbf{B}_i^T \\ \mathbf{I} \end{bmatrix} \mathbf{R}$$
$$\times \begin{bmatrix} \mathbf{A}_k^T \\ \mathbf{G}_l^T \mathbf{B}_k^T \\ \mathbf{I} \end{bmatrix}^T \begin{bmatrix} \mathbf{e}(t) \\ \mathbf{e}(t-\tau_s(t)) \\ \mathbf{m}_e(t) \end{bmatrix}$$
$$+ \mathbf{e}(t)^T \mathbf{P}_1 \mathbf{m}_e(t) + \mathbf{m}_e(t)^T \mathbf{P}_1 \mathbf{e}(t) + h_s \boldsymbol{\Upsilon}_{ij}^T \mathbf{R}^{-1} \boldsymbol{\Upsilon}_{kl}. \qquad (9.14)$$

where $\mathbf{Q}_{ij} = \mathbf{P}^T \begin{bmatrix} \mathbf{A}_i & \mathbf{B}_i\mathbf{G}_j \\ \mathbf{0} & \mathbf{0} \end{bmatrix} + \begin{bmatrix} \mathbf{A}_i & \mathbf{B}_i\mathbf{G}_j \\ \mathbf{0} & \mathbf{0} \end{bmatrix}^T \mathbf{P} + \begin{bmatrix} \mathbf{T}_{ij} \\ \mathbf{V}_{ij} \end{bmatrix} [\mathbf{I} \ -\mathbf{I}] + \begin{bmatrix} \mathbf{I} \\ -\mathbf{I} \end{bmatrix} [\mathbf{T}_{ij}^T \ \mathbf{V}_{ij}^T]$.

Denote $\mathbf{X} = \mathbf{P}^{-1} = \begin{bmatrix} \mathbf{X}_1 & \mathbf{0} \\ \mathbf{X}_2 & \mathbf{X}_3 \end{bmatrix} \in \Re^{2n \times 2n}$, $\mathbf{X}_2 = \varepsilon_2 \mathbf{X}_1 \in \Re^{n \times n}$, $\mathbf{X}_3 = \varepsilon_3 \mathbf{X}_1 \in \Re^{n \times n}$, $\mathbf{z}(t) = \begin{bmatrix} \mathbf{z}_1(t) \\ \mathbf{z}_2(t) \end{bmatrix} = \mathbf{X}^{-1} \begin{bmatrix} \mathbf{e}(t) \\ \mathbf{e}(t-\tau_s(t)) \end{bmatrix}$ and $\mathbf{s}(t) = \begin{bmatrix} \mathbf{z}(t) \\ \mathbf{m}_e(t) \end{bmatrix}$ where ε_2 and $\varepsilon_3 > 0$ are scalars.

Furthermore, consider the terms of $\sum_{i=1}^{p}\sum_{j=1}^{p} w_i m_j \mathbf{e}(t)^T \mathbf{P}_1 (\mathbf{R}_{ij} - \mathbf{R}_{ij}) \mathbf{P}_1 \mathbf{e}(t) = 0$ and $\sigma_1^2 \sum_{i=1}^{p}\sum_{j=1}^{p} w_i m_j \mathbf{m}_e(t)^T (\mathbf{R}_{ij} - \mathbf{R}_{ij}) \mathbf{m}_e(t) = 0$ where $\sigma_1 \neq 0$ is a scalar to be determined and $\mathbf{R}_{ij} = \mathbf{R}_{ij}^T \in \Re^{n \times n}$. Adding these two terms to (9.14), we have,

$$\dot{V}(t) \leq \sum_{i=1}^{p}\sum_{j=1}^{p} w_i m_j \mathbf{z}(t)^T \mathbf{X}^T \mathbf{Q}_{ij} \mathbf{X}\mathbf{z}(t) + h_s \sum_{i=1}^{p}\sum_{j=1}^{p}\sum_{k=1}^{p}\sum_{l=1}^{p} w_i m_j w_k m_l$$
$$\times \left(\mathbf{s}(t)^T \begin{bmatrix} \mathbf{X} & \mathbf{0} \\ \mathbf{0} & \mathbf{I} \end{bmatrix}^T \begin{bmatrix} \mathbf{A}_i^T \\ \mathbf{G}_j^T \mathbf{B}_i^T \\ \mathbf{I} \end{bmatrix} \mathbf{R} \begin{bmatrix} \mathbf{A}_k^T \\ \mathbf{G}_l^T \mathbf{B}_k^T \\ \mathbf{I} \end{bmatrix}^T \begin{bmatrix} \mathbf{X} & \mathbf{0} \\ \mathbf{0} & \mathbf{I} \end{bmatrix} \mathbf{s}(t) \right.$$
$$\left. + \mathbf{z}(t)^T \mathbf{X}^T \begin{bmatrix} \mathbf{T}_{ij} \\ \mathbf{V}_{ij} \end{bmatrix} \mathbf{X}_1 \mathbf{X}_1^{-1} \mathbf{R}^{-1} \mathbf{X}_1^{-1} \mathbf{X}_1 \begin{bmatrix} \mathbf{T}_{kl} \\ \mathbf{V}_{kl} \end{bmatrix}^T \mathbf{X}\mathbf{z}(t) \right)$$
$$- \sum_{i=1}^{p}\sum_{j=1}^{p} w_i m_j \mathbf{z}_1(t)^T \mathbf{R}_{ij} \mathbf{z}_1(t) + \sigma_1^2 \sum_{i=1}^{p}\sum_{j=1}^{p} w_i m_j \mathbf{m}_e(t)^T \mathbf{R}_{ij} \mathbf{m}_e(t)$$
$$+ \sum_{i=1}^{p}\sum_{j=1}^{p} w_i m_j \mathbf{s}(t)^T \begin{bmatrix} \mathbf{R}_{ij} & \mathbf{0} & \mathbf{I} \\ \mathbf{0} & \mathbf{0} & \mathbf{0} \\ \mathbf{I} & \mathbf{0} & -\sigma_1^2 \mathbf{R}_{ij} \end{bmatrix} \mathbf{s}(t). \tag{9.15}$$

Denote $\mathbf{U}_{ij} = \mathbf{X}_1 \mathbf{T}_{ij} \mathbf{X}_1 \in \Re^{n \times n}$ and $\mathbf{W}_{ij} = \mathbf{X}_1 \mathbf{V}_{ij} \mathbf{X}_1 \in \Re^{n \times n}$ for all i and j. Design the feedback gains as $\mathbf{G}_j = \mathbf{N}_j \mathbf{X}_1^{-1}$ where $\mathbf{N}_j \in \Re^{m \times n}$ for all j.

It follows from (9.15) that we have

$$\dot{V}(t) \leq \sum_{i=1}^{p}\sum_{j=1}^{p} w_i m_j \mathbf{s}(t)^T \mathbf{\Xi}_{ij} \mathbf{s}(t) + h_s \sum_{i=1}^{p}\sum_{j=1}^{p}\sum_{k=1}^{p}\sum_{l=1}^{p} w_i m_j w_k m_l$$
$$\times \mathbf{s}(t)^T (\mathbf{\Theta}_{ij}^T \mathbf{R} \mathbf{\Theta}_{kl} + \mathbf{\Phi}_{ij}^T \mathbf{X}_1^T \mathbf{R}^{-1} \mathbf{X}_1^T \mathbf{\Phi}_{kl}) \mathbf{s}(t)$$
$$+ \sum_{i=1}^{p}\sum_{j=1}^{p} w_i m_j (-\mathbf{z}_1(t)^T \mathbf{R}_{ij} \mathbf{z}_1(t) + \sigma_1^2 \mathbf{m}_e(t)^T \mathbf{R}_{ij} \mathbf{m}_e(t)) \tag{9.16}$$

where $\mathbf{\Xi}_{ij} = \begin{bmatrix} \mathbf{\Xi}_{ij}^{(11)} + \mathbf{\Xi}_{ij}^{(11)^T} + \mathbf{R}_{ij} & * & * \\ \mathbf{\Xi}_{ij}^{(21)} & -\varepsilon_3^2(\mathbf{W}_{ij} + \mathbf{W}_{ij}^T) & * \\ \mathbf{I} & \mathbf{0} & -\sigma_1^2 \mathbf{R}_{ij} \end{bmatrix}$, $\mathbf{\Xi}_{ij}^{(11)} = \mathbf{A}_i \mathbf{X}_1 + \varepsilon_2 \mathbf{B}_i \mathbf{N}_j + (1 - \varepsilon_2)\mathbf{U}_{ij} + \varepsilon_2(1 - \varepsilon_2)\mathbf{W}_{ij}$, $\mathbf{\Xi}_{ij}^{(21)} = \varepsilon_3 \mathbf{N}_j^T \mathbf{B}_i^T - \varepsilon_3 \mathbf{U}_{ij}^T + \varepsilon_3 \mathbf{W}_{ij} - \varepsilon_2 \varepsilon_3 (\mathbf{W}_{ij} + \mathbf{W}_{ij}^T)$, $\mathbf{\Theta}_{ij} = \begin{bmatrix} \mathbf{A}_i \mathbf{X}_1 + \varepsilon_2 \mathbf{B}_i \mathbf{N}_j & \varepsilon_3 \mathbf{B}_i \mathbf{N}_j & \mathbf{I} \end{bmatrix}$ and $\mathbf{\Phi}_{ij} = \begin{bmatrix} \mathbf{U}_{ij}^T + \varepsilon_2 \mathbf{W}_{ij}^T & \varepsilon_3 \mathbf{W}_{ij}^T & \mathbf{0} \end{bmatrix}$.

Consider that the following inequality holds:

$$\sum_{i=1}^{p}\sum_{j=1}^{p} w_i m_j \mathbf{\Xi}_{ij} + h_s \sum_{i=1}^{p}\sum_{j=1}^{p}\sum_{k=1}^{p}\sum_{l=1}^{p} w_i m_j w_k m_l$$
$$\times (\mathbf{\Theta}_{ij}^T \mathbf{R} \mathbf{\Theta}_{kl} + \mathbf{\Phi}_{ij}^T \mathbf{X}_1^T \mathbf{R}^{-1} \mathbf{X}_1^T \mathbf{\Phi}_{kl}) < 0 \tag{9.17}$$

It follows from (9.16) and (9.17), we have,

$$\dot{V}(t) \leq \sum_{i=1}^{p}\sum_{j=1}^{p} w_i m_j(-\mathbf{z}_1(t)^T\mathbf{R}_{ij}\mathbf{z}_1(t) + \sigma_1^2\mathbf{m}_e(t)^T\mathbf{R}_{ij}\mathbf{m}_e(t)). \tag{9.18}$$

Considering the termination time of control t_f [117], we take integration on both sides of (9.18) to achieve the following H_∞ performance.

$$\int_0^{t_f} \dot{V}(t)dt \leq \int_0^{t_f} \sum_{i=1}^{p}\sum_{j=1}^{p} w_i m_j(-\mathbf{z}_1(t)^T\mathbf{R}_{ij}\mathbf{z}_1(t) + \sigma_1^2\mathbf{m}_e(t)^T\mathbf{R}_{ij}\mathbf{m}_e(t))$$

$$\sigma_1^2 \geq \frac{V(t_f) - V(0) + \int_0^{t_f}\sum_{i=1}^{p}\sum_{j=1}^{p} w_i m_j \mathbf{z}_1(t)^T\mathbf{R}_{ij}\mathbf{z}_1(t)}{\int_0^{t_f}\sum_{i=1}^{p}\sum_{j=1}^{p} w_i m_j \mathbf{m}_e(t)^T\mathbf{R}_{ij}\mathbf{m}_e(t)} \tag{9.19}$$

If the inequality of (9.17) is satisfied, the sampled-data FMB model reference control system (9.1) is guaranteed to be asymptotically stable subject to the H_∞ performance (9.19). It can be seen that a good performance is ensured by a lower value of σ_1.

By Schur complement, the inequality (9.17) is equivalent to the one in the following.

$$\mathbf{Q} = \sum_{i=1}^{p}\sum_{j=1}^{p} w_i m_j \overline{\mathbf{Q}}_{ij} < 0 \tag{9.20}$$

where $\overline{\mathbf{Q}}_{ij} = \begin{bmatrix} \mathbf{\Xi}_{ij} & * & * \\ h_s\mathbf{\Theta}_{ij} & -h_s\mathbf{M} & * \\ h_s\mathbf{\Phi}_{ij} & \mathbf{0} & -h_s\mathbf{X}_1\mathbf{M}^{-1}\mathbf{X}_1 \end{bmatrix}$ and $\mathbf{M} = \mathbf{R}^{-1}$. It can be seen that $\overline{\mathbf{Q}}_{ij} < 0$ for all i and j, the inequality of (9.20) is satisfied. As there exists the term of $\mathbf{X}_1\mathbf{M}^{-1}\mathbf{X}_1$ in $\overline{\mathbf{Q}}_{ij}$, the condition of $\overline{\mathbf{Q}}_{ij} < 0$ is not an LMI of which the solution cannot be solved by convex programming techniques. To circumvent this difficulty, we consider the inequality (8.15). Consequently, the holding of the following inequality implies the holding of $\mathbf{Q} < 0$.

$$\mathbf{Q} = \sum_{i=1}^{p}\sum_{j=1}^{p} w_i m_j \overline{\mathbf{Q}}_{ij} \leq \sum_{i=1}^{p}\sum_{j=1}^{p} w_i m_j \hat{\mathbf{Q}}_{ij} < 0 \tag{9.21}$$

where $\hat{\mathbf{Q}}_{ij} = \begin{bmatrix} \mathbf{\Xi}_{ij} & * & * \\ h_s\mathbf{\Theta}_{ij} & -h_s\mathbf{M} & * \\ h_s\mathbf{\Phi}_{ij} & \mathbf{0} & -h_s(2\zeta\mathbf{X}_1 - \zeta^2\mathbf{M}) \end{bmatrix}$ and ζ is a scalar.

It can be seen that $\hat{\mathbf{Q}}_{ij} < 0$ implies the holding of the inequality (9.20) and guarantees the stability of the error system (9.7). However, it offers a conservative result if we do not consider the information of membership functions. To alleviate the conservativeness of the stability analysis, we consider the membership function satisfying $m_i - w_i + \delta_i \geq 0$ for all i, $\mathbf{x}(t)$ and $\mathbf{x}(t_\gamma)$

where δ_i are scalars. With these inequalities, some free matrices are introduced to relax the stability conditions. Rewrite (9.21) as follows.

$$\begin{aligned}\mathbf{Q} &\leq \sum_{i=1}^{p}\sum_{j=1}^{p} w_i(m_j - w_j + \delta_j + w_j - \delta_j)\hat{\mathbf{Q}}_{ij} \\ &= \sum_{i=1}^{p}\sum_{j=1}^{p} w_i(w_j - \delta_j)\hat{\mathbf{Q}}_{ij} + \sum_{j=1}^{p} w_i(m_j - w_j + \delta_j)\hat{\mathbf{Q}}_{ij} \\ &= \sum_{i=1}^{p}\sum_{j=1}^{p} w_i w_j(\hat{\mathbf{Q}}_{ij} - \sum_{k=1}^{p}\delta_k\hat{\mathbf{Q}}_{ik}) + \sum_{i=1}^{p}\sum_{j=1}^{p} w_i(m_j - w_j + \delta_j)\hat{\mathbf{Q}}_{ij}. \quad (9.22)\end{aligned}$$

Furthermore, considering the property of the membership functions, we have $\sum_{i=1}^{p}\sum_{j=1}^{p} w_i(m_j - w_j)\mathbf{\Lambda}_i = \mathbf{0}$ where $\mathbf{\Lambda}_i = \mathbf{\Lambda}_i^T \in \Re^{n\times n}$ for all i. It can be written as $\sum_{i=1}^{p}\sum_{j=1}^{p} w_i(m_j - w_j + \delta_j - \delta_j)\mathbf{\Lambda}_i = -\sum_{i=1}^{p}\sum_{j=1}^{p} w_i\delta_j\mathbf{\Lambda}_i + \sum_{i=1}^{p}\sum_{j=1}^{p} w_i(m_j - w_j + \delta_j)\mathbf{\Lambda}_i = \mathbf{0}$. Adding these inequalities to (9.22), we have

$$\begin{aligned}\mathbf{Q} \leq &\sum_{i=1}^{p}\sum_{j=1}^{p} w_i w_j\big(\hat{\mathbf{Q}}_{ij} - \sum_{k=1}^{p}\delta_k(\hat{\mathbf{Q}}_{ik} + \mathbf{\Lambda}_i)\big) \\ &+ \sum_{i=1}^{p}\sum_{j=1}^{p} w_i(m_j - w_j + \delta_j)(\hat{\mathbf{Q}}_{ij} + \mathbf{\Lambda}_i). \quad (9.23)\end{aligned}$$

Introducing matrices $\mathbf{S}_{ij} = \mathbf{S}_{ji}^T \in \Re^{n\times n}$, $i,\ j = 1,\ 2,\ \cdots,\ p$, we consider the following inequalities.

$$\mathbf{R}_{ii} > \hat{\mathbf{Q}}_{ii} - \sum_{k=1}^{p}\delta_k(\hat{\mathbf{Q}}_{ik} + \mathbf{\Lambda}_i)\ \forall\, i \quad (9.24)$$

$$\mathbf{R}_{ij} + \mathbf{R}_{ij}^T \geq \hat{\mathbf{Q}}_{ij} - \sum_{k=1}^{p}\delta_k(\hat{\mathbf{Q}}_{ik} + \mathbf{\Lambda}_i) + \hat{\mathbf{Q}}_{ji} - \sum_{k=1}^{p}\delta_k(\hat{\mathbf{Q}}_{jk} + \mathbf{\Lambda}_j)\ \forall\, j; i < j \quad (9.25)$$

From (9.23) to (9.25), we have

$$\begin{aligned}\mathbf{Q} &\leq \sum_{i=1}^{p} w_i^2 \mathbf{S}_{ii} + \sum_{j=1}^{p}\sum_{i<j} w_i w_j (\mathbf{S}_{ij} + \mathbf{S}_{ij}^T) \\ &\quad + \sum_{i=1}^{p}\sum_{j=1}^{p} w_i (m_j - w_j + \delta_j)(\hat{\mathbf{Q}}_{ij} + \mathbf{\Lambda}_i) \\ &= \mathbf{r}(t)^T \mathbf{S} \mathbf{r}(t) + \sum_{i=1}^{p}\sum_{j=1}^{p} w_i (m_j - w_j + \delta_j)(\hat{\mathbf{Q}}_{ij} + \mathbf{\Lambda}_i). \end{aligned} \tag{9.26}$$

where $\mathbf{r}(t) = \begin{bmatrix} w_1\mathbf{I} \\ w_2\mathbf{I} \\ \vdots \\ w_p\mathbf{I} \end{bmatrix}$, $\mathbf{S} = \begin{bmatrix} \mathbf{S}_{11} & \mathbf{S}_{12} & \cdots & \mathbf{S}_{1p} \\ \mathbf{S}_{21} & \mathbf{S}_{22} & \cdots & \mathbf{S}_{2p} \\ \vdots & \vdots & \vdots & \vdots \\ \mathbf{S}_{p1} & \mathbf{S}_{p2} & \cdots & \mathbf{S}_{pp} \end{bmatrix}$.

It can be seen from (9.26) that if the LMIs of $\mathbf{S} < 0$ and $\hat{\mathbf{Q}}_{ij} + \mathbf{\Lambda}_i < 0$ for all i and j are satisfied, $\mathbf{Q} < 0$ is achieved. As a result, the error system (9.6) is guaranteed to be asymptotically stable subject to the H_∞ performance (9.19). The analysis result is summarized in the following theorem.

Theorem 9.1. *The erorr system (9.6), formed by the nonlinear plant represented by the fuzzy model (2.2), the stable reference model (9.1) and the sampled-data fuzzy controller (9.3), is asymptotically stable subject to the H_∞ performance (9.19) if there exist predefined constant scalars $h_s > 0$, ζ, σ_1, ε_2, $\varepsilon_3 > 0$ and δ_i that satisfy $m_i(\mathbf{x}(t_\gamma)) - w_i(\mathbf{x}(t)) + \delta_i \geq 0$ for all i, $\mathbf{x}(t_\gamma)$ and $\mathbf{x}(t)$, and there exist matrices $\mathbf{M} = \mathbf{M}^T \in \Re^{n\times n}$, $\mathbf{N}_i \in \Re^{m\times n}$, $\mathbf{R}_{ij} = \mathbf{R}_{ij}^T \in \Re^{n\times n}$, $\mathbf{S}_{ij} = \mathbf{S}_{ji}^T \in \Re^{n\times n}$, $\mathbf{U}_{ij} \in \Re^{n\times n}$, $\mathbf{W}_{ij} \in \Re^{n\times n}$, $i = 1, 2, \cdots, p$, $\mathbf{X}_1 = \mathbf{X}_1^T \in \Re^{n\times n}$, $\mathbf{Y} = \mathbf{Y}^T \in \Re^{n\times n}$ such that the following LMIs are satisfied.*

$$\mathbf{X}_1 > 0;$$

$$\mathbf{M} > 0;$$

$$\mathbf{S} > 0;$$

$$\hat{\mathbf{Q}}_{ij} + \mathbf{\Lambda}_i < 0 \;\forall\, i,\, j;$$

and the feedback gains are designed as $\mathbf{G}_j = \mathbf{N}_j\mathbf{X}^{-1}$ for all j.

Remark 9.3. It should be noted that the stability analysis is valid if $\mathbf{X} = \mathbf{P}^{-1} = \begin{bmatrix} \mathbf{X}_1 & \mathbf{0} \\ \varepsilon_2\mathbf{X}_1 & \varepsilon_2\mathbf{X}_1 \end{bmatrix}$ is invertible. Referring to Theorem 9.1, ε_3 is a non-zero positive scalar and if there exists solution to the stability conditions, we have $\mathbf{X}_1 = \mathbf{X}_1^T > 0$ which is the sufficient condition to ensure that $\mathbf{X}$ is invertible.

Example 9.1. Consider the inverted pendulum on a cart in Example 3.2 as the nonlinear plant. It is assumed that the inverted pendulum is working in the operating domain of $x_1(t) \in \left[-\frac{\pi}{3}, \frac{\pi}{3}\right]$ (rad) and $x_1(t) \in \left[-15, 15\right]$

(rad/s). There are no limitations on $x_3(t)$ and $x_4(t)$. The proposed sampled-data fuzzy controller (9.3) is employed to control the inverted pendulum to track the system states of a stable linear reference model defined as follows.

$$\dot{\hat{\mathbf{x}}}(t) = \begin{bmatrix} 0 & 1 & 0 & 0 \\ -600 & -60 & -2 & -20 \\ 0 & 0 & 0 & 1 \\ 380 & 30 & 0 & 10 \end{bmatrix} \hat{\mathbf{x}}(t) + \begin{bmatrix} 0 \\ 0 \\ 0 \\ 1 \end{bmatrix} r(t) \tag{9.27}$$

where $r(t) = \sin(t/5)$.

A sampled-data fuzzy controller with two rules of the following format is employed for the control process.

$$\begin{aligned} &\text{Rule } j\text{: IF } x_1(t_\gamma) \text{ is } N_1^j \\ &\qquad \text{THEN } u(t) = \mathbf{G}_j \mathbf{e}(t_\gamma), t_\gamma \leq t < t_{\gamma+1} \end{aligned} \tag{9.28}$$

The inferred sampled-data fuzzy controller is defined as follows.

$$u(t) = \sum_{j=1}^{2} m_j(x_1(t_\gamma))\mathbf{G}_j \mathbf{e}(t_\gamma) \tag{9.29}$$

The membership functions of the sampled-data fuzzy controller are chosen to be $m_1(x_1(t_\gamma)) = \mu_{N_1^1}(x_1(t_\gamma)) = e^{-\frac{x_1(t_\gamma)^2}{2\times 2.2^2}}$ and $m_2(x_1(t_\gamma)) = \mu_{N_1^2}(x_1(t_\gamma)) = 1 - w_1(x_1(t_\gamma))$.

Based on Theorem 9.1, choosing the sample period $h_s = 0.015s$ (sampling frequency $\approx 66.6667Hz$), $\sigma_1 = 0.01$ and $\zeta = 2$, we obtain $\mathbf{G}_1 = \begin{bmatrix} 692.2467 & 54.9574 & 5.2048 & 52.1226 \end{bmatrix}$ and $\mathbf{G}_2 = \begin{bmatrix} 726.9190 & 57.1806 & 5.4344 & 53.5087 \end{bmatrix}$ with the MATLAB LMI toolbox. The sampled-data fuzzy controller (9.29) is employed to control the inverted pendulum.

In this example, the nonlinear plant is operating in the domain characterized by $\dot{x}_1(t) = x_2(t) \in [-15, 15]$. With this information and considering $t_\gamma \leq t \leq t_\gamma + h_s$, we have $x_1(t) = x_1(t_\gamma) + \int_{t_\gamma}^{t} x_2(t)dt$ which offers the lower and upper bounds as $x_1(t_\gamma) - 15\int_{t_\gamma}^{t_\gamma+h_s} dt = x_1(t_\gamma) - 15h_s = x_1(t_\gamma) - 0.225$ and $x_1(t_\gamma) + 15\int_{t_\gamma}^{t_\gamma+h_s} dt = x_1(t_\gamma) + 15h_s = x_1(t_\gamma) + 0.225$, respectively. Consequently, for any sampling instant t_γ, the value of $x_1(t)$ on or before the next sampling instant is in the range of $x_1(t_\gamma) - 0.225 \leq x_1(t) \leq x_1(t_\gamma) + 0.225$ for $t_\gamma \leq t \leq t_\gamma + h_s$. With the defined membership functions, it is found that $\delta_1 = 0.3420$ and $\delta_2 = 0.3380$ satisfy the inequalities of $m_i(x_1(t_\gamma)) - w_i(x_1(t)) + \delta_i > 0$ for all i.

The fuzzy controller (9.29) is employed to control the inverted pendulum. The system state responses and control signal of the sampled-data FMB model reference tracking control system under the initial system state

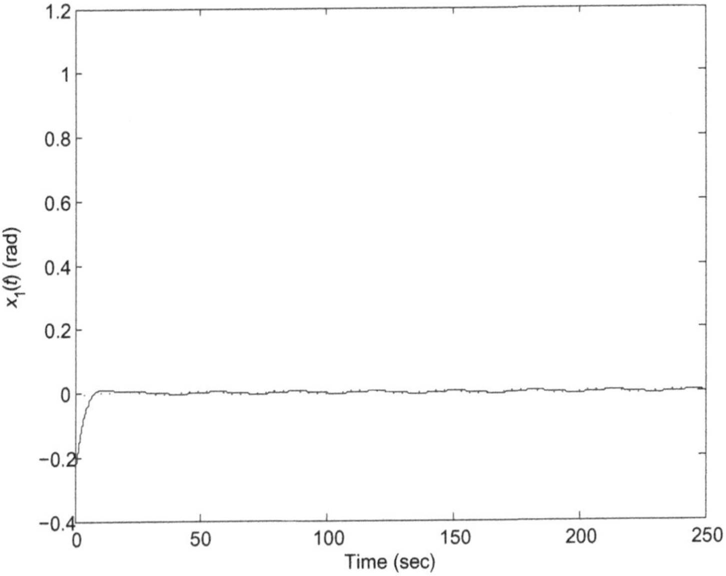

(a) $x_1(t)$.

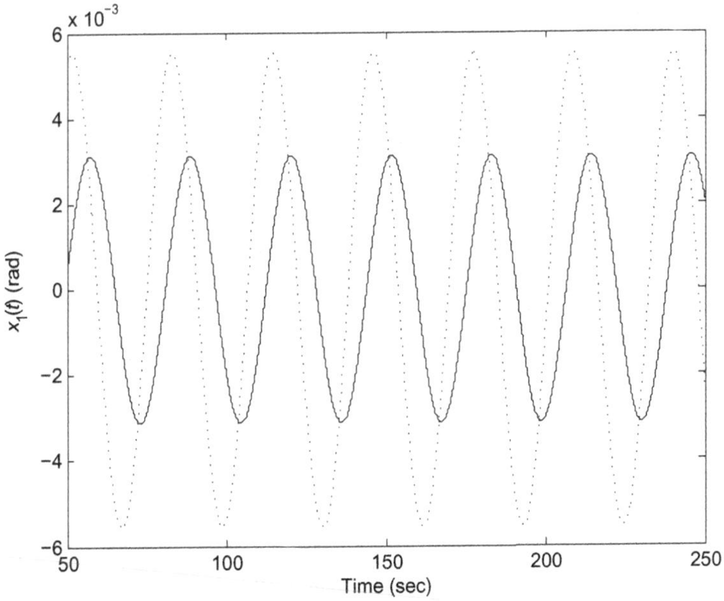

(b) $x_1(t)$ for $50s \leq t \leq 250s$.

Fig. 9.2 System state responses of the inverted pendulum with sampled-data fuzzy controller (Solid lines). Systems state responses of the linear reference model (Dotted lines).

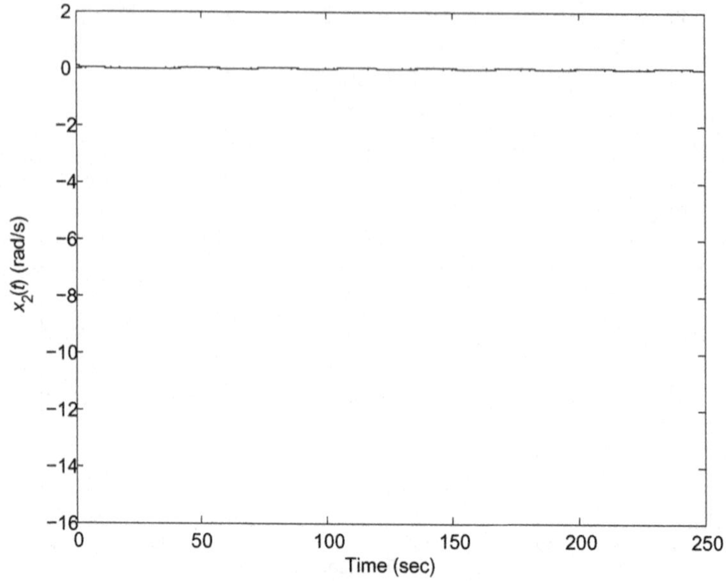

(c) $x_2(t)$.

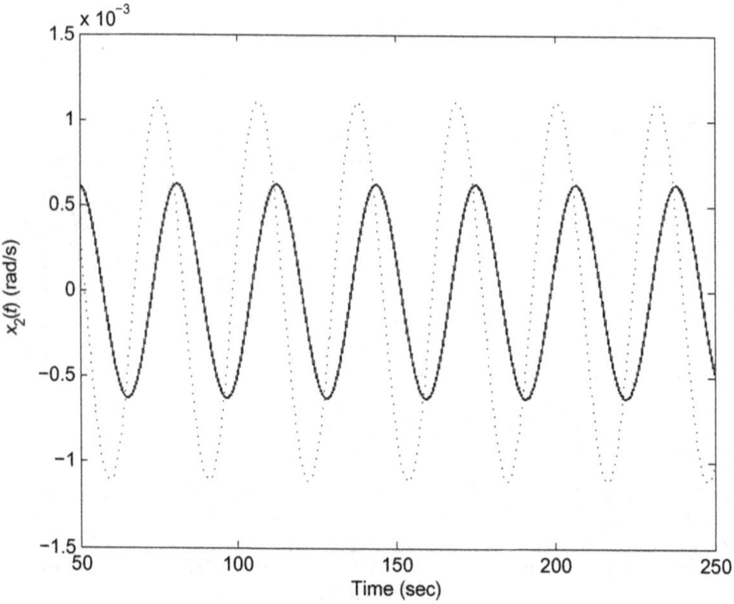

(d) $x_2(t)$ for $50s \leq t \leq 250s$.

Fig. 9.2 (*continued*)

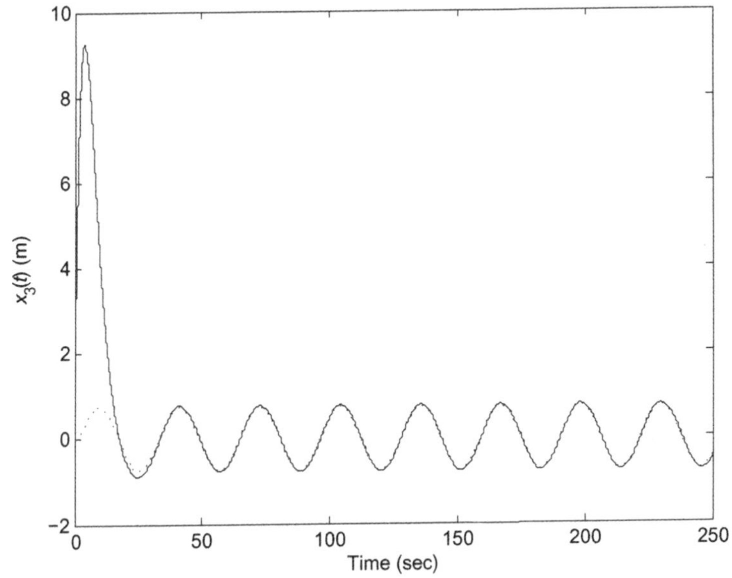

(e) $x_3(t)$.

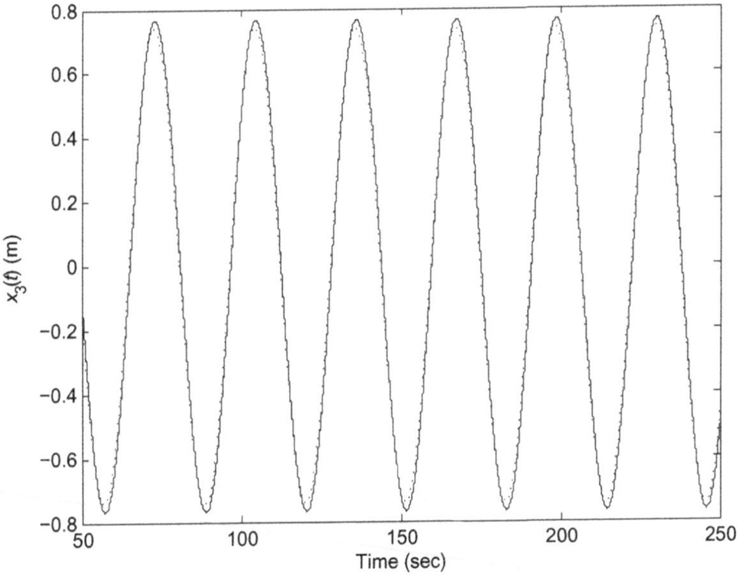

(f) $x_3(t)$ for $50s \leq t \leq 250s$.

Fig. 9.2 (*continued*)

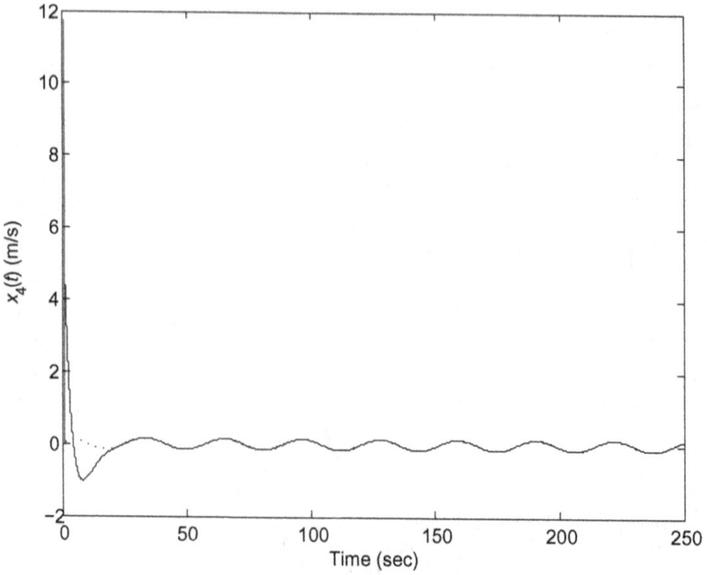

(g) $x_4(t)$.

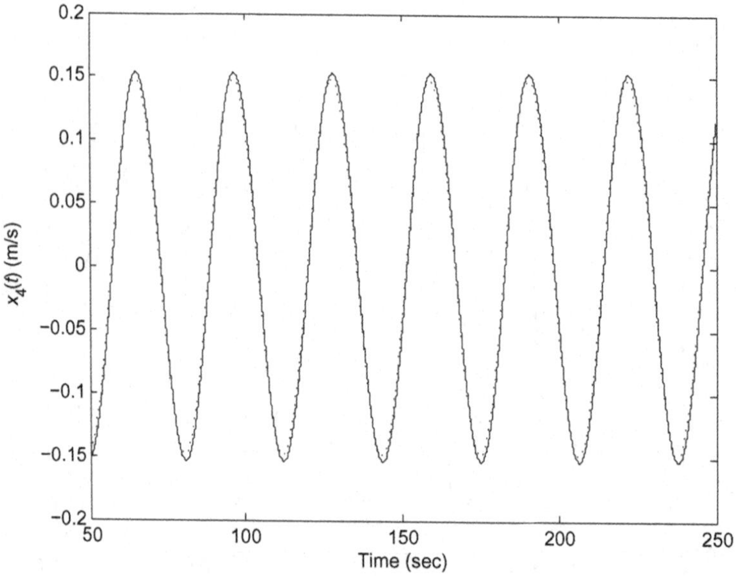

(h) $x_4(t)$ for $50s \leq t \leq 250s$.

Fig. 9.2 (*continued*)

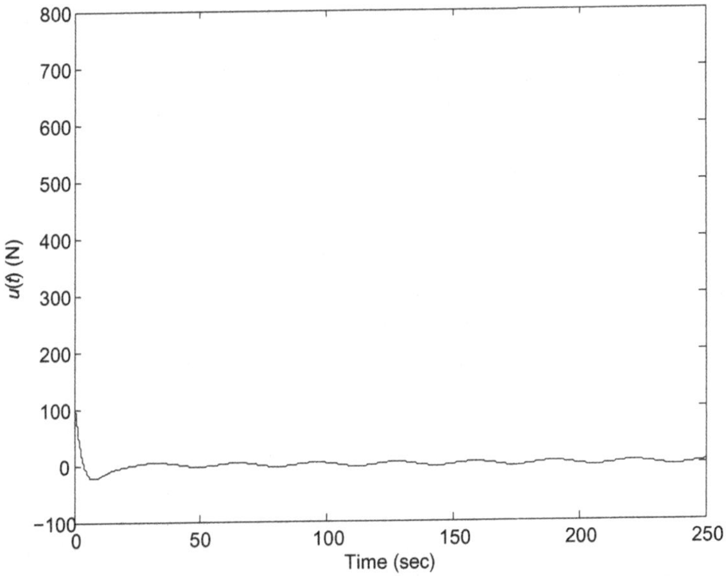

(a) $u(t)$.

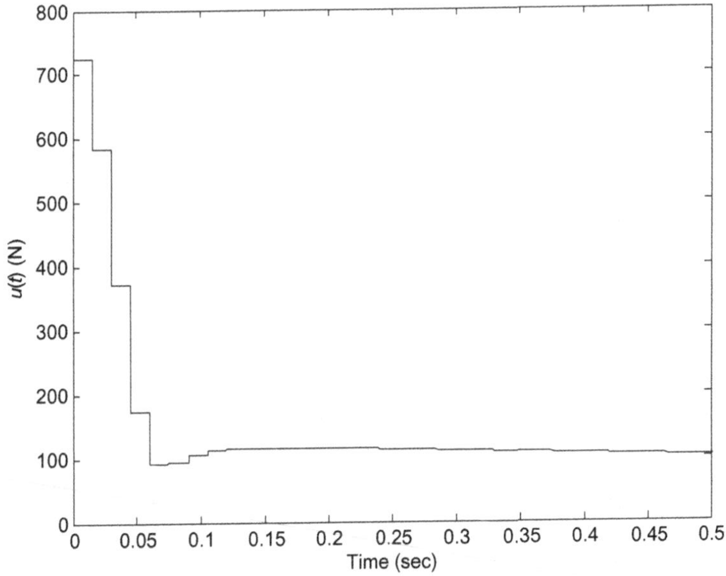

(b) $u(t)$ for $0s \leq t \leq 0.5s$.

Fig. 9.3 Control signal of the sampled-data fuzzy controller.

conditions of $\mathbf{x}(0) = \left[\frac{\pi}{3}\ 0\ 0\ 0\right]^T$ and $\hat{\mathbf{x}}(0) = \left[0\ 0\ 0\ 0\right]^T$ for $t \in \mathbf{x}(0) = \left[-h_s\ 0\right]$ are shown in Fig. 9.2 and Fig. 9.3, respectively. It can be seen that the system states of the inverted pendulum are driven to following those of the reference model of (9.27). It can be seen from Fig. 9.3 that control signal is a stepwise signal with its value being held constant during the sampling period.

9.4 Conclusion

The stability of the sampled-data FMB model reference tracking control systems has been investigated. A sampled-data fuzzy controller has been proposed to drive the system states of the nonlinear plant to follow those of a stable reference model. The sampled-data fuzzy controller can be realized by a microcontroller or digital computer to lower the implementation complexity and time. The information of the membership functions of both the TS fuzzy model and sampled-data controller has been employed to facilitate the stability analysis based on the Lyapunov stability theory. LMI-based stability conditions have been derived to design of the sampled-data FMB model reference tracking control systems. A simulation example has been given to illustrate the effectiveness of the proposed approach.

Chapter 10
IT2 FMB Control Systems

10.1 Introduction

Type-1 FMB control systems have drawn the attention of fuzzy control researchers for many years. The type-1 TS fuzzy model (2.2) offers a general framework for system analysis and controller synthesis. However, as the membership functions for the type-1 fuzzy sets have limited capability of capturing uncertainty information, the control problem cannot be handled directly if the nonlinear plant is subject to parameter uncertainties. In [18, 88, 89], a type-2 fuzzy logic system (FLS) was proposed to deal with uncertain grades of membership with type-2 fuzzy sets. The type-2 FLS can be regarded as a collection of theoretically infinite number of type-1 FLSs. As a result, additional information, including system parameter uncertainty, can be captured by the type-2 FLS [18, 88, 89]. Defuzzication and type reduction for general type-2 FLS are computational expensive. Instead, interval type-2 fuzzy (IT2) sets are employed to lower the computational demand that the grades of membership for the secondary membership functions are all one. Some theories of IT2 fuzzy sets [18, 88, 89] were developed to theoretically support the design and implementation of type-2 FLSs. The superiority of IT2 fuzzy sets over type-1 fuzzy sets on dealing with uncertain grades of membership for various applications have been shown in [20, 38, 42, 78, 80]. The applications include adaptive filtering [78], autonomous mobile robots [38], DC-DC power converter [80], intelligent fuzzy agent [20] and video streaming [42] etc.

The outstanding feature of the type-2 FLS on handling uncertainty offers a potential vehicle for carrying out stability analysis of FMB control systems subject to parameter uncertainties. In this chapter, we generalize and extend our preliminary analysis result in [55] to IT2 FMB control systems [78]. An IT2 fuzzy model is proposed to represent the nonlinear plant subject to parameter uncertainties captured by the lower and upper membership functions of the interval type-2 fuzzy sets. With the IT2 fuzzy sets, the IT2 TS fuzzy model can be regarded as a collection of a number of type-1 TS fuzzy models. This explains why the IT2 TS fuzzy model is superior to the type-1

H.-K. Lam and F.H.F. Leung: FMB Control Systems, STUDFUZZ 264, pp. 191–215.
springerlink.com

one. Although the IT2 TS fuzzy model demonstrates a stronger capability of capturing parameter uncertainties in the footprint of uncertainty (FOU), it eliminates the favourable property of the type-1 FMB analysis approach for stability analysis and thus leads to conservative stability analysis results. To facilitate the stability analysis, the lower and upper membership functions are employed to implement the proposed IT2 fuzzy controller [70] and their boundary information are considered. With such information, slack matrices are introduced to alleviate the conservativeness of the stability analysis results. Based on the Lyapunov stability theory [46, 100, 121], stability conditions in terms of LMIs are derived to achieve a stable IT2 FMB control system.

10.2 IT2 TS Fuzzy Model and Fuzzy Controller

An IT2 TS fuzzy model is employed to represent a nonlinear plant subject to parameter uncertainties to facilitate the stability analysis and controller synthesis. Based on the IT2 TS fuzzy model, an IT2 fuzzy controller is then proposed to close the feedback loop to form an IT2 FMB control system.

10.2.1 IT2 TS Fuzzy Model

Consider an IT2 TS fuzzy model [78] with p rules of the following format of which each rule the antecedent contains IT2 fuzzy sets and the consequent is a linear dynamical system.

$$\begin{aligned}&\text{Rule } i\text{: IF } f_1(\mathbf{x}(t)) \text{ is } \tilde{M}_1^i \text{ AND } \cdots \text{ AND } f_\Psi(\mathbf{x}(t)) \text{ is } \tilde{M}_\Psi^i\\ &\qquad \text{THEN } \dot{\mathbf{x}}(t) = \mathbf{A}_i\mathbf{x}(t) + \mathbf{B}_i\mathbf{u}(t)\end{aligned} \tag{10.1}$$

where $\tilde{M}_\alpha^i$ is an IT2 fuzzy set of rule i corresponding to the function $f_\alpha(\mathbf{x}(t))$, $\alpha = 1, 2, \cdots, \Psi$; $i = 1, 2, \cdots, p$; Ψ is a positive integer; $\mathbf{x}(t) \in \Re^n$ is the system state vector; $\mathbf{A}_i \in \Re^{n\times n}$ and $\mathbf{B}_i \in \Re^{n\times m}$ are the known system and input matrices, respectively; $\mathbf{u}(t) \in \Re^m$ is the input vector. The firing strength of the i-th rule is the following interval sets.

$$\tilde{w}_i(\mathbf{x}(t)) = \left[w_i^L(\mathbf{x}(t)),\, w_i^U(\mathbf{x}(t))\right], i = 1, 2, \cdots, p \tag{10.2}$$

where

$$w_i^L(\mathbf{x}(t)) = \prod_{\alpha=1}^{\Psi} \underline{\mu}_{\tilde{M}_\alpha^i}(f_\alpha(\mathbf{x}(t))) \geq 0, \tag{10.3}$$

$$w_i^U(\mathbf{x}(t)) = \prod_{\alpha=1}^{\Psi} \overline{\mu}_{\tilde{M}_\alpha^i}(f_\alpha(\mathbf{x}(t))) \geq 0, \tag{10.4}$$

in which $w_i^L(\mathbf{x}(t))$ and $w_i^U(\mathbf{x}(t))$ denote the lower and upper grades of membership, respectively; $\underline{\mu}_{\tilde{M}_\alpha^i}(f_\alpha(\mathbf{x}(t))) \geq 0$ and $\overline{\mu}_{\tilde{M}_\alpha^i}(f_\alpha(\mathbf{x}(t))) \geq 0$ denote the lower and upper membership functions, respectively. It exhibits the property that $\overline{\mu}_{\tilde{M}_\alpha^i}(f_\alpha(\mathbf{x}(t))) \geq \underline{\mu}_{\tilde{M}_\alpha^i}(f_\alpha(\mathbf{x}(t)))$ which leads to $w_i^U(\mathbf{x}(t)) \geq w_i^L(\mathbf{x}(t))$ for all i. The inferred IT2 TS fuzzy model is defined as follows.

$$\begin{aligned}\dot{\mathbf{x}}(t) &= \sum_{i=1}^{p} w_i^L(\mathbf{x}(t))\underline{v}_i(\mathbf{x}(t))(\mathbf{A}_i\mathbf{x}(t) + \mathbf{B}_i\mathbf{u}(t)) \\ &\quad + \sum_{i=1}^{p} w_i^U(\mathbf{x}(t))\overline{v}_i(\mathbf{x}(t))(\mathbf{A}_i\mathbf{x}(t) + \mathbf{B}_i\mathbf{u}(t)) \\ &= \sum_{i=1}^{p} w_i(\mathbf{x}(t))(\mathbf{A}_i\mathbf{x}(t) + \mathbf{B}_i\mathbf{u}(t)) \end{aligned} \tag{10.5}$$

where

$$w_i(\mathbf{x}(t)) = w_i^L(\mathbf{x}(t))\underline{v}_i(\mathbf{x}(t)) + w_i^U(\mathbf{x}(t))\overline{v}_i(\mathbf{x}(t)) \geq 0 \ \forall\, i, \sum_{i=1}^{p} w_i(\mathbf{x}(t)) = 1, \tag{10.6}$$

in which $\underline{v}_i(\mathbf{x}(t)) \geq 0$ and $\overline{v}_i(\mathbf{x}(t)) \geq 0$ are nonlinear functions and exhibit the property of $\underline{v}_i(\mathbf{x}(t)) + \overline{v}_i(\mathbf{x}(t)) = 1$ for all i. In [78], the functions of $\underline{v}_i(\mathbf{x}(t))$ and $\overline{v}_i(\mathbf{x}(t))$ are both defined as 0.5.

Remark 10.1. As the nonlinear plant is subject to parameter uncertainties, $\underline{v}_i(\mathbf{x}(t))$ and $\overline{v}_i(\mathbf{x}(t))$ may be dependent on the parameter uncertainties. However, they are not necessarily known in this investigation.

Remark 10.2. It should be noted that the IT2 TS fuzzy model serves as a mathematical tool to facilitate the design of the IT2 fuzzy controller, and is not necessarily implemented.

Remark 10.3. Referring to the IT2 TS fuzzy model (10.5), the unknown functions of $\underline{v}_i(\mathbf{x}(t))$ and $\overline{v}_i(\mathbf{x}(t))$ will lead to uncertain $w_i(\mathbf{x}(t))$. Consequently, the stability analysis for FMB control system subject to perfectly matched premise membership functions presented in Chapter 3, which requires $w_i(\mathbf{x}(t))$ to be known for all i, cannot be applied to investigate the system stability.

Example 10.1. A simple example is given below to illustrate the construction of IT2 TS fuzzy model for a nonlinear plant subject to parameter uncertainty. Considering the following simple scalar dynamical system subject to unknown/uncertain parameter of $a(x(t))$ with known lower and upper bounds,

$$\dot{x}(t) = \sin\big(a(x(t))x(t)\big)x(t) \tag{10.7}$$

where $x(t) \in \left[-2,\, 2\right]$ is the system state. For demonstration purposes, it is assumed that $a(x(t)) = \frac{b(t)x(t)^2+1}{10}$ and $b(t) \in \left[0,\, 1\right]$ is an unknown function. Hence, it can be seen that $\underline{a} \leq a(x(t)) \leq \overline{a}$ where $\underline{a} = 0.1$ and $\overline{a} = 0.5$ are the constant lower and upper bounds of $a(x(t))$, respectively.

We set the lower and upper bounds of $\sin\big(a(x(t))x(t)\big)$ to be 1 and -1, respectively, for constructing the type-1 TS fuzzy model using the sector nonlinearity technique [122]. Referring to the variables in Section 2.2 and denoting the membership functions as

$$\mu_{M_1^1}(x(t), a(x(t))) = \frac{1 - \sin\big(a(x(t))x(t)\big)}{2}$$

and

$$\mu_{M_1^2}(x(t), a(x(t))) = 1 - \mu_{M_1^1}(x(t), a(x(t))) = \frac{1 + \sin\big(a(x(t))x(t)\big)}{2},$$

the system (10.7) can be written as

$$\dot{x}(t) = \Big(\mu_{M_1^1}(x(t), a(x(t)))(-1) + \mu_{M_1^2}(x(t), a(x(t)))(1)\Big)x(t).$$

When $a(x(t))$ is considered as a constant, the following type-1 fuzzy rule is employed to describe the nonlinear system of (10.7).

$$\begin{aligned} &\text{Rule } i\text{: IF } x(t) \text{ is } M_1^i \\ &\qquad \text{THEN } \dot{x}(t) = A_i x(t), i = 1, 2 \end{aligned} \tag{10.8}$$

where M_1^1 and M_1^2 are type-1 fuzzy sets; $A_1 = -1$ and $A_2 = 1$. The type-1 TS fuzzy model is defined as follows.

$$\dot{x}(t) = \sum_{i=1}^{2} \hat{w}_i(x(t)) A_i x(t) \tag{10.9}$$

where the normalized grades of membership are defined as $\hat{w}_i(x(t)) = \frac{\mu_{M_1^i}(x(t))}{\mu_{M_1^1}(x(t)) + \mu_{M_1^2}(x(t))}$, $i = 1, 2$. It should be noted that $a(x(t))$ is assumed to be a constant; hence, it is not a variable parameter of the type-1 membership functions. The type-1 membership functions of $\mu_{M_1^1}(x(t)) = \frac{1-\sin\left(a(x(t))x(t)\right)}{2}$ with $a(x(t))$ taking various constant values are shown in Fig. 10.1(a).

Now, $a(x(t))$ is considered as an uncertain function with a value in the range of $\underline{a}$ to $\overline{a}$. Consequently, $\mu_{M_1^1}(x(t))$ is no longer a crisp membership function but characterized by the lower and upper memberships of $\underline{\mu}_{\tilde{M}_1^1}(x(t))$ and $\overline{\mu}_{\tilde{M}_1^2}(x(t))$, respectively. With the information of the type-1 membership functions, the lower and upper membership functions satisfying $\underline{\mu}_{\tilde{M}_1^1}(x(t)) \leq \mu_{M_1^1}(x(t)) \leq \overline{\mu}_{\tilde{M}_1^1}(x(t))$ are obtained as follows.

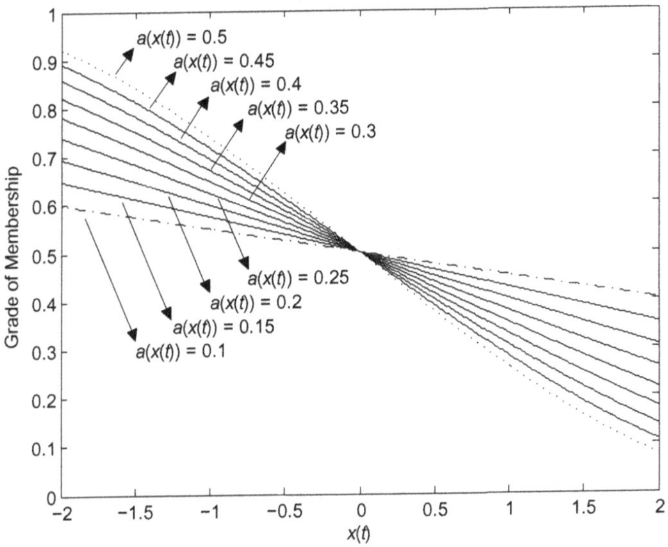

(a) $\mu_{M_1^1}(x(t))$ with various values of $a(x(t))$.

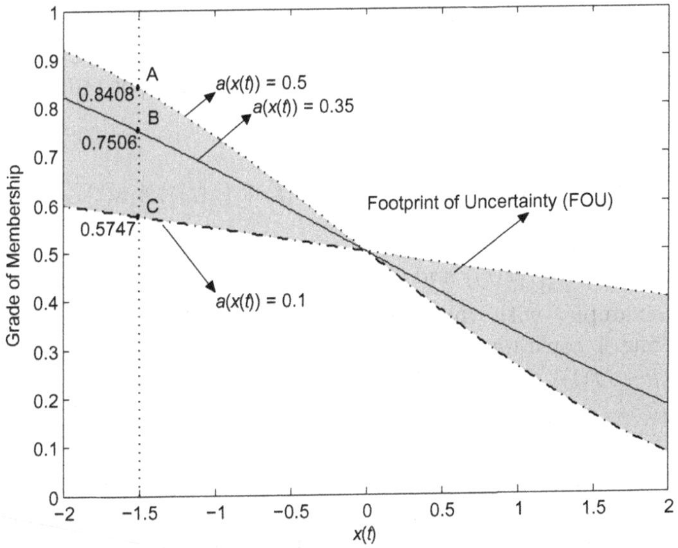

(b) Lower membership function $\underline{\mu}_{\tilde{M}_1^1}(x(t))$ (dash-dot line), upper membership function $\overline{\mu}_{\tilde{M}_1^1}(x(t))$ (dotted line), membership function $\mu_{M_1^1}(x(t))$ with $a(t) = 0.35$ (solid line) and footprint of uncertainty (grey area).

Fig. 10.1 Plot of $\mu_{M_1^1}(x(t)) = \frac{1-\sin\left(a(x(t))x(t)\right)}{2}$ with various values of $a(x(t))$, and illustration of footprint of uncertainty, lower and upper membership functions.

$$\underline{\mu}_{\tilde{M}_1^1}(x(t)) = \begin{cases} \frac{1-\sin(0.1x(t))}{2} & \text{for } x(t) < 0 \\ \frac{1-\sin(0.5x(t))}{2} & \text{for } x(t) \geq 0 \end{cases} \tag{10.10}$$

$$\overline{\mu}_{\tilde{M}_1^1}(x(t)) = \begin{cases} \frac{1-\sin(0.5x(t))}{2} & \text{for } x(t) < 0 \\ \frac{1-\sin(0.1x(t))}{2} & \text{for } x(t) \geq 0 \end{cases} \tag{10.11}$$

Similarly, we have $\underline{\mu}_{\tilde{M}_1^2}(x(t)) \leq \mu_{M_1^2}(x(t)) \leq \overline{\mu}_{\tilde{M}_1^2}(x(t))$ where

$$\underline{\mu}_{\tilde{M}_1^2}(x(t)) = \begin{cases} \frac{1+\sin(0.5x(t))}{2} & \text{for } x(t) < 0 \\ \frac{1+\sin(0.1x(t))}{2} & \text{for } x(t) \geq 0 \end{cases} \tag{10.12}$$

$$\overline{\mu}_{\tilde{M}_1^2}(x(t)) = \begin{cases} \frac{1+\sin(0.1x(t))}{2} & \text{for } x(t) < 0 \\ \frac{1+\sin(0.5x(t))}{2} & \text{for } x(t) \geq 0 \end{cases} \tag{10.13}$$

The following rules for the IT2 TS fuzzy model are employed to describe the nonlinear system (10.7).

$$\begin{aligned} &\text{Rule } i\text{: IF } x(t) \text{ is } \tilde{M}_1^i \\ &\qquad \text{THEN } \dot{x}(t) = A_i x(t), i = 1, 2 \end{aligned} \tag{10.14}$$

From (10.5), the IT2 TS fuzzy model for the nonlinear plant (10.7) is defined as follows.

$$\dot{\mathbf{x}}(t) = \sum_{i=1}^{2} \left(w_i^L(x(t))\underline{v}_i(x(t)) + w_i^U(x(t))\overline{v}_i(x(t)) \right) A_i x(t) \tag{10.15}$$

where $w_i^L(x(t)) = \underline{\mu}_{\tilde{M}_1^i}(x(t))$ and $w_i^U(x(t)) = \overline{\mu}_{\tilde{M}_1^i}(x(t))$ for all i.

In this example, with the type-1 membership function $\mu_{M_1^1}(x(t))$ for $a(x(t))$ taking a constant value of 0.35, the lower and upper membership functions ($\underline{\mu}_{\tilde{M}_1^i}(x(t))$ and $\overline{\mu}_{\tilde{M}_1^i}(x(t))$, $i = 1, 2$) are shown in Fig. 10.1(b). The area in between the lower and upper membership functions is defined as the FOU.

Based on the lower and upper membership functions, any type-1 membership functions in between can be reconstructed with the unknown nonlinear functions of $\underline{v}_i(x(t))$ and $\overline{v}_i(x(t))$. Referring to Fig. 10.1(b), taking $x(t) = -1.5$ for example, the grades of membership at points A, B and C are 0.8408, 0.7506 and 0.5747, respectively. The grade of membership at point B (0.7506) can be reconstructed by using the grades of lower and upper membership functions, i.e., $0.7506 = 0.5747(0.3390) + 0.8408(0.6610)$. By the same line of logic, every point of the membership function with $a(t) = 0.35$ can be determined based on the lower and upper membership functions. In general, the in-between membership functions can be reconstructed and represented in the form of a linear combination of the lower and upper membership functions, i.e., $w_i^L(x(t))\underline{v}_i(x(t)) + w_i^U(x(t))\overline{v}_i(x(t))$ where $\underline{v}_i(x(t)) + \overline{v}_i(x(t)) = 1$ for any $0.1 \leq a(x(t)) \leq 0.5$. In the above example, it can be seen that $\underline{v}_i(x(t)) = 0.3390$ and $\overline{v}_i(x(t)) = 0.6610$ at $a(t) = 0.35$.

With the IT2 fuzzy sets, the parameter uncertainties can be captured by the lower and upper membership functions. Consequently, an IT2 TS fuzzy model of (10.15) can be used to describe the nonlinear system (10.7) subject to the uncertain parameter of $a(x(t))$. The IT2 TS fuzzy model (10.15) can be regarded as a collection of type-1 TS fuzzy models with $a(x(t))$ taking various constant values.

Remark 10.4. It should be noted that the exact values of $\underline{v}_i(x(t))$ and $\overline{v}_i(x(t))$ are unknown as they depend on parameter uncertainty $a(x(t))$. However, it can be seen that the functions of $\underline{v}_i(x(t))$ and $\overline{v}_i(x(t))$ exist and not necessarily known in the stability analysis.

Remark 10.5. The lower and upper membership functions for an IT2 TS fuzzy model is not unique. Referring to Fig. 10.1(b), the footprint of uncertainty can be obtained by the area bounded by the lower and upper membership functions of $\underline{\mu}_{\tilde{M}_1^1}(x(t))$ and $\overline{\mu}_{\tilde{M}_1^1}(x(t))$, respectively. With the nonlinear functions of $\underline{v}_i(x(t))$ and $\overline{v}_i(x(t))$, any in-between membership functions can be reproduced. Based on this concept, by considering any arbitrary lower and upper membership functions denoted by $\underline{\eta}_{\tilde{M}_1^1}(x(t))$ and $\overline{\eta}_{\tilde{M}_1^1}(x(t))$ satisfying the conditions of $\underline{\eta}_{\tilde{M}_1^1}(x(t)) \leq \underline{\mu}_{\tilde{M}_1^1}(x(t))$ and $\overline{\eta}_{\tilde{M}_1^1}(x(t)) \geq \overline{\mu}_{\tilde{M}_1^1}(x(t))$, respectively, the actual membership functions can be reconstructed with the nonlinear functions of $\underline{v}_i(x(t))$ and $\overline{v}_i(x(t))$ in other forms. There is still little understanding on choosing appropriate lower and upper membership functions and more effort should be put on this topic to make the IT2 fuzzy control a powerful fuzzy control approach on handling nonlinear plants subject to parameter uncertainties.

10.2.2 IT2 Fuzzy Controller

An IT2 fuzzy controller with the following fuzzy rules is proposed to stabilize the nonlinear plant represented by the IT2 TS fuzzy model (10.5).

$$\begin{aligned}&\text{Rule } j\text{: IF } f_1(\mathbf{x}(t)) \text{ is } \tilde{M}_1^j \text{ AND } \cdots \text{ AND } f_\Psi(\mathbf{x}(t)) \text{ is } \tilde{M}_\Psi^j\\ &\qquad\text{THEN } \mathbf{u}(t) = \mathbf{G}_j\mathbf{x}(t)\end{aligned} \tag{10.16}$$

where $\mathbf{G}_j \in \Re^{m\times n}$, $j = 1, 2, \cdots, p$, are the feedback gains to be determined. The inferred IT2 fuzzy controller is defined as,

$$\mathbf{u}(t) = \sum_{j=1}^{p}(\underline{w}_j(\mathbf{x}(t)) + \overline{w}_j(\mathbf{x}(t)))\mathbf{G}_j\mathbf{x}(t) \tag{10.17}$$

where

$$\underline{w}_j(\mathbf{x}(t)) = \frac{w_j^L(\mathbf{x}(t))}{\sum_{k=1}^{p}(w_k^L(\mathbf{x}(t)) + w_k^U(\mathbf{x}(t)))} \geq 0$$

and

$$\overline{w}_j(\mathbf{x}(t)) = \frac{w_j^U(\mathbf{x}(t))}{\sum_{k=1}^{p}(w_k^L(\mathbf{x}(t)) + w_k^U(\mathbf{x}(t)))} \geq 0$$

which exhibits the property that $\sum_{k=1}^{p}(\underline{w}_j(\mathbf{x}(t)) + \overline{w}_j(\mathbf{x}(t))) = 1$. It can be seen from (10.17) that the type reduction for the proposed IT2 fuzzy controller is characterized by the average normalized membership grades of the lower and upper membership functions.

10.2.3 Stability Analysis of IT2 FMB Control Systems

The stability of the IT2 FMB control systems is investigated based on Lyapunov stability theory [46, 100, 121] in this section. The IT2 FMB control system formed by the nonlinear plant represented by the IT2 TS fuzzy model (10.5) and the type-2 fuzzy controller of (10.17) connected in a closed loop is defined as follows.

$$\begin{aligned}\dot{\mathbf{x}}(t) &= \sum_{i=1}^{p} w_i(\mathbf{x}(t))\big(\mathbf{A}_i\mathbf{x}(t) + \mathbf{B}_i\sum_{j=1}^{p}(\underline{w}_j(\mathbf{x}(t)) + \overline{w}_j(\mathbf{x}(t)))\mathbf{G}_j\mathbf{x}(t)\big) \\ &= \sum_{i=1}^{p}\sum_{j=1}^{p} w_i(\mathbf{x}(t))(\underline{w}_j(\mathbf{x}(t)) + \overline{w}_j(\mathbf{x}(t)))(\mathbf{A}_i + \mathbf{B}_i\mathbf{G}_j)\mathbf{x}(t)) \qquad (10.18)\end{aligned}$$

In the following analysis, for brevity, $w_i(\mathbf{x}(t))$, $\underline{w}_j(\mathbf{x}(t))$ and $\overline{w}_j(\mathbf{x}(t))$ are denoted as w_i, $\underline{w}_j$ and $\overline{w}_j$, respectively. Given by the property of the membership functions, the equality of $\sum_{i=1}^{p} w_i = \sum_{j=1}^{p}(\underline{w}_j + \overline{w}_j) = \sum_{i=1}^{p}\sum_{j=1}^{p} w_i(\underline{w}_j + \overline{w}_j) = 1$ is employed to facilitate the stability analysis. To investigate the system stability of IT2 FMB control system (10.18), we consider the quadratic Lyapunov function candidate (3.1).

From (3.1) and (10.18), we have,

$$\begin{aligned}\dot{V}(t) &= \dot{\mathbf{x}}(t)^T\mathbf{P}\mathbf{x}(t) + \mathbf{x}(t)^T\mathbf{P}\dot{\mathbf{x}}(t) \\ &= \sum_{i=1}^{p}\sum_{j=1}^{p} w_i(\underline{w}_j + \overline{w}_j)\mathbf{x}(t)^T\left((\mathbf{A}_i + \mathbf{B}_i\mathbf{G}_j)^T\mathbf{P} + \mathbf{P}(\mathbf{A}_i + \mathbf{B}_i\mathbf{G}_j)\right)\mathbf{x}(t).\end{aligned} \qquad (10.19)$$

Denote $\mathbf{X} = \mathbf{P}^{-1}$ and $\mathbf{z}(t) = \mathbf{X}^{-1}\mathbf{x}(t)$. Design the feedback gains as $\mathbf{G}_j = \mathbf{N}_j\mathbf{X}^{-1}$ where $\mathbf{N}_j \in \Re^{m\times n}$ for all j. From (10.19), we have,

$$\begin{aligned}\dot{V}(t) &= \sum_{i=1}^{p}\sum_{j=1}^{p} w_i(\underline{w}_j + \overline{w}_j)\mathbf{z}(t)^T\mathbf{Q}_{ij}\mathbf{z}(t)\\ &= \mathbf{z}(t)^T\mathbf{\Theta}\mathbf{z}(t)\end{aligned} \tag{10.20}$$

where

$$\mathbf{\Theta} = \sum_{i=1}^{p}\sum_{j=1}^{p} w_i(\underline{w}_j + \overline{w}_j)\mathbf{Q}_{ij} \tag{10.21}$$

and $\mathbf{Q}_{ij} = \mathbf{A}_i\mathbf{X} + \mathbf{X}\mathbf{A}_i^T + \mathbf{B}_i\mathbf{N}_j + \mathbf{N}_j^T\mathbf{B}_i^T$.

In the following, we derive the stability conditions that $\dot{V}(t) < 0$ for $\mathbf{z}(t) \neq \mathbf{0}$ ($\mathbf{x}(t) \neq \mathbf{0}$), which implies the asymptotic stability of the IT2 FMB control system of (10.18). It can be seen from (10.20) that $\dot{V}(t) < 0$ is satisfied for $\mathbf{\Theta} < 0$ that can be simply achieved by $\mathbf{Q}_{ij} < 0$ for all i and j. It is expected that such set of stability conditions is very conservative as the membership functions are not considered in stability analysis. For relaxing the stability analysis result, we consider the equality $\sum_{i=1}^{p}(w_i - \underline{w}_i - \overline{w}_i) = 0$ given by the property of the membership functions that $\sum_{i=1}^{p} w_i = \sum_{i=1}^{p}(\underline{w}_i + \overline{w}_i) = 1$. By considering MFSI slack matrices, $\mathbf{M}_j = \mathbf{M}_j^T \in \Re^{n\times n}$, $\mathbf{V}_j = \mathbf{V}_j^T \in \Re^{n\times n}$ and $\mathbf{\Lambda}_j = \mathbf{\Lambda}_j^T \in \Re^{n\times n}$ for all j, we have the following MFSI equality.

$$\mathbf{\Xi} = \sum_{i=1}^{p}\sum_{j=1}^{p}(w_i - \underline{w}_i - \overline{w}_i)(w_j\mathbf{M}_j + \underline{w}_j\mathbf{V}_j + \overline{w}_j\mathbf{\Lambda}_j) = \mathbf{0} \tag{10.22}$$

Remark 10.6. The slack matrices $\mathbf{M}_j$, $\mathbf{V}_j$, and $\mathbf{\Lambda}_j$ in (10.22) are MFSI as they are independent of the membership functions of the IT2 TS fuzzy model or fuzzy controller.

Expanding the terms in the right hand side of (10.22), we have,

$$\begin{aligned}\mathbf{\Xi} = \sum_{i=1}^{p}\sum_{j=1}^{p} \Big(& w_iw_j\mathbf{M}_j + w_i\underline{w}_j(\mathbf{V}_j - \mathbf{M}_i) + w_i\overline{w}_j(\mathbf{\Lambda}_j - \mathbf{M}_i)\\ &- \underline{w}_i\underline{w}_j\mathbf{V}_j - \underline{w}_i\overline{w}_j(\mathbf{\Lambda}_j + \mathbf{V}_i) - \overline{w}_i\overline{w}_j\mathbf{\Lambda}_j\Big)\end{aligned} \tag{10.23}$$

Similar to the MFSD analysis approach in Chapter 3, we further introduce some MFSD slack matrices in the stability analysis. Consider the scalars ρ_{i1}, σ_{i1}, γ_{i1}, ρ_{i2}, σ_{i2}, γ_{i2} such that the membership functions satisfy the following inequalities for all i and $\mathbf{x}(t)$.

$$-w_i + \rho_{i1}\underline{w}_i + \sigma_{i1}\overline{w}_i + \gamma_{i1} \geq 0 \tag{10.24}$$

$$w_i - \rho_{i2}\underline{w}_i - \sigma_{i2}\overline{w}_i + \gamma_{i2} \geq 0 \tag{10.25}$$

Introduce the MFSD slack matrices $\mathbf{R}_{ij} = \mathbf{R}_{ij}^T \in \Re^{n\times n}$ satisfying $\mathbf{R}_{ij} \geq 0$; from (10.24) and (10.25), we have,

$$\mathbf{\Phi} = \sum_{i=1}^{p}\sum_{j=1}^{p}(-w_i + \rho_{i1}\underline{w}_i + \sigma_{i1}\overline{w}_i + \gamma_{i1})(w_j - \rho_{j2}\underline{w}_j - \sigma_{j2}\overline{w}_j + \gamma_{j2})\mathbf{R}_{ij} \geq 0 \tag{10.26}$$

Remark 10.7. The slack matrices $\mathbf{R}_{ij}$ in (10.26) are MFSD as they depend on both the membership functions of the IT2 TS fuzzy model and the fuzzy controller.

Expanding the terms in the right hand side of (10.26), we have,

$$\begin{aligned}
\mathbf{\Phi} = \sum_{i=1}^{p}\sum_{j=1}^{p}&(-w_iw_j + \rho_{j2}w_i\underline{w}_j + \sigma_{j2}w_i\overline{w}_j - w_i\gamma_{j2} \\
&+ \rho_{i1}\underline{w}_iw_j - \rho_{i1}\rho_{j2}\underline{w}_i\underline{w}_j - \rho_{i1}\sigma_{j2}\underline{w}_i\overline{w}_j + \rho_{i1}\underline{w}_i\gamma_{j2} \\
&+ \sigma_{i1}\overline{w}_iw_j - \sigma_{i1}\rho_{j2}\overline{w}_i\underline{w}_j - \sigma_{i2}\sigma_{j2}\overline{w}_i\overline{w}_j + \sigma_{i1}\overline{w}_i\gamma_{j2} \\
&+ \gamma_{i1}w_j - \gamma_{i1}\rho_{j2}\underline{w}_j - \gamma_{i1}\sigma_{j2}\overline{w}_j + \gamma_{i1}\gamma_{j2})\mathbf{R}_{ij} \\
= -&\sum_{i=1}^{p}\sum_{j=1}^{p} w_iw_j\Big(\mathbf{R}_{ij} + \sum_{k=1}^{p}(\gamma_{k2}\mathbf{R}_{ik} - \gamma_{k1}\mathbf{R}_{kj}) - \sum_{k=1}^{p}\sum_{l=1}^{p}\gamma_{k1}\gamma_{l2}\mathbf{R}_{kl}\Big) \\
-&\sum_{i=1}^{p}\sum_{j=1}^{p}\underline{w}_i\underline{w}_j\rho_{i1}\rho_{j2}\mathbf{R}_{ij} - \sum_{i=1}^{p}\sum_{j=1}^{p}\overline{w}_i\overline{w}_j\sigma_{i1}\sigma_{j2}\mathbf{R}_{ij} \\
+&\sum_{i=1}^{p}\sum_{j=1}^{p} w_i\underline{w}_j\Big(\rho_{j2}\mathbf{R}_{ij} + \rho_{j1}\mathbf{R}_{ji} - \sum_{k=1}^{p}(\rho_{j2}\gamma_{k1}\mathbf{R}_{kj} - \rho_{j1}\gamma_{k2}\mathbf{R}_{jk})\Big) \\
+&\sum_{i=1}^{p}\sum_{j=1}^{p} w_i\overline{w}_j\Big(\sigma_{j2}\mathbf{R}_{ij} + \sigma_{j1}\mathbf{R}_{ji} - \sum_{k=1}^{p}(\sigma_{j2}\gamma_{k1}\mathbf{R}_{kj} - \sigma_{j1}\gamma_{k2}\mathbf{R}_{jk})\Big) \\
-&\sum_{i=1}^{p}\sum_{j=1}^{p}\underline{w}_i\overline{w}_j\Big(\rho_{i1}\sigma_{j2}\mathbf{R}_{ij} + \sigma_{j1}\rho_{i2}\mathbf{R}_{ji}\Big).
\end{aligned} \tag{10.27}$$

From (10.21), (10.23) and (10.27), we have

$$\begin{aligned}
\boldsymbol{\Theta} &\leq \boldsymbol{\Theta} + \boldsymbol{\Xi} + \boldsymbol{\Phi} \\
&= -\sum_{i=1}^{p}\sum_{j=1}^{p} w_i w_j \Big(\mathbf{R}_{ij} + \sum_{k=1}^{p}(\gamma_{k2}\mathbf{R}_{ik} - \gamma_{k1}\mathbf{R}_{kj}) - \sum_{k=1}^{p}\sum_{l=1}^{p}\gamma_{k1}\gamma_{l2}\mathbf{R}_{kl} - \mathbf{M}_j\Big) \\
&\quad - \sum_{i=1}^{p}\sum_{j=1}^{p} \underline{w}_i \underline{w}_j \Big(\rho_{i1}\rho_{j2}\mathbf{R}_{ij} + \mathbf{V}_j\Big) - \sum_{i=1}^{p}\sum_{j=1}^{p} \overline{w}_i \overline{w}_j \Big(\sigma_{i1}\sigma_{j2}\mathbf{R}_{ij} + \boldsymbol{\Lambda}_j\Big) \\
&\quad + \sum_{i=1}^{p}\sum_{j=1}^{p} w_i \underline{w}_j \Big(\rho_{j2}\mathbf{R}_{ij} + \rho_{j1}\mathbf{R}_{ji} - \sum_{k=1}^{p}(\rho_{j2}\gamma_{k1}\mathbf{R}_{kj} - \rho_{j1}\gamma_{k2}\mathbf{R}_{jk}) \\
&\quad + \mathbf{Q}_{ij} + \mathbf{V}_j - \mathbf{M}_i\Big) + \sum_{i=1}^{p}\sum_{j=1}^{p} w_i \overline{w}_j \Big(\sigma_{j2}\mathbf{R}_{ij} + \sigma_{j1}\mathbf{R}_{ji} \\
&\quad - \sum_{k=1}^{p}(\sigma_{j2}\gamma_{k1}\mathbf{R}_{kj} - \sigma_{j1}\gamma_{k2}\mathbf{R}_{jk}) + \mathbf{Q}_{ij} + \boldsymbol{\Lambda}_j - \mathbf{M}_i\Big) \\
&\quad - \sum_{i=1}^{p}\sum_{j=1}^{p} \underline{w}_i \overline{w}_j \Big(\rho_{i1}\sigma_{j2}\mathbf{R}_{ij} + \sigma_{j1}\rho_{i2}\mathbf{R}_{ji} + \boldsymbol{\Lambda}_j + \mathbf{V}_i\Big). \qquad (10.28)
\end{aligned}$$

Introduce the slack matrices $\mathbf{S}_{ij} = \mathbf{S}_{ji}^T \in \Re^{n\times n}$, $\mathbf{T}_{ij} \in \Re^{n\times n}$, $\mathbf{U}_{ij} \in \Re^{n\times n}$, $\mathbf{W}_{ij} \in \Re^{n\times n}$, $\mathbf{Y}_{ij} = \mathbf{Y}_{ji}^T \in \Re^{n\times n}$ and $\mathbf{Z}_{ij} = \mathbf{Z}_{ji}^T \in \Re^{n\times n}$ for all i and j. We consider the following LMIs:

$$\mathbf{S}_{ii} < \mathbf{R}_{ii} + \sum_{k=1}^{p}(\gamma_{k2}\mathbf{R}_{ik} - \gamma_{k1}\mathbf{R}_{ki}) - \sum_{k=1}^{p}\sum_{l=1}^{p}\gamma_{k1}\gamma_{l2}\mathbf{R}_{kl} - \mathbf{M}_i \ \forall\, i; \qquad (10.29)$$

$$\begin{aligned}
\mathbf{S}_{ij} + \mathbf{S}_{ij}^T &\leq \mathbf{R}_{ij} + \sum_{k=1}^{p}(\gamma_{k2}\mathbf{R}_{ik} - \gamma_{k1}\mathbf{R}_{kj}) - 2\sum_{k=1}^{p}\sum_{l=1}^{p}\gamma_{k1}\gamma_{l2}\mathbf{R}_{kl} - \mathbf{M}_j \\
&\quad + \mathbf{R}_{ji} + \sum_{k=1}^{p}(\gamma_{k2}\mathbf{R}_{jk} - \gamma_{k1}\mathbf{R}_{ki}) - \mathbf{M}_i \ \forall\, j; i < j; \qquad (10.30)
\end{aligned}$$

$$\begin{aligned}
\mathbf{T}_{ij} + \mathbf{T}_{ij}^T &\geq \rho_{j2}\mathbf{R}_{ij} + \rho_{j1}\mathbf{R}_{ji} - \sum_{k=1}^{p}(\rho_{j2}\gamma_{k1}\mathbf{R}_{kj} - \rho_{j1}\gamma_{k2}\mathbf{R}_{jk}) \\
&\quad + \mathbf{Q}_{ij} + \mathbf{V}_j - \mathbf{M}_i \ \forall\, i,\ j; \qquad (10.31)
\end{aligned}$$

$$\begin{aligned}
\mathbf{U}_{ij} + \mathbf{U}_{ij}^T &\geq \sigma_{j2}\mathbf{R}_{ij} + \sigma_{j1}\mathbf{R}_{ji} - \sum_{k=1}^{p}(\sigma_{j2}\gamma_{k1}\mathbf{R}_{kj} \\
&\quad - \sigma_{j1}\gamma_{k2}\mathbf{R}_{jk}) + \mathbf{Q}_{ij} + \boldsymbol{\Lambda}_j - \mathbf{M}_i \ \forall\, i,\ j; \qquad (10.32)
\end{aligned}$$

$$\mathbf{W}_{ij} + \mathbf{W}_{ij}^T \le \rho_{i1}\sigma_{j2}\mathbf{R}_{ij} + \sigma_{j1}\rho_{i2}\mathbf{R}_{ji} + \mathbf{\Lambda}_j + \mathbf{V}_i \ \forall\, i,\ j; \tag{10.33}$$

$$\mathbf{Y}_{ii} < \rho_{i1}\rho_{j2}\mathbf{R}_{ii} + \mathbf{V}_i \ \forall\, i; \tag{10.34}$$

$$\mathbf{Y}_{ij} + \mathbf{Y}_{ij}^T \le \rho_{i1}\rho_{j2}\mathbf{R}_{ij} + \mathbf{V}_j + \rho_{j1}\rho_{i2}\mathbf{R}_{ji} + \mathbf{V}_i \ \forall\, j; i < j; \tag{10.35}$$

$$\mathbf{Z}_{ii} < \sigma_{i1}\sigma_{i2}\mathbf{R}_{ii} + \mathbf{\Lambda}_i \ \forall\, i; \tag{10.36}$$

$$\mathbf{Z}_{ij} + \mathbf{Z}_{ij}^T \le \sigma_{i1}\sigma_{j2}\mathbf{R}_{ij} + \mathbf{\Lambda}_j + \sigma_{j1}\sigma_{i2}\mathbf{R}_{ji} + \mathbf{\Lambda}_i \ \forall\, j; i < j; \tag{10.37}$$

From (10.29) to (10.37), we have

$$\begin{aligned}
\dot{V}(t) \le & -\sum_{i=1}^{p} w_i^2 \mathbf{z}(t)^T \mathbf{S}_{ii} \mathbf{z}(t) - \sum_{j=1}^{p}\sum_{i<j} w_i w_j \mathbf{z}(t)^T (\mathbf{S}_{ij} + \mathbf{S}_{ij}^T)\mathbf{z}(t) \\
& - \sum_{i=1}^{p} \underline{w}_i^2 \mathbf{z}(t)^T \mathbf{Y}_{ii} \mathbf{z}(t) - \sum_{j=1}^{p}\sum_{i<j} \underline{w}_i \underline{w}_j \mathbf{z}(t)^T (\mathbf{Y}_{ij} + \mathbf{Y}_{ij}^T)\mathbf{z}(t) \\
& - \sum_{i=1}^{p} \overline{w}_i^2 \mathbf{z}(t)^T \mathbf{Z}_{ii} \mathbf{z}(t) - \sum_{j=1}^{p}\sum_{i<j} \overline{w}_i \overline{w}_j \mathbf{z}(t)^T (\mathbf{Z}_{ij} + \mathbf{Z}_{ij}^T)\mathbf{z}(t) \\
& + \sum_{i=1}^{p}\sum_{j=1}^{p} w_i \underline{w}_j \mathbf{z}(t)^T (\mathbf{T}_{ij} + \mathbf{T}_{ij}^T)\mathbf{z}(t) \\
& + \sum_{i=1}^{p}\sum_{j=1}^{p} w_i \overline{w}_j \mathbf{z}(t)^T (\mathbf{U}_{ij} + \mathbf{U}_{ij}^T)\mathbf{z}(t) \\
& - \sum_{i=1}^{p}\sum_{j=1}^{p} \underline{w}_i \overline{w}_j \mathbf{z}(t)^T (\mathbf{W}_{ij} + \mathbf{W}_{ij}^T)\mathbf{z}(t) \\
= & \begin{bmatrix} \mathbf{r}(t) \\ \mathbf{s}(t) \\ \mathbf{t}(t) \end{bmatrix}^T \begin{bmatrix} -\mathbf{S} & \mathbf{T} & \mathbf{U} \\ \mathbf{T}^T & -\mathbf{Y} & -\mathbf{W} \\ \mathbf{U}^T & -\mathbf{W}^T & -\mathbf{Z} \end{bmatrix} \begin{bmatrix} \mathbf{r}(t) \\ \mathbf{s}(t) \\ \mathbf{t}(t) \end{bmatrix}.
\end{aligned} \tag{10.38}$$

where $\mathbf{r}(t) = \begin{bmatrix} w_1\mathbf{z}(t) \\ w_2\mathbf{z}(t) \\ \vdots \\ w_p\mathbf{z}(t) \end{bmatrix}$, $\mathbf{s}(t) = \begin{bmatrix} \underline{w}_1\mathbf{z}(t) \\ \underline{w}_2\mathbf{z}(t) \\ \vdots \\ \underline{w}_p\mathbf{z}(t) \end{bmatrix}$, $\mathbf{t}(t) = \begin{bmatrix} \overline{w}_1\mathbf{z}(t) \\ \overline{w}_2\mathbf{z}(t) \\ \vdots \\ \overline{w}_p\mathbf{z}(t) \end{bmatrix}$,
$\mathbf{S}= \begin{bmatrix} \mathbf{S}_{11} & \mathbf{S}_{12} & \cdots & \mathbf{S}_{1p} \\ \mathbf{S}_{21} & \mathbf{S}_{22} & \cdots & \mathbf{S}_{2p} \\ \vdots & \vdots & \vdots & \vdots \\ \mathbf{S}_{p1} & \mathbf{S}_{p2} & \cdots & \mathbf{S}_{pp} \end{bmatrix}$, $\mathbf{T}= \begin{bmatrix} \mathbf{T}_{11} & \mathbf{T}_{12} & \cdots & \mathbf{T}_{1p} \\ \mathbf{T}_{21} & \mathbf{T}_{22} & \cdots & \mathbf{T}_{2p} \\ \vdots & \vdots & \vdots & \vdots \\ \mathbf{T}_{p1} & \mathbf{T}_{p2} & \cdots & \mathbf{T}_{pp} \end{bmatrix}$, $\mathbf{U}= \begin{bmatrix} \mathbf{U}_{11} & \mathbf{U}_{12} & \cdots & \mathbf{U}_{1p} \\ \mathbf{U}_{21} & \mathbf{U}_{22} & \cdots & \mathbf{U}_{2p} \\ \vdots & \vdots & \vdots & \vdots \\ \mathbf{U}_{p1} & \mathbf{U}_{p2} & \cdots & \mathbf{U}_{pp} \end{bmatrix}$,
$\mathbf{W} = \begin{bmatrix} \mathbf{W}_{11} & \mathbf{W}_{12} & \cdots & \mathbf{W}_{1p} \\ \mathbf{W}_{21} & \mathbf{W}_{22} & \cdots & \mathbf{W}_{2p} \\ \vdots & \vdots & \vdots & \vdots \\ \mathbf{W}_{p1} & \mathbf{W}_{p2} & \cdots & \mathbf{W}_{pp} \end{bmatrix}$, $\mathbf{Y} = \begin{bmatrix} \mathbf{Y}_{11} & \mathbf{Y}_{12} & \cdots & \mathbf{Y}_{1p} \\ \mathbf{Y}_{21} & \mathbf{Y}_{22} & \cdots & \mathbf{Y}_{2p} \\ \vdots & \vdots & \vdots & \vdots \\ \mathbf{Y}_{p1} & \mathbf{Y}_{p2} & \cdots & \mathbf{Y}_{pp} \end{bmatrix}$ and
$\mathbf{Z} = \begin{bmatrix} \mathbf{Z}_{11} & \mathbf{Z}_{12} & \cdots & \mathbf{Z}_{1p} \\ \mathbf{Z}_{21} & \mathbf{Z}_{22} & \cdots & \mathbf{Z}_{2p} \\ \vdots & \vdots & \vdots & \vdots \\ \mathbf{Z}_{p1} & \mathbf{Z}_{p2} & \cdots & \mathbf{Z}_{pp} \end{bmatrix}$.

From (3.1) and (10.38), based on the the Lyapunov stability theory, $V(t) > 0$ and $\dot{V}(t) < 0$ for $\mathbf{z}(t) \neq \mathbf{0}$ ($\mathbf{x}(t) \neq \mathbf{0}$) implying the asymptotic stability of the FMB control system (10.18), i.e., $\mathbf{x}(t) \to 0$ when time $t \to \infty$, can be achieved if the stability conditions summarized in the following theorem are satisfied.

Theorem 10.1. *The IT2 FMB control system (10.18), formed by the nonlinear plant represented by the IT2 fuzzy model (10.5) and the IT2 TS fuzzy controller (10.17) connected in a closed loop, is asymptotically stable if there exist predefined constant scalars* ρ_{i1}, σ_{i1}, γ_{i1}, ρ_{i2}, σ_{i2}, γ_{i2} i, $j = 1, 2, \cdots, p$, *satisfying the inequalities (10.24) to (10.25) for all* i, j *and* $\mathbf{x}(t)$, *and there exist matrices* $\mathbf{M}_j = \mathbf{M}_j^T \in \Re^{n\times n}$, $\mathbf{N}_j \in \Re^{m\times n}$, $\mathbf{R}_{ij} = \mathbf{R}_{ij}^T \in \Re^{n\times n}$, $\mathbf{S}_{ij} = \mathbf{S}_{ji}^T \in \Re^{n\times n}$, $\mathbf{T}_{ij} \in \Re^{n\times n}$, $\mathbf{U}_{ij} \in \Re^{n\times n}$, $\mathbf{V}_i = \mathbf{V}_i^T \in \Re^{n\times n}$, $\mathbf{W}_{ij} \in \Re^{n\times n}$, $\mathbf{X} = \mathbf{X}^T \in \Re^{n\times n}$, $\mathbf{Y}_{ij} = \mathbf{Y}_{ji}^T \in \Re^{n\times n}$, $\mathbf{Z}_{ij} = \mathbf{Z}_{ji}^T \in \Re^{n\times n}$ *and* $\mathbf{\Lambda}_i = \mathbf{\Lambda}_i^T \in \Re^{n\times n}$ *such that the following LMIs are satisfied.*

$$\mathbf{X} > 0;$$

$$\mathbf{R}_{ij} \geq 0 \,\forall\, i,\, j;$$

(10.29) to (10.37);

$$\begin{bmatrix} -\mathbf{S} & \mathbf{T} & \mathbf{U} \\ \mathbf{T}^T & -\mathbf{Y} & -\mathbf{W} \\ \mathbf{U}^T & -\mathbf{W}^T & -\mathbf{Z} \end{bmatrix} < 0;$$

and the feedback gains are designed as $\mathbf{G}_j = \mathbf{N}_j\mathbf{X}^{-1}$ *for all* j.

Example 10.2. Consider an IT2 TS fuzzy model with 3 fuzzy rules in the following format.

$$\text{Rule } i\text{: IF } x_1(t) \text{ is } \tilde{M}_1^i$$
$$\text{THEN } \dot{\mathbf{x}}(t) = \mathbf{A}_i\mathbf{x}(t) + \mathbf{B}_i u(t), i = 1, 2, 3 \qquad (10.39)$$

where $\mathbf{x}(t) = \begin{bmatrix} x_1(t) \; x_2(t) \end{bmatrix}^T$ $\mathbf{A}_1 = \begin{bmatrix} 2.78 & -5.63 \\ 0.01 & 0.33 \end{bmatrix}$, $\mathbf{A}_2 = \begin{bmatrix} 0.2 & -3.22 \\ 0.35 & 0.12 \end{bmatrix}$, $\mathbf{A}_3 = \begin{bmatrix} -a & -6.63 \\ 0.45 & 0.15 \end{bmatrix}$, $\mathbf{B}_1 = \begin{bmatrix} 2 \\ -1 \end{bmatrix}$, $\mathbf{B}_2 = \begin{bmatrix} 8 \\ 0 \end{bmatrix}$, $\mathbf{B}_3 = \begin{bmatrix} -b+6 \\ -1 \end{bmatrix}$ where $22 \le a \le 30$ and $20 \le b \le 25$ are constant parameters.

It is assumed that the IT2 TS fuzzy model is working in the operating domain $x_1(t) \in \begin{bmatrix} -10, 10 \end{bmatrix}$. The lower and upper membership functions are given in Table 10.1. The actual membership functions subject to parameter uncertainties are inside the region (footprint of uncertainty) bounded by the lower and upper membership functions.

Table 10.1 Lower and upper membership functions for Example 10.2.

Lower Membership Functions	Upper Membership Functions
$w_1^L(x_1(t)) = 0.95 - \frac{0.925}{1+e^{-\frac{x_1(t)+3.5}{8}}}$	$w_1^U(x_1(t)) = 0.95 - \frac{0.925}{1+e^{-\frac{x_1(t)+4.5}{8}}}$
$w_2^L(x_1(t)) = 1 - w_1^L(x_1(t)) - w_3^L(x_1(t))$	$w_2^U(x_1(t)) = 1 - w_1^U(x_1(t)) - w_3^U(x_1(t))$
$w_3^L(x_1(t)) = 0.025 + \frac{0.925}{1+e^{-\frac{x_1(t)-3.5}{8}}}$	$w_3^U(x_1(t)) = 0.025 + \frac{0.925}{1+e^{-\frac{x_1(t)-4.5}{8}}}$

For demonstration purposes, with the IT2 fuzzy controller defined in (10.17), it is assumed that $\rho_{i1} = \sigma_{i1} = 2$, $\rho_{i2} = \sigma_{i2} = 0.01$, $\gamma_{i1} = -0.025$ and $\gamma_{i2} = -0.024$ satisfy the inequalities of (10.24) to (10.25). The stability conditions in Theorem 10.1 are employed to check for the stability region using the MATLAB LMI toolbox. The stability region is shown in Fig. 10.2(a) indicated by '$\circ$'. To investigate the influence of the slack matrices, $\mathbf{M}_i$, $\mathbf{V}_i$ and $\mathbf{\Lambda}_i$, to the stability conditions, we remove them from Theorem 10.1. The stability region without these slack matrices is shown in Fig. 10.2(a) indicated by '$\times$'. Referring to Fig. 10.2(a), it can be seen that the stability conditions can be relaxed with these slack matrices as evidenced by a larger stability region.

To investigate the influence of the bounding coefficients of the inequalities of (10.24) to (10.25) to the stability region, we choose γ_{i1} as -0.048 and -0.012 with other coefficients unchanged. The stability region is shown in Fig. 10.2(b). It can be seen that different bounding coefficients (due to different lower and upper membership functions) will lead to different sizes of stability regions.

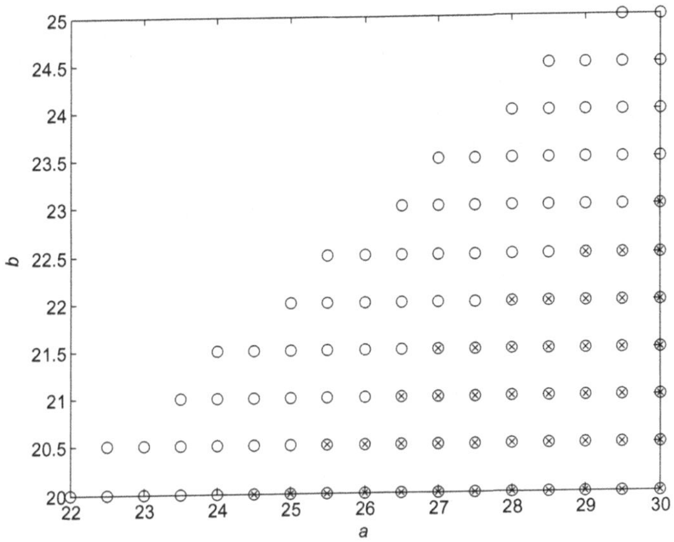

(a) Stability regions with ('o') and without ('×') $\mathbf{M}_i$, $\mathbf{V}_i$ and $\mathbf{\Lambda}_i$.

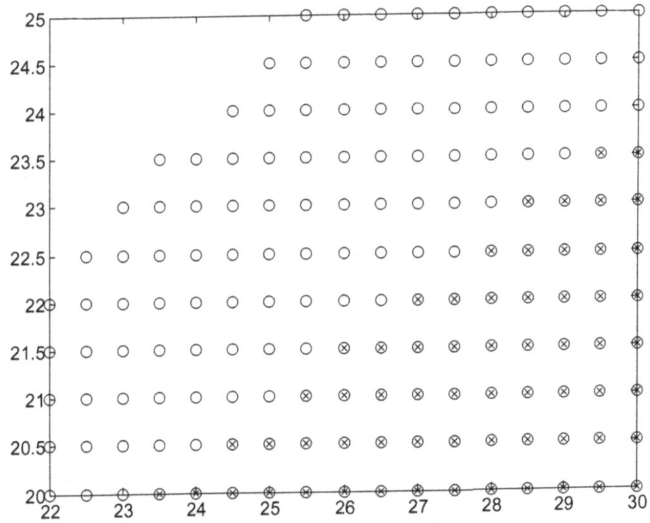

(b) Stability region with different bounding coefficients of $\gamma_{i1} = -0.048$ ('o') and $\gamma_{i1} = -0.012$ ('×').

Fig. 10.2 Stability regions for the stability conditions in Theorem 10.1.

Example 10.3. Consider the inverted pendulum in Example 7.1 subject to parameter uncertainty as the nonlinear plant in this example. We consider the mass of the pole m_p and the mass of the cart M_c are uncertain. An IT2 TS fuzzy model is first constructed to represent the inverted pendulum. Different sets of lower and upper membership functions are considered to capture the parameter uncertainties. The proposed IT2 fuzzy controller is then employed to stabilize the inverted pendulum.

The dynamics of the inverted pendulum is described by the dynamical equations (7.33). It is considered that the mass of the inverted pendulum $m_p \in \left[m_{p_{min}}, m_{p_{max}}\right] = \left[2, 3\right]$ and the mass of the cart $M_c \in \left[M_{c_{min}}, M_{c_{max}}\right] = \left[8, 16\right]$ are uncertain. It was reported in [56] that the uncertainty-free inverted pendulum, of which m_p and M_c take constant values, can be exactly represented by the type-1 TS fuzzy model (7.35). The fuzzy model is restated as follows.

$$\dot{\mathbf{x}}(t) = \sum_{i=1}^{4} \hat{w}_i(\mathbf{x}(t))(\mathbf{A}_i\mathbf{x}(t) + \mathbf{B}_i u(t)) \tag{10.40}$$

The inverted pendulum is considered working in the operating domain $x_1(t) = \theta(t) \in \left[-\frac{5\pi}{12}, \frac{5\pi}{12}\right]$ and $x_2(t) = \dot{\theta}(t) \in \left[-5, 5\right]$. Consequently, we have $f_{1_{min}} = 10.0078$ and $f_{1_{max}} = 18.4800$, $f_{2_{min}} = -0.1765$ and $f_{2_{max}} = -0.0261$. The rest variables are defined in Example 7.1.

It can be seen from the type-1 TS fuzzy model (10.40) that the grades of membership become uncertain in value when m_p and M_c are uncertain. Consequently, the stability conditions subject to perfectly matched premise membership functions in Chapter 2 cannot be applied to design a stable fuzzy controller. An IT2 TS fuzzy model with 4 rules of the following format is employed to describe the inverted pendulum subject to parameter uncertainties.

$$\begin{aligned} &\text{Rule } i\text{: IF } x_1(t) \text{ is } \tilde{M}_1^i \text{ AND } x_1(t) \text{ is } \tilde{M}_2^i \\ &\qquad \text{THEN } \dot{\mathbf{x}}(t) = \mathbf{A}_i\mathbf{x}(t) + \mathbf{B}_i u(t), i = 1, 2, 3, 4 \end{aligned} \tag{10.41}$$

The IT2 TS fuzzy model is defined as follows.

$$\begin{aligned} \dot{\mathbf{x}}(t) &= \sum_{i=1}^{4} \left(w_i^L(x_1(t))\underline{v}_i(\mathbf{x}(t)) + w_i^U(x_1(t))\overline{v}_i(\mathbf{x}(t))\right)\left(\mathbf{A}_i\mathbf{x}(t) + \mathbf{B}_i\ u(t)\right) \\ &= \sum_{i=1}^{4} \tilde{w}_i(\mathbf{x}(t))\left(\mathbf{A}_i\mathbf{x}(t) + \mathbf{B}_i u(t)\right) \end{aligned} \tag{10.42}$$

where $\tilde{w}_i(\mathbf{x}(t)) = w_i^L(x_1(t))\underline{v}_i(\mathbf{x}(t)) + w_i^U(x_1(t))\overline{v}_i(\mathbf{x}(t))$.

The lower and upper membership functions are required to satisfy the following inequalities.

$$\underline{\mu}_{\tilde{M}_1^i}(x_1(t)) \leq \mu_{M_1^i}(f_1(\mathbf{x}(t)) \leq \overline{\mu}_{\tilde{M}_1^i}(x_1(t)), i = 1, 2, 3, 4 \tag{10.43}$$

$$\underline{\mu}_{\tilde{M}_2^i}(x_1(t)) \leq \mu_{M_2^i}(f_2(x_1(t)) \leq \overline{\mu}_{\tilde{M}_2^i}(x_1(t)), i = 1, 2, 3, 4 \tag{10.44}$$

It can be seen from (10.43) and (10.44) that the lower and upper membership functions form the FOU that captures the parameter uncertainties m_p and M_c. Any type-1 membership functions in between can be reconstructed based on the lower and upper membership functions. As a result, the IT2 TS fuzzy model is equivalent to have infinite number of type-1 fuzzy models in the form of (10.40) where each one of them takes different constant values of m_p and M_c within the predefined lower and upper bounds.

By taking $x_2(t)$, m_p and M_c as constant values in their operating ranges of $f_1(\mathbf{x}(t))$ and $f_2(x_1(t))$ and considering all combinations of them, we can determine the lower and upper bounds of $\mu_{M_1^i}(f_1(\mathbf{x}(t))$ and $\mu_{M_2^i}(f_2(x_1(t))$ that they are taken as the lower and upper membership functions of the IT2 TS fuzzy model. The lower and upper membership functions are defined in Table 10.2 and Table 10.3, respectively. It should be noted that $\underline{\mu}_{\tilde{M}_1^i}(x_1(t))$ for all i depend on $x_1(t)$ only due to $x_2(t)$ is taken as a constant value. The lower and upper normalized grades of membership for each rule are defined as $w_i^L(x_1(t)) = \underline{\mu}_{\tilde{M}_1^i}(x_1(t))\underline{\mu}_{\tilde{M}_2^i}(x_1(t))$ and $w_i^U(x_1(t)) = \overline{\mu}_{\tilde{M}_1^i}(x_1(t))\overline{\mu}_{\tilde{M}_2^i}(x_1(t))$, respectively, for all i. It can be shown that $w_i^L(x_1(t)) \leq \tilde{w}_i(\mathbf{x}(t)) \leq w_i^U(x_1(t))$. The plots of the lower and upper normalized membership functions of the IT2 TS fuzzy model are shown in Fig. 10.3.

Table 10.2 Lower membership functions for Example 10.3.

$$
\begin{aligned}
\underline{\mu}_{\tilde{M}_1^1}(x_1(t)) &= \tfrac{-f_1(\mathbf{x}(t))+f_{1_{max}}}{f_{1_{max}}-f_{1_{min}}} \text{ with } x_2(t)=0,\ m_p=m_{p_{max}},\ M_c=M_{c_{min}}\\
\underline{\mu}_{\tilde{M}_1^2}(x_1(t)) &= \tfrac{-f_1(\mathbf{x}(t))+f_{1_{max}}}{f_{1_{max}}-f_{1_{min}}} \text{ with } x_2(t)=0,\ m_p=m_{p_{max}},\ M_c=M_{c_{min}}\\
\underline{\mu}_{\tilde{M}_1^3}(x_1(t)) &= \tfrac{f_1(\mathbf{x}(t))-f_{1_{min}}}{f_{1_{max}}-f_{1_{min}}} \text{ with } x_2(t)=x_{2_{max}},\ m_p=m_{p_{max}},\ M_c=M_{c_{min}}\\
\underline{\mu}_{\tilde{M}_1^4}(x_1(t)) &= \tfrac{f_1(\mathbf{x}(t))-f_{1_{min}}}{f_{1_{max}}-f_{1_{min}}} \text{ with } x_2(t)=x_{2_{max}},\ m_p=m_{p_{max}},\ M_c=M_{c_{min}}\\
\underline{\mu}_{\tilde{M}_2^1}(x_1(t)) &= \tfrac{-f_2(x_1(t))+f_{2_{max}}}{f_{2_{max}}-f_{2_{min}}} \text{ with } m_p=m_{p_{max}},\ M_c=M_{c_{max}}\\
\underline{\mu}_{\tilde{M}_2^2}(x_1(t)) &= \tfrac{f_2(x_1(t))-f_{2_{min}}}{f_{2_{max}}-f_{2_{min}}} \text{ with } m_p=m_{p_{min}},\ M_c=M_{c_{min}}\\
\underline{\mu}_{\tilde{M}_2^3}(x_1(t)) &= \tfrac{-f_2(x_1(t))+f_{2_{max}}}{f_{2_{max}}-f_{2_{min}}} \text{ with } m_p=m_{p_{max}},\ M_c=M_{c_{max}}\\
\underline{\mu}_{\tilde{M}_2^4}(x_1(t)) &= \tfrac{f_2(x_1(t))-f_{2_{min}}}{f_{2_{max}}-f_{2_{min}}} \text{ with } m_p=m_{p_{min}},\ M_c=M_{c_{min}}
\end{aligned}
$$

Based on the IT2 TS fuzzy model, an IT2 fuzzy controller with four rules of the following format is employed to stabilize the inverted pendulum (7.33), i.e., $\mathbf{x}(t) \to \mathbf{0}$ as time $t \to \infty$.

Table 10.3 Upper membership functions for Example 10.3.

$\overline{\mu}_{\tilde{M}_1^1}(x_1(t)) = \frac{-f_1(\mathbf{x}(t))+f_{1_{max}}}{f_{1_{max}}-f_{1_{min}}}$ with $x_2(t) = x_{2_{max}}$, $m_p = m_{p_{max}}$, $M_c = M_{c_{min}}$
$\overline{\mu}_{\tilde{M}_1^2}(x_1(t)) = \frac{-f_1(\mathbf{x}(t))+f_{1_{max}}}{f_{1_{max}}-f_{1_{min}}}$ with $x_2(t) = x_{2_{max}}$, $m_p = m_{p_{max}}$, $M_c = M_{c_{min}}$
$\overline{\mu}_{\tilde{M}_1^3}(x_1(t)) = \frac{f_1(\mathbf{x}(t))-f_{1_{min}}}{f_{1_{max}}-f_{1_{min}}}$ with $x_2(t) = 0$, $m_p = m_{p_{max}}$, $M_c = M_{c_{min}}$
$\overline{\mu}_{\tilde{M}_1^4}(x_1(t)) = \frac{f_1(\mathbf{x}(t))-f_{1_{min}}}{f_{1_{max}}-f_{1_{min}}}$ with $x_2(t) = 0$, $m_p = m_{p_{max}}$, $M_c = M_{c_{min}}$
$\overline{\mu}_{\tilde{M}_2^1}(x_1(t)) = \frac{-f_2(x_1(t))+f_{2_{max}}}{f_{2_{max}}-f_{2_{min}}}$ with $m_p = m_{p_{min}}$, $M_c = M_{c_{min}}$
$\overline{\mu}_{\tilde{M}_2^2}(x_1(t)) = \frac{f_2(x_1(t))-f_{2_{min}}}{f_{2_{max}}-f_{2_{min}}}$ with $m_p = m_{p_{max}}$, $M_c = M_{c_{max}}$
$\overline{\mu}_{\tilde{M}_2^3}(x_1(t)) = \frac{-f_2(x_1(t))+f_{2_{max}}}{f_{2_{max}}-f_{2_{min}}}$ with $m_p = m_{p_{min}}$, $M_c = M_{c_{min}}$
$\overline{\mu}_{\tilde{M}_2^4}(x_1(t)) = \frac{f_2(x_1(t))-f_{2_{min}}}{f_{2_{max}}-f_{2_{min}}}$ with $m_p = m_{p_{max}}$, $M_c = M_{c_{max}}$

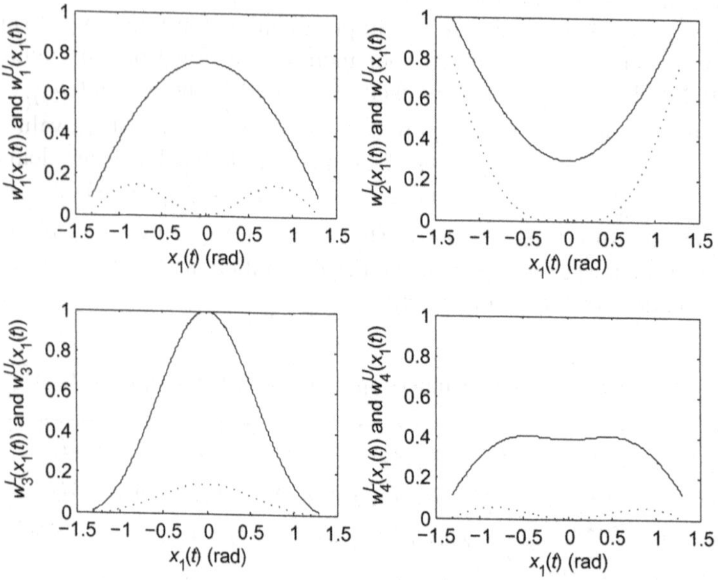

Fig. 10.3 Lower membership functions $w_i^L(x_1(t))$ (dotted lines) and upper membership functions $w_i^U(x_1(t))$ (solid lines), $i = 1, 2, 3, 4$ for Example 10.3.

$$\begin{aligned} &\text{Rule } j\text{: IF } x_1(t) \text{ is } \tilde{M}_1^j \text{ AND } x_1(t) \text{ is } \tilde{M}_\Psi^j \\ &\qquad \text{THEN } u(t) = \mathbf{G}_j\mathbf{x}(t), j = 1, 2, 3, 4 \end{aligned} \tag{10.45}$$

Referring to (10.17), the inferred IT2 fuzzy controller is defined as,

$$u(t) = \frac{\sum_{j=1}^{p}(w_j^L(x_1(t)) + w_j^U(x_1(t)))\mathbf{G}_j\mathbf{x}(t)}{\sum_{k=1}^{p}(w_k^L(x_1(t)) + w_k^U(x_1(t))} \tag{10.46}$$

Considering the chosen lower and upper membership functions, it can be shown numerically that $\rho_{ik} = 1$, $\sigma_{ik} = 0.001$, $\gamma_{11} = 0.441$, $\gamma_{21} = 0.191$, $\gamma_{31} = 0.525$, $\gamma_{41} = 0.2$, $\gamma_{i2} = 0.001$, $i = 1, 2, 3, 4$; $k = 1, 2$, satisfy the inequalities (10.24) and (10.25). Based on the stability conditions in Theorem 10.1, with the MATLAB LMI toolbox, the feedback gains are found numerically and shown in Table 10.4.

Table 10.4 Feedback gains $\mathbf{G}_j$ for Example 10.3.

$\mathbf{G}_1 =$	$\begin{bmatrix} 934.7822 & 238.7590 \end{bmatrix}$
$\mathbf{G}_2 =$	$\begin{bmatrix} 1027.3843 & 260.9691 \end{bmatrix}$
$\mathbf{G}_3 =$	$\begin{bmatrix} 1000.4873 & 254.0293 \end{bmatrix}$
$\mathbf{G}_4 =$	$\begin{bmatrix} 1138.3668 & 289.6253 \end{bmatrix}$

The IT2 fuzzy controller (10.46) with the feedback gains in Table 10.4 is employed to control the inverted pendulum (7.33). Considering the initial system states $\mathbf{x}(0) = \begin{bmatrix} \frac{5\pi}{12} & 0 \end{bmatrix}^T$, $\mathbf{x}(0) = \begin{bmatrix} -\frac{5\pi}{12} & 0 \end{bmatrix}^T$, $\mathbf{x}(0) = \begin{bmatrix} \frac{\pi}{6} & 0 \end{bmatrix}^T$ and $\mathbf{x}(0) = \begin{bmatrix} -\frac{\pi}{6} & 0 \end{bmatrix}^T$, the system state responses and control signals of the IT2 FMB control system with $m_p = m_{p_{min}}$ and $M_c = M_{c_{min}}$ are shown in Fig. 10.4. To check for the robustness of the IT2 fuzzy controller, we choose $m_p = m_{p_{max}}$ and $M_c = M_{c_{max}}$. It can be seen from Fig. 10.4 that the IT2 fuzzy controller is able to stabilize the inverted pendulum with the chosen values of m_p and M_c.

As discussed in Remark 10.5, any functions in the range of 0 to 1 satisfying the inequalities (10.43) and (10.44) can be taken as the lower and upper membership functions. As a result, the IT2 TS fuzzy model for a nonlinear system is not unique and it offers a design flexibility to the IT2 fuzzy controller. However, different sets of lower and upper membership functions will lead to different values of ρ_{i1}, σ_{i1}, γ_{i1}, ρ_{i2}, σ_{i2}, γ_{i2}. As shown in Example 10.1, it will lead to different stability result.

For demonstration purposes, another set of lower and upper membership functions for the IT2 TS fuzzy model with the rules in (10.41) are chosen and shown in Table 10.5. The plots of the lower and upper membership functions are shown in Fig. 10.5. It can be shown that the normalized membership functions satisfy the inequality of $w_i^L(x_1(t)) \leq \tilde{w}_i(\mathbf{x}(t)) \leq w_i^U(x_1(t))$ for all i.

With other membership functions in simple forms being employed for the construction of the IT2 TS fuzzy model, it is able to lower the structural complexity of the IT2 fuzzy controller (10.46) and reduce the implementation cost. Considering the lower and upper membership functions in Table 10.5, it can be shown numerically that $\rho_{ik} = 1$, $\sigma_{ik} = 0.001$, $\gamma_{11} = 0.456$, $\gamma_{21} = 0.27$, $\gamma_{31} = 0.565$, $\gamma_{41} = 0.195$, $\gamma_{i2} = 0.001$, $i = 1, 2, 3, 4$; $k = 1, 2$, satisfy the inequalities (10.24) and (10.25). It should be noted that different coefficients for the inequalities (10.24) and (10.25) will give different feedback gains for the IT2 fuzzy controller (10.46). Hence, different IT2 fuzzy

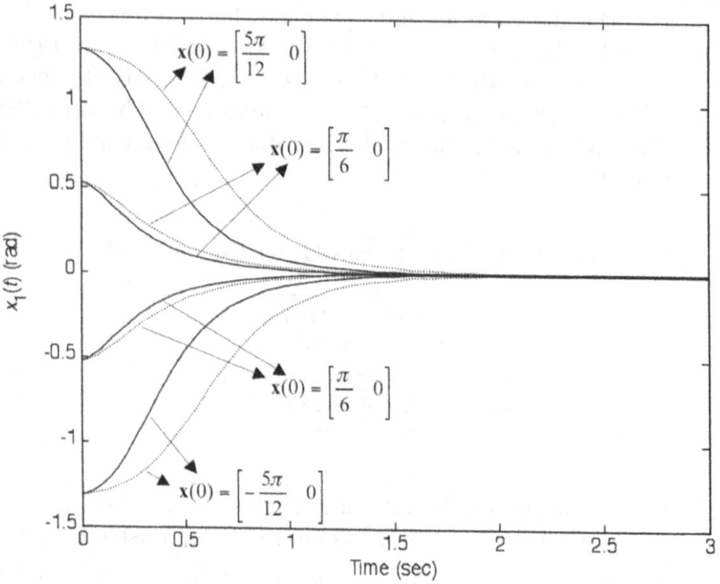

(a) $x_1(t)$.

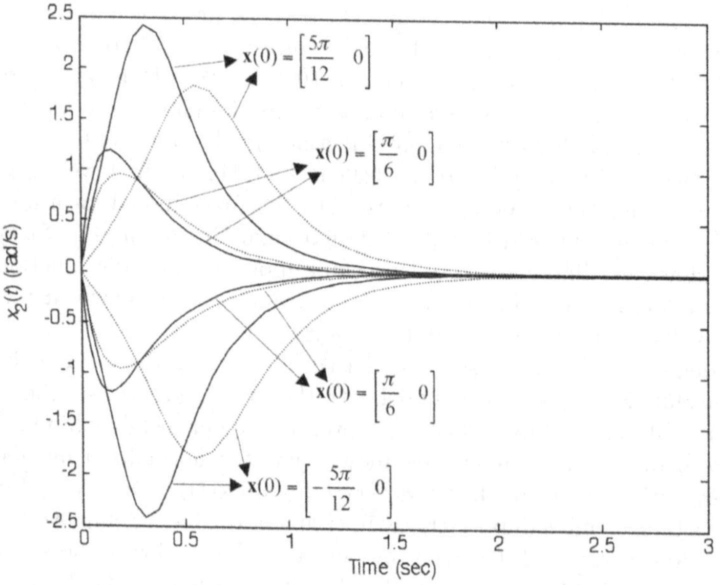

(b) $x_2(t)$.

Fig. 10.4 System state responses and control signals for the inverted pendulum in Example 10.3. Solid lines: $m_p = m_{p_{min}}$ and $M_c = M_{c_{min}}$. Dotted lines: $m_p = m_{p_{max}}$ and $M_c = M_{c_{max}}$.

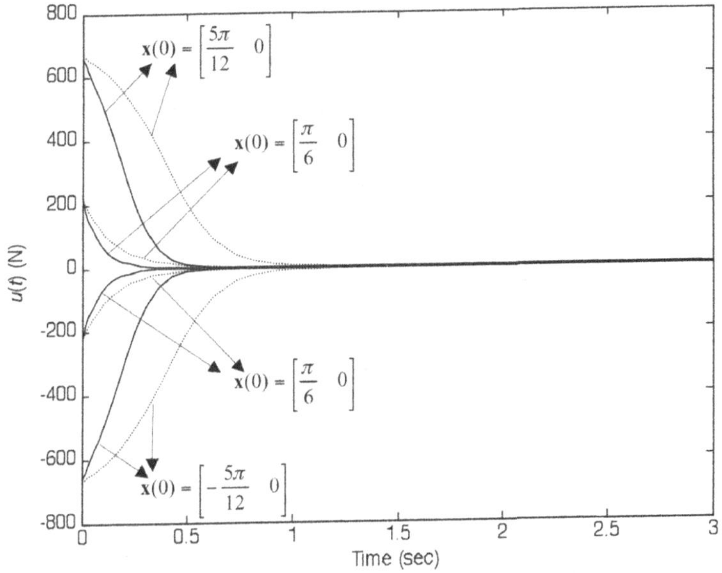

Fig. 10.4 *(continued)*

Table 10.5 Another set of lower and upper membership functions for the IT2 TS fuzzy model for Example 10.3.

Lower membership functions	Upper membership functions
$\underline{\mu}_{\tilde{M}_1^1}(x_1(t)) = 1 - e^{-\frac{x_1(t)^2}{1.2}}$	$\overline{\mu}_{\tilde{M}_1^1}(x_1(t)) = 1 - 0.23e^{-\frac{x_1(t)^2}{0.25}}$
$\underline{\mu}_{\tilde{M}_1^2}(x_1(t)) = 1 - e^{-\frac{x_1(t)^2}{1.2}}$	$\overline{\mu}_{\tilde{M}_1^2}(x_1(t)) = 1 - 0.23e^{-\frac{x_1(t)^2}{0.25}}$
$\underline{\mu}_{\tilde{M}_1^3}(x_1(t)) = 0.23e^{-\frac{x_1(t)^2}{0.25}}$	$\overline{\mu}_{\tilde{M}_1^3}(x_1(t)) = e^{-\frac{x_1(t)^2}{1.2}}$
$\underline{\mu}_{\tilde{M}_1^4}(x_1(t)) = 0.23e^{-\frac{x_1(t)^2}{0.25}}$	$\overline{\mu}_{\tilde{M}_1^4}(x_1(t)) = e^{-\frac{x_1(t)^2}{1.2}}$
$\underline{\mu}_{\tilde{M}_2^1}(x_1(t)) = 0.5e^{-\frac{x_1(t)^2}{0.25}}$	$\overline{\mu}_{\tilde{M}_2^1}(x_1(t)) = e^{-\frac{x_1(t)^2}{1.5}}$
$\underline{\mu}_{\tilde{M}_2^2}(x_1(t)) = 1 - e^{-\frac{x_1(t)^2}{1.5}}$	$\overline{\mu}_{\tilde{M}_2^2}(x_1(t)) = 1 - 0.5e^{-\frac{x_1(t)^2}{0.25}}$
$\underline{\mu}_{\tilde{M}_2^3}(x_1(t)) = 0.5e^{-\frac{x_1(t)^2}{0.25}}$	$\overline{\mu}_{\tilde{M}_2^3}(x_1(t)) = e^{-\frac{x_1(t)^2}{1.5}}$
$\underline{\mu}_{\tilde{M}_2^4}(x_1(t)) = 1 - e^{-\frac{x_1(t)^2}{1.5}}$	$\overline{\mu}_{\tilde{M}_2^4}(x_1(t)) = 1 - 0.5e^{-\frac{x_1(t)^2}{0.25}}$

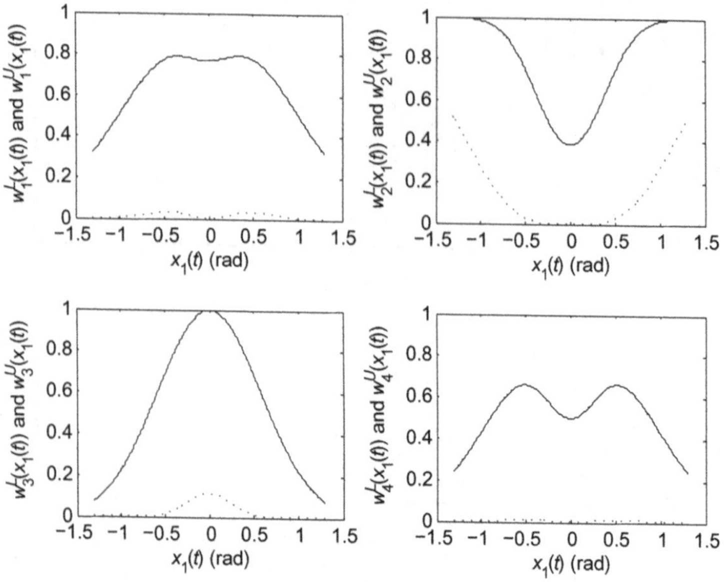

Fig. 10.5 Another set of lower membership functions $w_i^L(x_1(t))$ (dotted lines) and upper membership functions $w_i^U(x_1(t))$ (solid lines), $i = 1, 2, 3, 4$ for Example 10.3.

controllers designed based on different IT2 TS fuzzy models will offer different system performance. The performance analysis given in Chapter 3 can be considered to realize the system performance.

Based on the stability conditions in Theorem 10.1, with the MATLAB LMI toolbox, we have the feedback gains listed in Table 10.6.

Table 10.6 Another set of feedback gains $\mathbf{G}_j$ for Example 10.3.

$\mathbf{G}_1 = \begin{bmatrix} 885.2658 & 227.9417 \end{bmatrix}$
$\mathbf{G}_2 = \begin{bmatrix} 951.8250 & 243.5480 \end{bmatrix}$
$\mathbf{G}_3 = \begin{bmatrix} 942.8617 & 241.3597 \end{bmatrix}$
$\mathbf{G}_4 = \begin{bmatrix} 1072.0733 & 275.0160 \end{bmatrix}$

The IT2 fuzzy controller with the lower and upper membership functions in Table 10.5, and the feedback gains in Table 10.6 is employed to control the inverted pendulum. Considering the same initial system states as above, the system state responses and control signals for the IT2 FMB control system with $m_p = m_{p_{min}}$ and $M_c = M_{c_{min}}$, and $m_p = m_{p_{max}}$ and $M_c = M_{c_{max}}$ are shown in Fig. 10.6. Referring to Fig. 10.6, it can be seen that the inverted pendulum subject to different parameter values can be stabilized by the IT2

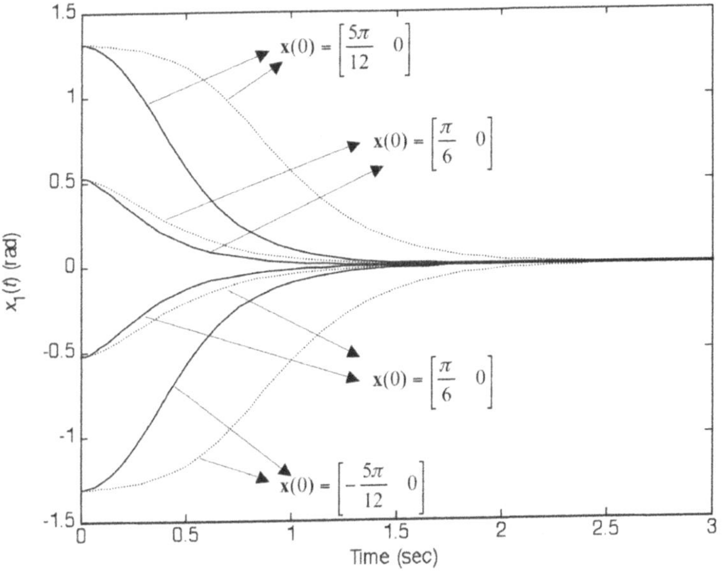

(a) $x_1(t)$.

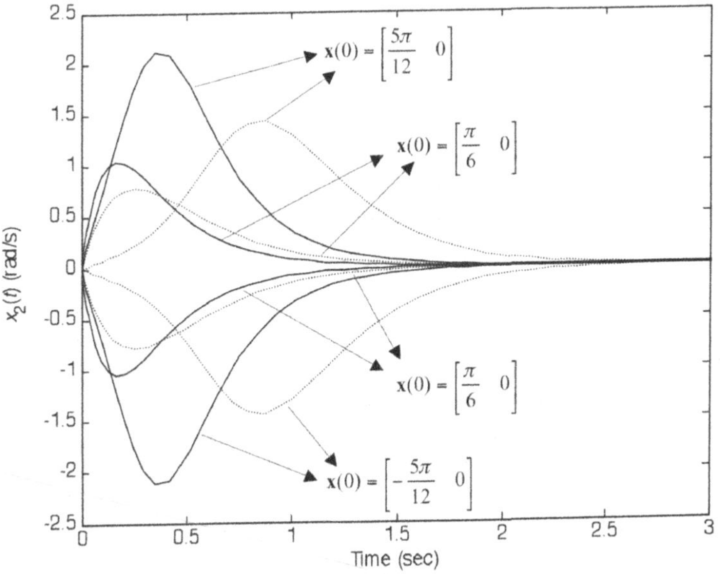

(b) $x_2(t)$.

Fig. 10.6 System state responses and control signals for the inverted pendulum in Example 10.3 with another set of feedback gains. Solid lines: $m_p = m_{p_{min}}$ and $M_c = M_{c_{min}}$. Dotted lines: $m_p = m_{p_{max}}$ and $M_c = M_{c_{max}}$.

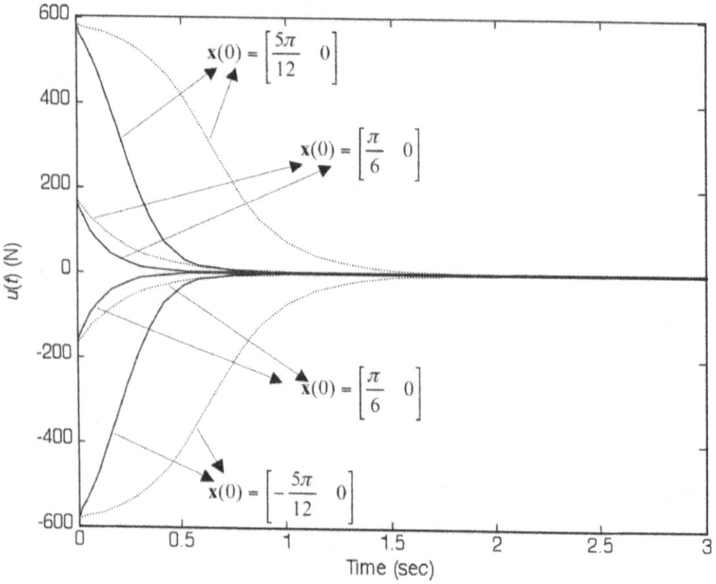

(c) $u(t)$.

Fig. 10.6 (*continued*)

fuzzy controller of (10.46). Comparing with the IT2 fuzzy controllers taking different sets of membership functions and feedback gains, it can be seen from Fig. 10.5 and Fig. 10.6 that both IT2 fuzzy controller are able to stabilize the inverted pendulum subject to different values of system parameters.

10.3 Conclusion

An IT2 TS fuzzy model has been proposed to represent the nonlinear plant subject to parameter uncertainties. The parameter uncertainties of the nonlinear plant are captured by the IT2 fuzzy sets characterized by the lower and upper membership functions to facilitate the stability analysis and controller synthesis. An IT2 fuzzy controller has been proposed to close the feedback loop to form an IT2 FMB control system. As the actual grades of the membership are uncertain in value due to the parameter uncertainties, the MFSI stability analysis approach in Chapter 3 will lead to very conservative stability conditions. To circumvent the difficulty, the nice property of the lower and upper membership functions offering the boundary information of uncertain parameters is utilized to introduce some MFSI and MFSD slack matrices to the stability analysis. By considering the MFSI and MFSD slack matrices, the stability of the IT2 FMB control systems has been investigated using the Lyapunov stability theory. LMI-based stability conditions have been derived to guarantee the system stability and design the IT2 fuzzy controller. Some simulation examples have been given to illustrate the effectiveness of the proposed IT2 FMB control approach.

References

1. Anderson, B.D.O., Moore, J.B., Naidu, D.S.: Optimal Control: Linear Quadratic Methods. Prentice-Hall, Englewood Cliffs (1990)
2. Ariño, C., Sala, A.: Extensions to stability analysis of fuzzy control systems subject to uncertain grades of membership. IEEE Trans. Syst., Man and Cybern, Part B: Cybernetics 38(2), 558–563 (2008)
3. Besheer, A.H., Emara, H.M., Aziz, M.M.A.: Fuzzy-based output-feedback H 1 control for uncertain nonlinear systems: an LMI approach. IET Control Theory & Applications 1(4), 1176–1185 (2007)
4. Biglarbegian, M., Melek, W.W., Mendel, J.M.: On the Stability of Interval Type-2 TSK Fuzzy Logic Control Systems. IEEE Trans. Syst., Man and Cybern., - Part B: Cybernetics 40(3), 798–818 (2010)
5. Blažič, S., Škrjanc, I., Matko, D.: Globally stable direct fuzzy model reference adaptive control. Fuzzy Sets and Systems 139(1), 3–33 (2003)
6. Boyd, S.P.: Linear matrix inequalities in system and control theory. Society for Industrial and Applied Mathematics (SIAM) (1994)
7. Cao, Y.Y., Frank, P.M.: Analysis and synthesis of nonlinear time-delay systems via fuzzy control approach. IEEE Trans. on Fuzzy Systems 8(2), 200–211 (2000)
8. Cao, Y.Y., Frank, P.M.: Stability analysis and synthesis of nonlinear time-delay systems via linear Takagi-Sugeno fuzzy models. Fuzzy Sets and Systems 124(2), 213–229 (2001)
9. Chang, W., Park, J.B., Joo, Y.H.: GA-based intelligent digital redesign of fuzzy-model-based controllers. IEEE Trans. on Fuzzy Systems 11(1), 35–44 (2003)
10. Chen, B., Liu, X.: Delay-Dependent Robust H_infty Control for T-S Fuzzy Systems With Time Delay. IEEE Trans. on Fuzzy Systems 13(4), 544–556 (2005)
11. Chen, B., Liu, X., Tong, S.: Delay-dependent stability analysis and control synthesis of fuzzy dynamic systems with time delay. Fuzzy Sets and Systems 157(16), 2224–2240 (2006)
12. Chen, B., Liu, X., Tong, S.: New delay-dependent stabilization conditions of TS fuzzy systems with constant delay. Fuzzy sets and systems 158(20), 2209–2224 (2007)
13. Chen, B.S., Lee, B.K., Guo, L.B.: Optimal tracking design for stochastic fuzzy systems. IEEE Trans. on Fuzzy Systems 11(6), 796–813 (2003)
14. Chen, B.S., Tseng, C.S., Uang, H.J.: Mixed H_2/H_∞ fuzzy output feedback control design for nonlinear dynamic systems: an LMI approach. IEEE Trans. Fuzzy Systems 8(3), 249–265 (2000)
15. Chen, C.L., Chen, P.C., Chen, C.K.: Analysis and design of fuzzy control system. Fuzzy Sets and Systems 57(2), 125–140 (1993)
16. Choi, H.H.: LMI-based nonlinear fuzzy observer-controller design for uncertain MIMO nonlinear systems. IEEE Trans. Fuzzy Systems 15(5), 956–971 (2007)
17. Chung, H.Y., Wu, S.M., Yu, F.M., Chang, W.J.: Evolutionary design of static output feedback controller for Takagi-Sugeno fuzzy systems. IET Control Theory Appl. 1(4), 1096–1103 (2007)

18. Coupland, S., John, R.: Geometric type-1 and type-2 fuzzy logic systems. IEEE Trans. Fuzzy Systems 15(1), 3–15 (2007)
19. Ding, B.C., Sun, H.X., Yang, P.: Further studies on LMI-based relaxed stabilization conditions for nonlinear systems in Takagi-Sugeno's form. Automatica 42(3), 503–508 (2006)
20. Doctor, F., Hagras, H., Callaghan, V.: An intelligent fuzzy agent approach for realising ambient intelligence in intelligent inhabited environments. IEEE Transactions on Systems, Man and Cybernetics, Part A: Systems and Humans 35(1), 55–65 (2005)
21. Dong, J., Yang, G.H.: Dynamic output feedback control synthesis for continuous-time T-S fuzzy systems via a switched fuzzy control scheme. IEEE Trans. Syst., Man and Cybern., Part B: Cybernetics 38(4), 1166–1175 (2008)
22. Dong, J., Yang, G.H.: State feedback control of continuous-time T-S fuzzy systems via switched fuzzy controllers. Information Sciences 178(6), 1680–1695 (2008)
23. EI Ghaoui, L., Niculescu, S.I.: Advances in linear matrix inequality methods in control. Society for Industrial Mathematics (2000)
24. Fang, C.H., Liu, Y.S., Kau, S.W., Hong, L., Lee, C.H.: A new LMI-based approach to relaxed quadratic stabilization of Takagi-Sugeno fuzzy control systems. IEEE Trans. Fuzzy Systems 14(3), 386–397 (2006)
25. Feng, G.: Controller synthesis of fuzzy dynamic systems based on piecewise Lyapunov functions. IEEE Trans. Fuzzy Systems 11(5), 605–612 (2003)
26. Feng, G.: H_∞ controller design of fuzzy dynamic systems based on piecewise Lyapunov functions. IEEE Trans. Syst., Man and Cybern., Part B: Cybernetics 34(1), 283–292 (2004)
27. Feng, G.: A survey on analysis and design of model-based fuzzy control systems. IEEE Trans. Fuzzy Systems 14(5), 676–697 (2006)
28. Feng, G., Chen, C.L., Sun, D., Zhu, Y.: H_∞ controller synthesis of fuzzy dynamic systems based on piecewise Lyapunov functions and bilinear matrix inequalities. IEEE Trans. on Fuzzy Systems 13(1), 94–103 (2005)
29. Feng, M., Harris, C.J.: Piecewise Lyapunov stability conditions of fuzzy systems. IEEE Trans. Syst., Man and Cybern., Part B: Cybernetics 31(2), 259–262 (2001)
30. Fridman, E., Seuret, A., Richard, J.P.: Robust sampled-data stabilization of linear systems: an input delay approach* 1. Automatica 40(8), 1441–1446 (2004)
31. Fridman, E., Shaked, U.: Delay-dependent stability and H control: constant and time-varying delays. International Journal of Control 76(1), 48–60 (2003)
32. Gao, H., Chen, T.: Stabilization of nonlinear systems under variable sampling: a fuzzy control approach. IEEE Transactions on Fuzzy Systems 15(5), 972–983 (2007)
33. Gao, Z., Shi, X., Ding, S.X.: Fuzzy State/Disturbance Observer Design for T-S Fuzzy Systems With Application to Sensor Fault Estimation. IEEE Trans. Syst., Man and Cybern., - Part B: Cybernetics 38(3), 875 (2008)
34. Guan, X.P., Chen, C.L.: Delay-dependent guaranteed cost control for TS fuzzy systems with time delays. IEEE Trans. Fuzzy Systems 12(2), 236–249 (2004)
35. Guerra, T.M., Delmotte, F., Vermeiren, L., Tirmant, H.: Compensation and division control law for fuzzy models. In: The 10th IEEE International Conference on Fuzzy Systems, 2001, vol. 1, pp. 521–524 (2001)

36. Guerra, T.M., Vermeiren, L.: LMI-based relaxed nonquadratic stabilization conditions for nonlinear systems in the Takagi-Sugeno's form* 1. Automatica 40(5), 823–829 (2004)
37. Guo, Y., Woo Woo, P.Y.: An adaptive fuzzy sliding mode controller for robotic manipulators. IEEE Transactions on Systems, Man, and Cybernetics. Part A, Systems and Humans 33(2), 149–159 (2003)
38. Hagras, H.: A hierarchical type-2 fuzzy logic control architecture for autonomous mobile robots. IEEE Trans. on Fuzzy Systems 12(4), 524–539 (2004)
39. Holmblad, L.P., Ostergaard, J.J.: Control of a cement kiln by fuzzy logic techniques. In: Proc. of Conf. 8-th IFAC, Kyoto, Japan, pp. 809–814 (1981)
40. Huang, D., Nguang, S.K.: Static output feedback controller design for fuzzy systems: An ILMI approach. Information Sciences 177(14), 3005–3015 (2007)
41. Ioannou, P.A., Sun, J.: Robust Adaptive Control. Prentice-Hall, Englewood Cliffs (1996)
42. Jammeh, E.A., Fleury, M., Wagner, C., Hagras, H., Ghanbari, M.: Interval type-2 fuzzy logic congestion control for video streaming across IP networks. IEEE Trans. on Fuzzy Systems 17(5), 1123–1142 (2009)
43. Johansson, M., Rantzer, A., Arzen, K.E.: Piecewise quadratic stability of fuzzy systems. IEEE Trans. Fuzzy Systems 7(6), 713–722 (1999)
44. Katayama, H., Ichikawa, A.: H_∞ control for sampled-data nonlinear systems described by Takagi-Sugeno fuzzy systems. Fuzzy Sets and Systems 148(3), 431–452 (2004)
45. Kau, S.W., Lee, H.J., Yang, C.M., Lee, C.H., Hong, L., Fang, C.H.: Robust H $_\infty$ fuzzy static output feedback control of T-S fuzzy systems with parametric uncertaintiesif. Fuzzy Sets and Systems 158(2), 135–146 (2007)
46. Khalil, H.K., Grizzle, J.W.: Nonlinear Systems. Prentice-Hall, Englewood Cliffs (1996)
47. Kickert, W.J.M., Mamdani, E.H.: Analysis of a fuzzy logic controller. Fuzzy Sets and Systems 1(1), 29–44 (1978)
48. Kim, E., Lee, H.: New approaches to relaxed quadratic stability condition of fuzzy control systems. IEEE Trans. Fuzzy Systems 8(5), 523–534 (2000)
49. Kim, E., Park, M., Ji, S., Park, M.: A new approach to fuzzy modeling. IEEE Trans. Fuzzy Systems 5(3), 328–337 (1997)
50. Kim, J.H., Hyun, E., Park, M.: Adaptive synchronization of uncertain chaotic systems based on T-S fuzzy model. IEEE Trans. Fuzzy Systems 15(3), 359–369 (2007)
51. Lam, H.K.: Sampled-data fuzzy-model-based control systems: stability analysis with consideration of analogue-to-digital converter and digital-to-analogue converter. Control Theory & Applications, IET 4(7), 1131–1144 (2010)
52. Lam, H.K., Leung, F.H.F.: Fuzzy combination of linear state-feedback and switching controllers. Electronics Letters 40(7), 410 (2004)
53. Lam, H.K., Leung, F.H.F.: Fuzzy combination of fuzzy and switching state-feedback controllers for nonlinear systems subject to parameter uncertainties. IEEE Trans. Syst., Man and Cybern., Part B: Cybernetics 35(2), 269–281 (2005)
54. Lam, H.K., Leung, F.H.F.: Fuzzy rule-based combination of linear and switching state-feedback controllers. Fuzzy Sets and Systems 156(2), 153–184 (2005)

55. Lam, H.K., Leung, F.H.F.: Stability analysis of fuzzy control systems subject to uncertain grades of membership. IEEE Trans. Syst., Man and Cybern, Part B: Cybernetics 35(6), 1322–1325 (2005)
56. Lam, H.K., Leung, F.H.F.: Design and stabilization of sampled-data neural-network-based control systems. IEEE Trans. on Systems, Man, and Cybern., Part B 36(5), 995–1005 (2006)
57. Lam, H.K., Leung, F.H.F.: Synchronization of uncertain chaotic systems based on the fuzzy-model-based approach. International Journal of Bifurcation and Chaos 16(5), 1435–1444 (2006)
58. Lam, H.K., Leung, F.H.F.: Fuzzy controller with stability and performance rules for nonlinear systems. Fuzzy Sets and Systems 158(2), 147–163 (2007)
59. Lam, H.K., Leung, F.H.F.: LMI-Based Stability and Performance Conditions for Continuous-Time Nonlinear Systems in Takagi–Sugeno's Form. IEEE Trans. Syst., Man and Cybern., Part B: Cybernetics 37(5), 1396–1406 (2007)
60. Lam, H.K., Leung, F.H.F.: Sampled-data fuzzy controller for time-delay nonlinear system: LMI-based and fuzzy-model-based approaches. IEEE Trans. Syst., Man and Cybern., Part B: Cybernetics 37(3), 617–629 (2007)
61. Lam, H.K., Leung, F.H.F.: Stability analysis and performance design for fuzzy model-based control systems using a BMI-based approach. Foundations of Generic Optimization 2, 261–281 (2008)
62. Lam, H.K., Leung, F.H.F., Lai, J.C.Y.: Fuzzy-model-based control systems using fuzzy combination techniques. International Journal of Fuzzy Systems 9(3), 123–132 (2007)
63. Lam, H.K., Leung, F.H.F., Lee, Y.S.: Design of a switching controller for nonlinear systems with unknown parameters based on a fuzzy logic approach. IEEE Trans. Syst., Man and Cybern., Part B: Cybernetics 34(2), 1068–1074 (2004)
64. Lam, H.K., Leung, F.H.F., Tam, P.K.S.: Stable and robust fuzzy control for uncertain nonlinear systems based on a grid-point approach. In: Proc. 6th IEEE Int. Conf. Fuzzy Syst. (FUZZ-IEEE 1997), pp. 87–92. IEEE, Los Alamitos (1997)
65. Lam, H.K., Leung, F.H.F., Tam, P.K.S.: A switching controller for uncertain nonlinear systems. IEEE Control Systems Magazine 22(1), 7–14 (2002)
66. Lam, H.K., Ling, W.K.: Sampled-data fuzzy controller for continuous nonlinear systems. IET Control Theory Appl. 2(1), 32–39 (2008)
67. Lam, H.K., Narimani, M.: Stability analysis and performance deign for fuzzy-model-based control system under imperfect premise matching. IEEE Trans. Fuzzy Systems 17(4), 949–961 (2009)
68. Lam, H.K., Narimani, M.: Quadratic stability analysis of fuzzy-model-based control systems using staircase membership functions. IEEE Trans. Fuzzy Systems 18(1), 125–137 (2010)
69. Lam, H.K., Seneviratne, L.D.: BMI-based stability and performance design for fuzzy-model-based control systems subject to parameter uncertainties. IEEE Trans. Syst., Man and Cybern., Part B: Cybernetics 37(3), 502–514 (2007)
70. Lam, H.K., Seneviratne, L.D.: Stability analysis of interval type-2 fuzzy-model-based control systems. IEEE Trans. Syst., Man and Cybern., Part B: Cybernetics 38(3), 617–628 (2008)
71. Lam, H.K., Seneviratne, L.D.: Tracking control of sampled-data fuzzy-model-based control systems. IET Control Theory Appl. 3(1), 56–67 (2009)

72. Lee, H.J., Kim, H., Joo, Y.H., Chang, W., Park, J.B.: A new intelligent digital redesign for TS fuzzy systems: global approach. IEEE Trans. on Fuzzy Systems 12(2), 274–284 (2004)
73. Lee, H.J., Park, J.B., Joo, Y.H.: Digitalizing a fuzzy observer-based output-feedback control: intelligent digital redesign approach. IEEE Trans. on Fuzzy Systems 13(5), 701–716 (2005)
74. Leung, F.H.F., Lam, H.K., Ling, S.H., Tam, P.K.S.: Optimal and stable fuzzy controllers for nonlinear systems based on an improved genetic algorithm. IEEE Trans. on Industrial Electronics 51(1), 172–182 (2004)
75. Li, C., Wang, H., Liao, X.: Delay-dependent robust stability of uncertain fuzzy systems with time-varying delays. In: IEE Proceedings-Control Theory and Applications, vol. 151(4), pp. 417–421 (2004)
76. Li, J., Wang, H., Niemann, D., Tanaka, K.: Dynamic parallel distributed compensation for Takagi-Sugeno fuzzy systems: An LMI approach. Information Sciences 123(3-4), 201–221 (2000)
77. Lian, K.Y., Tu, H.W., Liou, J.J.: Stability conditions for LMI-based fuzzy control from viewpoint of membership functions. IEEE Trans. Fuzzy Systems 14(6), 874–884 (2006)
78. Liang, Q., Mendel, J.M.: Equalization of nonlinear time-varying channels using type-2 fuzzyadaptive filters. IEEE Trans. Fuzzy Systems 8(5), 551–563 (2000)
79. Lien, C.: Stabilization for uncertain Takagi-Sugeno fuzzy systems with time-varying delays and bounded uncertainties. Chaos, Solitons & Fractals 32(2), 645–652 (2007)
80. Lin, P.Z., Lin, C.M., Hsu, C.F., Lee, T.T.: Type-2 fuzzy controller design using a sliding-mode approach for application to DC-DC converters. In: IEE Proceedings-Electric Power Applications, vol. 152(6), pp. 1482–1488 (2005)
81. Liu, X., Zhang, Q.: Approaches to quadratic stability conditions and H_∞ control designs for Takagi-Sugeno fuzzy systems. IEEE Trans. Fuzzy Systems 11(6), 830–839 (2003)
82. Liu, X., Zhang, Q.: New approaches to H_∞ controller designs based on fuzzy observers for Takagi-Sugeno fuzzy systems via LMI. Automatica 39(9), 1571–1582 (2003)
83. Lo, J.C., Kuo, Y.H.: Decoupled fuzzy sliding-mode control. IEEE Transactions on Fuzzy Systems 6(3), 426–435 (1998)
84. Ma, X.J., Sun, Z.Q.: Analysis and design of fuzzy reduced-dimensional observer and fuzzy functional observer. Fuzzy Sets and Systems 120(1), 35–63 (2001)
85. Ma, X.J., Sun, Z.Q., He, Y.Y.: Analysis and design of fuzzy controller and fuzzy observer. IEEE Trans. Fuzzy Systems 6(1), 41–51 (1998)
86. Mamdani, E.H.: Advances in the linguistic synthesis of fuzzy controllers. International Journal of Man-Machine Studies 8(6), 669–678 (1976)
87. Mamdani, E.H., Assilian, S.: An experiment in linguistic synthesis with a fuzzy logic controller. International Journal of Man-Machine Studies 7(1), 1–13 (1975)
88. Mendel, J., Liu, F.: Super-exponential convergence of the Karnik–Mendel algorithms for computing the centroid of an interval type-2 fuzzy set. IEEE Trans. Fuzzy Systems 15(2), 309–320 (2007)

89. Mendel, J.M., John, R.I., Liu, F.: Interval type-2 fuzzy logic systems made simple. IEEE Trans. Fuzzy Systems 14(6), 808–821 (2006)
90. Michalewicz, Z.: Genetic algorithms+ data structures. Springer, Heidelberg (1996)
91. Narimani, M., Lam, H.K.: Relaxed LMI-based stability conditions for Takagi-Sugeno fuzzy control systems using regional-membership-function-shape-dependent analysis approach. IEEE Trans. Fuzzy Systems 17(5), 1221–1228 (2009)
92. Nguang, S.K., Shi, P.: Fuzzy H_∞ output feedback control of nonlinear systems under sampled measurements. Automatica 39(12), 2169–2174 (2003)
93. Ohtake, H., Tanaka, K., Wang, H.O.: Switching fuzzy controller design based on switching Lyapunov function for a class of nonlinear systems. IEEE Trans. Syst., Man and Cybern., Part B: Cybernetics 36(1), 13–23 (2006)
94. Park, C.W., Cho, Y.W.: Adaptive tracking control of flexible joint manipulator based on fuzzy model reference approach. In: IEE Proceedings-Control Theory and Applications, vol. 150(2), pp. 198–204 (2003)
95. Park, C.W., Cho, Y.W.: TS model based indirect adaptive fuzzy control using online parameter estimation. IEEE Trans. Systems, Man, and Cybern., Part B 34(6), 2293–2302 (2004)
96. Rhee, B.J., Won, S.: A new fuzzy Lyapunov function approach for a Takagi-Sugeno fuzzy control system design. Fuzzy Sets and Systems 157(9), 1211–1228 (2006)
97. Sala, A., Ariño, C.: Asymptotically necessary and sufficient conditions for stability and performance in fuzzy control: Applications of Polya's theorem. Fuzzy Sets Syst. 158(24), 2671–2686 (2007)
98. Sala, A., Ariño, C.: Relaxed stability and performance conditions for Takagi-Sugeno fuzzy systems with knowledge on membership function overlap. IEEE Trans. Syst., Man and Cybern., Part B: Cybernetics 37(3), 727–732 (2007)
99. Sala, A., Ariño, C.: Relaxed stability and performance LMI conditions for Takagi-Sugeno fuzzy systems with polynomial constraints on membership function shapes. IEEE Trans. Fuzzy Systems 16(5), 1328–1336 (2008)
100. Slotine, J.J.E., Li, W.: Applied Nonlinear Control. Prentice-Hall, Englewood Cliffs (1991)
101. Sugeno, M., Kang, G.T.: Structure identification of fuzzy model. Fuzzy sets and systems 28(1), 15–33 (1988)
102. Takagi, T., Sugeno, M.: Fuzzy identification of systems and its applications to modelling and control. IEEE Trans. Sys., Man., Cybern. smc-15(1), 116–132 (1985)
103. Tanaka, K., Hori, T., Wang, H.O.: A multiple Lyapunov function approach to stabilization of fuzzy control systems. IEEE Trans. Fuzzy Systems 11(4), 582–589 (2003)

104. Tanaka, K., Ikeda, T., Wang, H.O.: Robust stabilization of a class of uncertain nonlinear systems via fuzzy control: Quadratic stability, H_∞ control theory, and linear matrix inequalities. IEEE Trans. on Fuzzy Systems 4(1), 1–13 (1996)
105. Tanaka, K., Ikeda, T., Wang, H.O.: Fuzzy regulators and fuzzy observers: Relaxed stability conditions and LMI-based designs. IEEE Trans. Fuzzy Systems 6(2), 250–265 (1998)
106. Tanaka, K., Iwasaki, M., Wang, H.O.: Switching control of an R/C hovercraft: stabilization and smoothswitching. IEEE Trans. Syst., Man and Cybern., Part B: Cybernetics 31(6), 853–863 (2001)
107. Tanaka, K., Ohtake, H., Wang, H.O.: A descriptor system approach to fuzzy control system design via fuzzy Lyapunov functions. IEEE Trans. Fuzzy Systems 15(3), 333–341 (2007)
108. Tanaka, K., Ohtake, H., Wang, H.O.: Guaranteed cost control of polynomial fuzzy systems via a sum of squares approach. IEEE Trans. Systems, Man and Cybern. - Part B: Cybern. 39(2), 561–567 (2009)
109. Tanaka, K., Sugeno, M.: Stability analysis and design of fuzzy control systems. Fuzzy sets and systems 45(2), 135–156 (1992)
110. Tanaka, K., Wang, H.O.: Fuzzy control systems design and analysis: a linear matrix inequality approach. Wiley Interscience, Hoboken (2001)
111. Tanaka, K., Yoshida, H., Ohtake, H., Wang, H.O.: A sum of squares approach to modeling and control of nonlinear dynamical systems with polynomial fuzzy systems. IEEE Trans. Fuzzy Systems 17(4), 911–922 (2009)
112. Teixeira, M.C.M., Assuncão, E., Avellar, R.G.: On relaxed LMI-based designs for fuzzy regulators and fuzzy observers. IEEE Trans. Fuzzy Systems 11(5), 613–623 (2003)
113. Tian, E., Peng, C.: Delay-dependent stability analysis and synthesis of uncertain TS fuzzy systems with time-varying delay. Fuzzy Sets and Systems 157(4), 544–559 (2006)
114. Ting, C., Li, T., Kung, F.: An approach to systematic design of the fuzzy control system. Fuzzy sets and Systems 77(2), 151–166 (1996)
115. Tong, R.M., Beck, M.B., Latten, A.: Fuzzy control of the activated sludge wastewater treatment process. Automatica 16(6), 695–701 (1980)
116. Tong, S., Wang, T., Li, H.X.: Fuzzy robust tracking control for uncertain nonlinear systems. International Journal of Approximate Reasoning 30(2), 73–90 (2002)
117. Tseng, C.S.: Model-reference output feedback fuzzy tracking control design for nonlinear discrete-time systems with time-delay. IEEE Trans. Fuzzy Systems 14(1), 58–70 (2006)
118. Tseng, C.S.: A novel approach to H_∞ decentralized fuzzy-observer-based fuzzy control design for nonlinear interconnected systems. IEEE Trans. Fuzzy Systems 16(5), 1337–1350 (2008)
119. Tseng, C.S., Chen, B.S., Uang, H.J.: Fuzzy tracking control design for nonlinear dynamic systems via T-S fuzzy model. IEEE Trans. Fuzzy Systems 9(3), 381–392 (2001)
120. VanAntwerp, J.G., Braatz, R.D.: A tutorial on linear and bilinear matrix inequalities. Journal of Process Control 10(4), 363–385 (2000)
121. Vidyasagar, M.: Nonlinear Systems Analysis. Society for Industrial Mathematics (2002)

122. Wang, H.O., Tanaka, K., Griffin, M.F.: An approach to fuzzy control of nonlinear systems: Stability and design issues. IEEE Trans. Fuzzy Systems 4(1), 14–23 (1996)
123. Wang, L.X.: Adaptive Fuzzy Systems and Control - Design and Stability Analysis. Prentice-Hall, Englewood Cliffs (1994)
124. Wang, R.J., Lin, W.W., Wang, W.J.: Stabilizability of linear quadratic state feedback for uncertain fuzzy time-delay systems. IEEE Trans. Syst., Man and Cybern., Part B: Cybernetics 34(2), 1288–1292 (2004)
125. Wang, W.J., Yan, S.F., Chiu, C.H.: Flexible stability criteria for a linguistic fuzzy dynamic system. Fuzzy Sets and Systems 105(1), 63–80 (1999)
126. Wang, Z., Ho, D.W.C., Liu, X.: A note on the robust stability of uncertain stochastic fuzzy systems with time-delays. IEEE Trans. Syst., Man and Cybern., - Part A 34(4), 570–576 (2004)
127. Wu, M., He, Y., She, J.H., Liu, G.P.: Delay-dependent criteria for robust stability of time-varying delay systems. Automatica 40(8), 1435–1439 (2004)
128. Xiu, Z., Ren, G.: Stability analysis and systematic design of Takagi-Sugeno fuzzy control systems. Fuzzy sets and systems 151(1), 119–138 (2005)
129. Yoneyama, J.: New robust stability conditions and design of robust stabilizing controllers for Takagi-Sugeno fuzzy time-delay systems. IEEE Transactions on Fuzzy Systems 15(5), 828–839 (2007)
130. Yoneyama, J., Nishikawa, M., Katayama, H., Ichikawa, A.: Output stabilization of Takagi-Sugeno fuzzy systems. Fuzzy Sets and Systems 111(2), 253–266 (2000)
131. Yoneyama, J., Nishikawa, M., Katayama, H., Ichikawa, A.: Design of output feedback controllers for Takagi-Sugeno fuzzy systems. Fuzzy Sets and Systems 121(1), 127–148 (2001)
132. Yu, W., Sun, C.: Fuzzy model based adaptive control for a class of nonlinear systems. IEEE Trans. Fuzzy Systems 9(3), 413–425 (2001)
133. Zhang, X., Wu, M., She, J., He, Y.: Delay-dependent stabilization of linear systems with time-varying state and input delays. Automatica 41(8), 1405–1412 (2005)

Index